Computational Intelligent Techniques in Mechatronics

Scrivener Publishing
100 Cummings Center, Suite 541J
Beverly, MA 01915-6106

Next-Generation Computing and Communication Engineering

Series Editor: Dr. G. R. Kanagachidambaresan (kanagachidambaresan@gmail.com)
and Dr. Kolla Bhanu Prakash (drkbp1981@gmail.com)

The series concentrates on next-generation computing and communication methodologies involving smart and ambient environment design. It is an effective publishing platform for monographs, handbooks, and edited volumes on Industry 4.0, agriculture, smart city development, new computing and communication paradigms. Although the series mainly focuses on design, it also addresses analytics and investigation of industry-related real-time problems.

Publishers at Scrivener
Martin Scrivener (martin@scrivenerpublishing.com)
Phillip Carmical (pcarmical@scrivenerpublishing.com)

Computational Intelligent Techniques in Mechatronics

Edited by

Kolla Bhanu Prakash

Department of Computer Science & Engineering, K. L. University, Vaddeswaram, India

Satish Kumar Peddapelli

Osmania University, Hyderabad, India

Ivan C. K. Tam

Director of Innovation & Engagement at the University of Newcastle in Singapore

Wai Lok Woo

Department: Computer and Information Sciences, Northumbria University, Newcastle, UK

and

Vishal Jain

Department of Computer Science and Engineering, Sharda University, Greater Noida, India

Scrivener Publishing

WILEY

This edition first published 2024 by John Wiley & Sons, Inc., 111 River Street, Hoboken, NJ 07030, USA and Scrivener Publishing LLC, 100 Cummings Center, Suite 541J, Beverly, MA 01915, USA

For more information about Scrivener publications please visit www.scrivenerpublishing.com.

Wiley Global Headquarters
111 River Street, Hoboken, NJ 07030, USA

For details of our global editorial offices, customer services, and more information about Wiley products visit us at www.wiley.com.

Library of Congress Cataloging-in-Publication Data

ISBN 978-1-394-17464-5

Cover image: Pixabay.Com
Cover design by Russell Richardson

Set in size of 11pt and Minion Pro by Manila Typesetting Company, Makati, Philippines

Printed in the USA

10 9 8 7 6 5 4 3 2 1

Contents

Preface

Mechatronics has been a revolutionary force in engineering and medical robotics over the past decade. It will lead to a major industrial revolution and affect research in every field of engineering. This book covers the basics of mechatronics, computational intelligence approaches, simulation and modeling concepts, architectures, nanotechnology, real-time monitoring and control, different actuators, and sensors. This resource is important for research and development departments in academia, government, and industry, and as a reference source in university libraries. This book explains clearly and comprehensively the engineering design process at different stages. As the historical divisions between the various branches of engineering and computer science become less clearly defined, mechatronics may provide a roadmap for nontraditional engineering students studying within the traditional university structure. This book covers all the algorithms and techniques found in mechatronics engineering, well explained with real-time examples, especially lab experiments that will be very informative to students and scholars.

Chapter 1 of this book provides an overview of promising AI approaches for modern mechatronics, along with the key benefits. The second chapter discusses classical thermodynamics and heat transfer principles that can optimize the blueprint, performance, and regulation of mechatronic apparatus.

Chapter 3 explores the role of data acquisition, sensors, and actuators in the mechatronics industry, while Chapter 4 discusses the challenges and opportunities in employing these techniques. Furthermore, it proposes future research directions to optimize mechatronic systems.

Chapter 5 highlights the potential of RL for the future of adaptive mechatronics, fostering a new era of intelligent and versatile systems that can adapt, learn, and excel in diverse and dynamic scenarios. The sixth chapter shows a study of how important PLCs are to advancing the mechatronics sector and enabling more advanced forms of industrial automation.

Chapter 7 discusses fuzzy logic and its applications in mechatronic control systems, while Chapter 8 highlights the need for addressing ethical considerations and ensuring responsible deployment for the widespread acceptance and success of autonomous systems in the real world. Chapter 9 explores this intersection by considering the key components, challenges, and benefits of integrating AI in self-driving transportation systems. This section also highlights the advancements made in the field, current state of the art, and potential future directions.

Chapter 10 illustrates how mechatronics-based power quality solutions have increased industrial productivity, and may result in smart and strong industrial infrastructures. The eleventh chapter delves into prospects, challenges, and emerging trends, underscoring the potential of neural networks and fuzzy logic to shape the future of autonomous electric transportation.

Chapter 12 explores the transformative impact of integrating artificial intelligence (AI) techniques within the field of mechatronics, showcasing the synergistic potential that emerges when advanced AI technologies are harnessed to enhance the functionality, adaptability, and intelligence of mechatronic systems. Then, Chapter 13 delves into the intriguing area of human-robot interaction and emotional intelligence, showcasing how AI enhances interaction and emotional understanding between humans and machines.

Chapter 14 proposes a way by which a driver can be alerted to drowsiness in order to reduce the numbers of road accidents, while Chapter 15 proposes the model of a multi-featured robot that can send live surveillance and notifications if human motion is detected. Finally, Chapter 16 highlights the latency of mechatronics in altering the current agriculture practices including productivity, and sustainability.

In conclusion, we offer our heartfelt thanks to all the contributors who have made this book a reality. Your expertise, dedication, and passion have enriched every page, creating a comprehensive and insightful collection. Without your valuable insights and commitment to excellence, this project would not have been possible. Your contributions have not only elevated the quality of this book but have also left a lasting impact on the field. Additionally, we have greatly appreciated the dedicated support and valuable assistance rendered by Martin Scrivener and the Scrivener Publishing team during the publication of this volume.

Editors
Kolla Bhanu Prakash
Satish Kumar Peddapelli
Ivan C. K. Tam
Wai Lok Woo
Vishal Jain

1

AI in Mechatronics

Vansh Gehlot* **and Prashant Singh Rana**

Thapar Institute of Engineering and Technology, Patiala, Punjab, India

Abstract

The chapter "AI in Mechatronics" provides a comprehensive overview of how artificial intelligence (AI) techniques can enhance mechatronic systems; it covers core AI approaches, including machine learning, computer vision, soft computing methods, natural language processing, AI planning, and intelligent system design tools. Key real-world applications across areas like manufacturing automation, robotics, vehicles, and human–machine interaction are analyzed. The integration of data-driven and knowledge-based AI methods can endow mechatronics with advanced capabilities around autonomy, environmental perception, reasoning, control, and human collaboration. This creates intelligent self-optimizing systems that continue learning and adapting. Current challenges around model accuracy, data availability, security, and algorithmic transparency are discussed along with emerging opportunities in embedded intelligence and bio-inspired robotics. Overall, the synergistic fusion of AI and mechatronics promises to transform technological systems across industrial and societal domains by enhancing automation, augmenting human capabilities, and enabling next-generation smart mechanized assistants.

***Keywords*:** Machine learning, human–robot collaboration, computer vision, robotic systems, natural language processing, manufacturing automation, self-driving vehicles

**Corresponding author*: gehlotvansh111@gmail.com

Kolla Bhanu Prakash, Satish Kumar Peddapelli, Ivan C. K. Tam, Wai Lok Woo and Vishal Jain (eds.) Computational Intelligent Techniques in Mechatronics, (1–40) © 2024 Scrivener Publishing LLC

1.1 Introduction to AI Techniques for Mechatronics

Mechatronics refers to the synergistic integration of mechanical, electrical, and computer engineering to design smart systems and products capable of exhibiting both information-processing and physical-action capabilities. It combines the core disciplines of mechanics, electronics, control systems, and computing to create high-performance systems with built-in intelligence.

Although mechatronics has its origins in industrial robotics and manufacturing automation, it has grown tremendously in scope and adoption over the past few decades. Today, mechatronic systems and principles are utilized in a diverse range of applications including automotive systems, aerospace systems, industrial production systems, defense systems, material handling systems, healthcare systems, and consumer product design. Figure 1.1 shows the flowchart depicting the design methodology for mechatronics systems, mechatronics integrates various engineering disciplines.

Modern mechatronic systems incorporate sensors, actuators, microprocessors, control systems, and mathematical modeling into product design. This facilitates features like automation, precision, efficiency, adaptability, reconfigurability, safety, reliability, and productivity. For example, an automated robotic manufacturing system may use servo motors and drives for motion control, various sensors for inspection and feedback, PLCs or microcontrollers for regulated actuation and monitoring, and sophisticated control software for supervision, analysis, and coordination [1].

With rapid advances in artificial intelligence (AI) and machine learning, integrating AI techniques into mechatronic systems has become an active area of innovation and research. AI can endow mechatronics with higher levels of intelligence, autonomy, perception, and decision-making capabilities to create smart, self-optimizing systems. This introductory chapter provides an overview of promising AI approaches for modern mechatronics along with the key benefits.

1.1.1 Overview of Key AI Approaches

Several AI techniques hold promise for advancing mechatronics.

- **Machine Learning:** This allows systems to automatically learn and improve from data without being explicitly programmed. Algorithms identify patterns and build models based on training data. Supervised, unsupervised, and reinforcement learning methods can be used for tasks like prediction, classification, and control optimization in mechatronic systems.

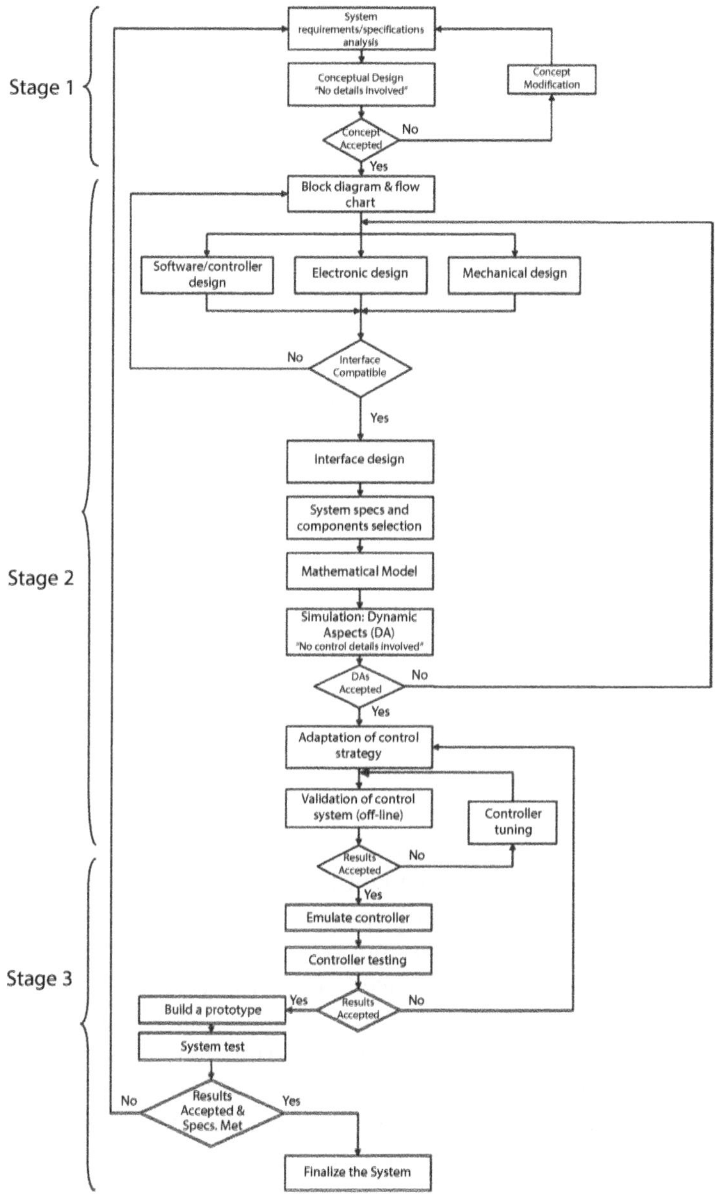

Figure 1.1 Design methodology for a mechatronic system [15].

- **Neural Networks:** Inspired by biological neurons, artificial neural networks consist of interconnected nodes that process and transmit data signals. The strengths of connections are adjusted during training to build robust models. Types like convolutional and recurrent neural nets are ideal for perception, analysis, and decision-making in mechatronic applications.
- **Fuzzy Logic:** This enables representing and reasoning with vague, imprecise information using degrees of truth rather than binary logic. Fuzzy rule–based systems can model complex nonlinear behavior for control and inference in mechatronic systems with uncertainty.

1.1.2 Benefits of Integrating AI in Mechatronic Systems

Integrating appropriate AI techniques enhances mechatronic systems with the following. Figure 1.2 depicts a block diagram illustrating various AI approaches in mechatronics, from machine learning to computer vision and soft computing.

- **Increased Autonomy:** With machine learning, systems can independently analyze data, identify patterns, and make optimal decisions for a given task. This reduces reliance on constant human oversight.
- **Improved Adaptability:** AI approaches like neural networks allow systems to adapt in real-time to changes in the environment, tasks, and operating conditions. This facilitates reconfiguration without reprogramming.
- **Smarter Control:** Advanced control strategies enabled by fuzzy logic and reinforcement learning lead to more efficient, stable, and optimal system performance.
- **Enhanced Perception:** Sensors coupled with AI-based perception and analysis enable mechatronic systems to accurately interpret ambiguous, complex data from the real world.
- **Higher-Level Cognition:** With neural networks and machine learning, mechatronic systems can demonstrate capabilities like scene understanding, forecasting, diagnosis, and strategy optimization.

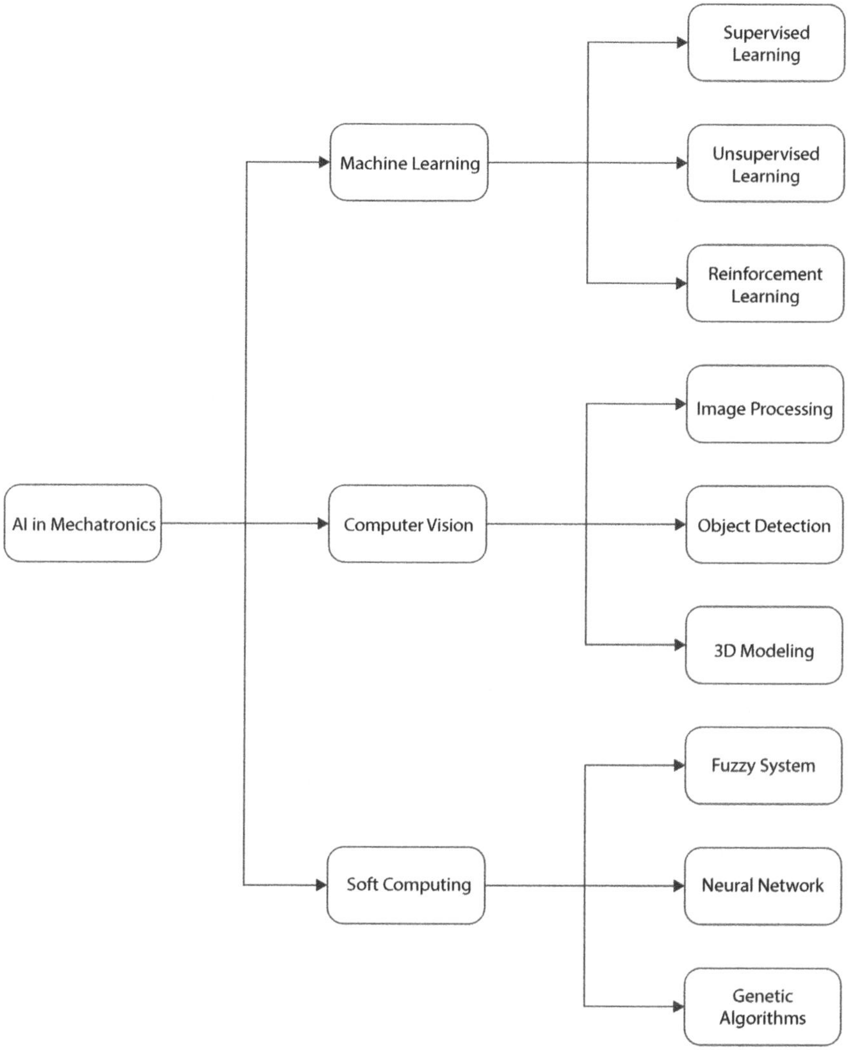

Figure 1.2 Different AI approaches in mechatronics.

1.2 Machine Learning for Mechatronic Systems

A wide repertoire of machine learning techniques can impart different types of learning abilities.

Supervised learning algorithms train predictive models by learning from labeled example inputs and expected outputs. Classification techniques like support vector machines enable tasks like image recognition and fault diagnosis. Regression methods like neural networks can model complex control systems and be used for predictive analytics [2].

Unsupervised learning techniques identify hidden patterns and relationships within unlabeled data. Clustering methods can detect anomalies and group similar data points. Dimensionality reduction can be used for simplifying complex high-dimensional data from diverse mechatronic sensors to enable efficient processing. Association rule mining reveals interesting correlations and connections in large datasets.

Reinforcement learning focuses on optimizing sequential decision-making via trial-and-error interactions with a dynamic environment. The system learns an optimal policy for taking actions to maximize cumulative reward. Dynamic programming, Monte Carlo methods, and temporal difference learning are key techniques applicable to the optimization of mechatronic control policies. Figure 1.3 shows the different parts of machine learning.

Key applications of machine learning in mechatronic systems include adaptive control, predictive maintenance, fault diagnosis, sensor fusion, motion planning, forecasting, multidimensional optimization, natural language interfaces, computer vision for inspection and navigation, and noise filtering. Learning capabilities enhance automation, precision, adaptability, productivity, and reliability [3].

Researchers have successfully demonstrated machine learning use for industrial robots, reusable launch vehicles, autonomous cars, smart prosthetics, and advanced manufacturing systems. As computing power grows exponentially, machine learning will become further ingrained in mechatronic system design.

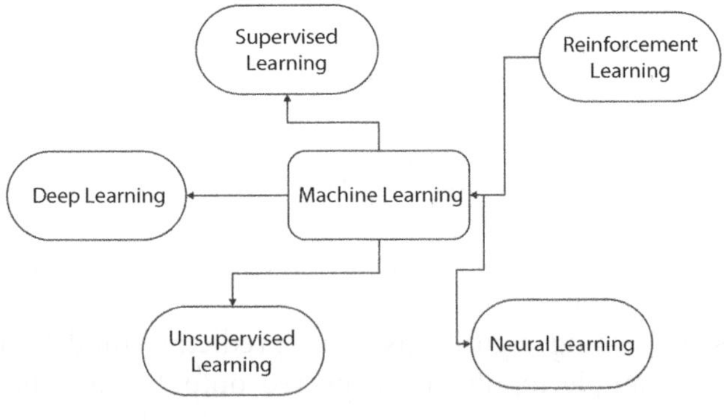

Figure 1.3 Machine learning map.

1.2.1 Supervised, Unsupervised, and Reinforcement Learning Techniques

Supervised Learning: Algorithms like linear regression, logistic regression, decision trees, support vector machines, and neural networks can be trained on labeled input–output data to learn prediction and classification tasks for mechatronic systems like quality evaluation, fault diagnosis, sensor fusion, trajectory modeling, and process optimization. Figure 1.4 illustrates the accuracy of different machine learning models on relevant mechatronics models.

Unsupervised Learning: Algorithms like k-means clustering, hierarchical clustering, principal component analysis, and autoencoders can find hidden patterns and groupings within complex unlabeled mechatronic data. This enables anomaly detection, novel insights, data compression, noise removal, etc., without explicit training.

Reinforcement Learning: Based on dynamic programming techniques like temporal difference and Q-learning, reinforcement learning agents interact with an environment, select actions, and receive rewards or penalties to automatically determine optimal policies, maximizing long-term reward. Applicable to mechatronic control optimization, maintenance scheduling, robot navigation, etc. [14].

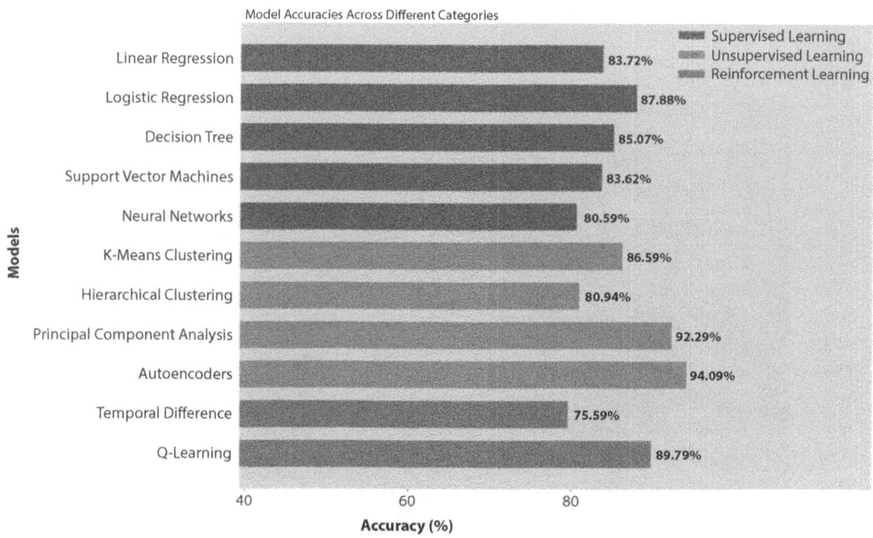

Figure 1.4 Accuracy of different machine learning models on relevant mechatronics models.

1.2.2 Applications in Control, Prediction, Optimization, and Diagnostics

Control: Reinforcement learning for adaptive servo motor tuning, proportional-integral-derivative (PID) controller optimization, and robotic arm manipulation; supervised learning for modeling drone flight dynamics.
Prediction: Classifiers predicting machine component failure from sensor data; regression forecasting manufacturing defects; neural networks for early fault diagnosis.
Optimization: Metaheuristics like genetic algorithms used for design optimization of mechatronic systems; reinforcement learning optimizes production scheduling.
Diagnostics: Unsupervised learning reveals outliers in quality testing data indicative of issues; supervised networks classify x-ray images for automated defect analysis.

1.2.3 Case Studies of Machine Learning in Robotics, Vehicles, and Automation

1. **NASA JPL** trained reinforcement learning models using deep neural networks to allow the Curiosity mars rover to autonomously navigate unknown rocky terrains and learn optimal driving policies. Figure 1.5 exhibits the automated manufacturing unit at Tesla.

Figure 1.5 Tesla manufacturing unit [18].

2. **Boston Dynamics** used deep reinforcement learning along with simulation to train the Atlas humanoid robot to perform complex locomotion and jumping maneuvers with flexibility and adaptability. Figure 1.6 pictures Atlas robot from Boston Scientific.
3. **OpenAI** developed a highly dexterous robotic hand capable of solving Rubik's cubes using deep reinforcement learning and automatic domain randomization to train on simulated environments.
4. **Waymo** applied deep neural networks trained on real-world driving data for automated object detection and situation analysis to enable reliable self-driving technology.
5. **Siemens** leveraged machine learning for predictive maintenance on high-speed rail infrastructure, reducing downtime over 50% and achieving 90% accuracy in predicting failures.
6. **Fanuc** used reinforcement learning to train collaborative robots to perform tasks like surface grinding and cable routing more quickly and precisely than human operators.
7. **Mercedes-Benz** successfully trained convolutional neural networks to interpret complex real-world driving scenes and enable automated vehicle control systems to self-drive up to 130 km/h on highways.
8. **Intel's** self-learning neuromorphic chip Loihi used policy gradient reinforcement learning to quickly learn optimal locomotion gaits for a hexapod robot.
9. **Google-owned DeepMind** created AlphaGo Zero, which used deep reinforcement learning to master the complex game Go from scratch, beating all human expert players.
10. **IBM Watson** leveraged natural language processing and machine learning for cognitive automation applications ranging from chatbots to supply chain optimization to predictive maintenance.

1.3 Computer Vision for Mechatronic Perception

The integration of computer vision techniques is pivotal for enabling mechatronic systems to perceive and understand the visual world around them much like human vision. Computer vision refers to the automated extraction of meaningful semantic information from digital images and videos. It focuses on developing methods and algorithms that can process,

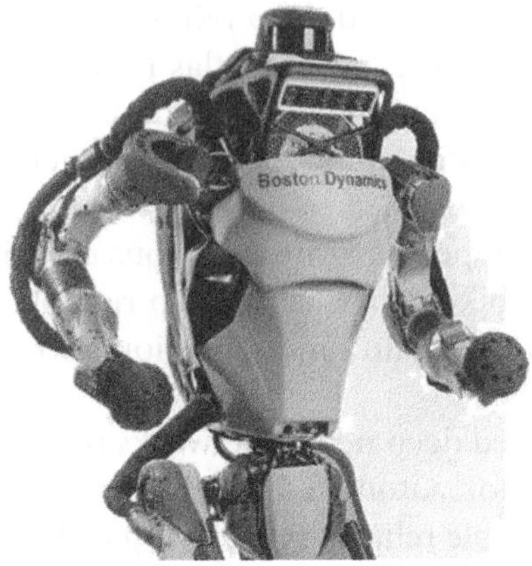

Figure 1.6 Boston dynamics atlas robot [19].

analyze, and comprehend 2D and 3D visual data acquired from cameras and imaging sensors to identify scenes, objects, motions, and environmental context.

By mimicking the mammalian visual cortex, computer vision seeks to empower machines with the exceptional visual perception abilities that humans possess. Sophisticated computer vision capabilities will allow future mechatronic systems to operate autonomously and efficiently in unstructured, real-world environments where uncertainty, variability, and dynamic changes abound.

Key computer vision techniques applicable to mechatronic systems include image classification, object recognition and detection, motion tracking, 3D mapping, simultaneous localization and mapping (SLAM), image segmentation, anomaly detection, image registration, and augmented reality. When integrated thoughtfully, these techniques provide mechatronic systems with crucial perception skills like understanding scenes and spatial relationships, identifying objects of interest, detecting obstacles, tracking moving objects, inspecting environments, determining position and orientation, reconstructing 3D structures, and monitoring changes.

This section provides an overview of promising computer vision approaches for modern mechatronics along with applications in navigation, control, inspection, modeling, and human–machine interfacing. Computer vision unlocks superior situational awareness, adaptable behaviors, precision manipulation, and autonomous decision-making for next-generation intelligent mechatronic systems.

1.3.1 Image Processing and Computer Vision Techniques

A wide range of image processing and computer vision techniques can extract meaningful information from digital images to understand scene contents:

- pre-processing techniques like image filtering, contrast enhancement, thresholding, morphological operations, and segmentation partition images into regions of interest and prepare them for further analysis;
- feature detection algorithms like scale-invariant feature (SIFT), speeded-up robust (SURF), oriented FAST and rotated BRIEF (ORB), features from accelerated segment test (FAST), and Harris corner detector identify distinct image interest points and descriptors that serve as robust visual features for tasks like image matching, registration, and object recognition;
- object classification based on machine learning assigns a class label like "person," "car," or "road" to image regions. Deep convolutional neural networks excel at visual classification due to their hierarchical feature learning capabilities;
- object detection locates and delineates objects in images via bounding boxes using region-proposal methods like region-based convolutional neural networks (R-CNN) and regression models like you only look once (YOLO). This enables detecting multiple object instances;
- semantic segmentation classifies each pixel in an image, providing per-pixel semantic understanding of a scene. fully convolutional networks (FCN) and Mask R-CNN are leading segmentation network architectures; and
- visual SLAM constructs a 3D map of an unknown environment while simultaneously determining the system's location within it based on image data.

1.3.2 Enabling Environmental Perception and Scene Understanding

By applying the above techniques, mechatronic systems can acquire capabilities like Figure 1.7 demonstrates the different parts of computer vision:

- recognizing objects like cars, roads, signs, and people critical for navigation and interaction;

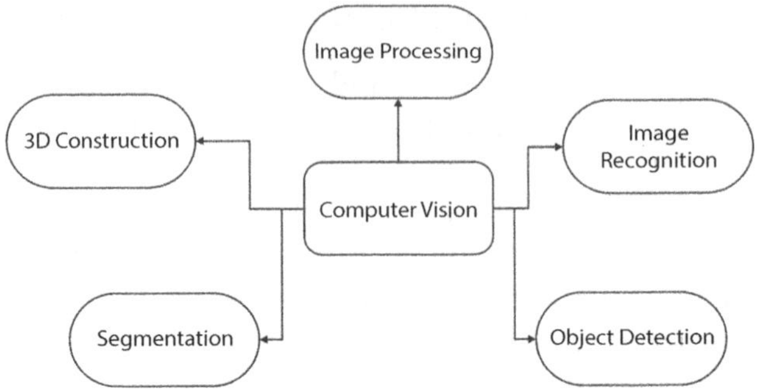

Figure 1.7 Computer vision map.

- detecting product defects and anomalies on manufacturing assembly lines for quality control;
- tracking multiple moving objects in the environment to maintain awareness of surroundings;
- mapping unfamiliar environments as a mobile robot navigates through them;
- determining position precisely within a visually mapped area for accurate localization; and
- reconstructing 3D structures from captured images through photogrammetry.

This enables mechatronic systems to perceive, comprehend and operate seamlessly within unstructured, changing environments.

1.3.3 Vision-Based Control, Inspection, and Monitoring

Key applications of computer vision in mechatronics are as follows:

- visual servo control uses live camera feedback enabling precise manipulation and maneuvering without relying on sensors;
- intelligent robotic inspection for product quality testing, replacing tedious and error-prone human inspection;
- surveillance systems detecting anomalies and incidents in public areas for safety and security;
- self-driving vehicles planning routes and detecting obstacles using computer vision for navigation; and
- augmented reality interfaces project virtual elements into operators' real-world views for teleoperation of mechatronic systems.

In summary, computer vision delivers powerful visual intelligence for automation, manipulation, navigation, and control, unlocking the true potential of mechatronic systems to tackle real-world challenges.

1.4 Soft Computing Techniques

Soft computing refers to a set of computational techniques that provide adaptive, flexible information processing capabilities for handling real-world problems that often involve uncertainty, imprecise inputs, noise, and complexity. Soft computing methods can model and analyze complex phenomena using an approach of approximation and partial truth to achieve robustness, tractability, and low solution cost. In contrast to conventional hard computing based on binary logic, crisp systems, and precise numerical analysis, soft computing exploits the given tolerance for imprecision, uncertainty, and partial truth to achieve efficient, cost-effective solutions.

Soft computing has emerged as an integral part of designing intelligent mechatronic systems that operate in ambiguous, unpredictable real-world environments. It provides nonlinear modeling capabilities, knowledge extraction techniques, and learning algorithms that enable mechatronic systems to make rational decisions, respond adaptively, infer intelligently, describe complex behaviors, and perform robust control under uncertainty and vague objectives [11].

Key soft computing approaches include fuzzy logic, artificial neural networks, evolutionary computation, Bayesian networks, swarm intelligence, and hybrid neuro-fuzzy techniques. This section provides an overview of major soft computing paradigms along with their applications in mechatronic system modeling, control, optimization, decision-making, and knowledge extraction. The strengths and benefits of soft computing in dealing with real-world complexity, uncertainty, and noise make it indispensable for developing next-generation intelligent mechatronic systems. Figure 1.8 exhibits the core parts of soft computing.

1.4.1 Fuzzy Logic Systems for Knowledge Representation and Control

Fuzzy logic systems are based on fuzzy set theory, which represents uncertainty and vagueness using membership functions, with degrees of truth between 0 and 1, instead of binary logic. This allows modeling subjective, linguistic knowledge from domain experts through fuzzy IF-THEN rules using linguistic variables.

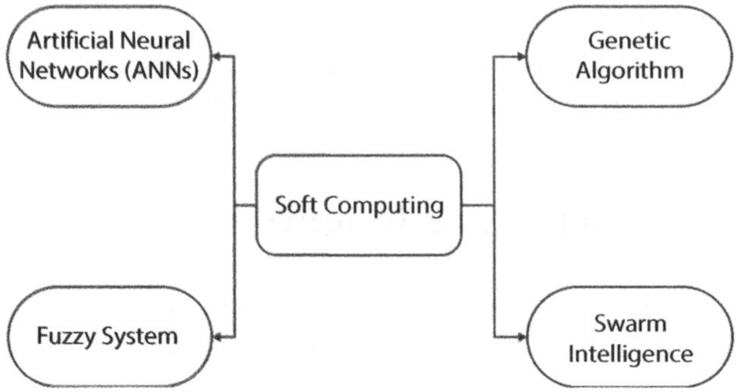

Figure 1.8 Soft computing map.

A fuzzy logic controller's knowledge base comprises these fuzzy rule sets along with membership functions representing fuzzy sets like "high temperature" and "fast speed." Fuzzy inference engines then combine and apply these fuzzy rules to determine optimal control actions. Defuzzification converts the fuzzy outputs back into crisp values.

Key benefits of fuzzy logic control for mechatronics include the following:

- handles nonlinear systems and complex behaviors difficult to model mathematically;
- smooth control outputs instead of abrupt switching;
- incorporates human operator knowledge and expertise; and
- robust to noise and disturbances

Fuzzy logic has been successfully applied for control of mechatronic systems like inverted pendulum balancing, autonomous vehicle steering, subway operation, washing machines, elevators, and camera focusing. Overall, fuzzy logic facilitates knowledge-driven control under uncertainty.

1.4.2 Bio-Inspired Algorithms Like Neural Networks and Genetic Algorithms

Artificial neural networks composed of interconnected neurons can learn from data to solve problems like pattern recognition, classification, prediction, function approximation, and optimal control. With enough layers and nodes, they can represent any nonlinear function. Key benefits include adaptive learning, self-organization, real-time operation, and resilience to noise. Recurrent NNs are effective for modeling dynamic mechatronic systems.

Evolutionary algorithms like genetic algorithms simulate natural selection to evolve optimal solutions. They maintain a population of candidate solutions, evaluate their fitness, and perform crossover and mutation to produce new generations. The fittest solutions survive, whereas the least fit are discarded. The advantages include global search, adaptation and obtaining multiple solutions. They are widely used for multi-objective and constrained optimization in mechatronic design.

Swarm intelligence algorithms like ant colony optimization and particle swarm optimization mimic social swarms for distributed optimization and collective intelligence. Bio-inspired algorithms provide powerful, flexible approaches for learning, modeling, and optimization in mechatronics [12].

1.4.3 Hybrid Intelligent Systems

Integrating different soft computing approaches via ensemble, sequential, or fused architectures combines their individual strengths synergistically. Key examples include the following:

- neuro-fuzzy systems incorporating neural learning into a fuzzy logic framework to tune membership functions and rules; this enhances model interpretability and adaptation;
- genetic algorithms used to optimize fuzzy rule bases and neural network topologies for mechatronic control; this automates development of optimal soft computing models;
- fuzzy logic combined with reinforcement learning for adaptive dynamic programming and optimal mechatronic control under uncertainty;
- neural network assists fuzzy controller by providing predictive system models for model-based control and compensation of uncertainties; and
- ensemble models comprising multiple networks or fuzzy systems provide more robust decisions compared to single models.

Hybridization facilitates developing flexible, self-learning systems with capabilities like explainable decisions, human-like reasoning, and noise tolerance. This makes them highly suited for mechatronic applications requiring intelligence, adaptability, and transparency.

Hybrid intelligent systems amalgamate the strengths of different soft computing techniques in a synergistic manner.

1.5 AI Planning and Decision-Making

AI planning and decision-making techniques empower mechatronic systems to operate autonomously in complex real-world environments by intelligently planning actions and making rational choices. These algorithms enable systems to reason about optimal sequences of actions to achieve specified goals, even with dynamic changes. Planning methods allow the systematic exploration of different action scenarios to synthesize an efficient plan considering constraints and resources.

Sophisticated decision-making capabilities under uncertainty are crucial for mechatronics to handle incomplete knowledge and noisy sensor observations while selecting optimal actions based on deduced probabilities and utility trade-offs. AI planning relies on informed search in decision spaces, whereas decision-making requires probabilistic reasoning and utility maximization.

Key techniques include automated planning algorithms like A*, heuristic search, hierarchical task planning, Markov decision processes (MDPs), partially observable MDPs, decision trees, Bayesian networks, utility theory, fuzzy logic, and reinforcement learning. By effectively applying such techniques, mechatronic systems can exhibit goal-driven behaviors, situational awareness, adaptable plans, and reliable decisions for tasks like manufacturing automation, dynamic navigation, predictive maintenance, disaster rescue, robotic manipulation, and process control under uncertainty [4].

This section covers the foundations of AI planning and decision-making for mechatronics along with sample applications across manufacturing, transportation, healthcare, and human–machine interaction. With planning and decision capabilities, mechatronic systems can operate resiliently and efficiently in uncontrolled, real-world environments.

1.5.1 Automated Planning Algorithms for Sequencing Actions

Automated AI planning algorithms systematically search for and organize sequences of actions to achieve specified goals. Key approaches include the following:

- heuristic search methods like A* that leverage heuristics to guide optimal path finding; this enables efficient navigation planning for mobile robots;
- hierarchical task planning decomposes tasks recursively into sub-tasks; this facilitates complex, multi-step reasoning and scheduling;
- partial order planning determines ordering constraints between actions rather than total sequence; this is suited for dynamic, reactive planning;
- constraint satisfaction formulations using finite domain variables, constraints, and search techniques; this handles resource limits;
- planning graphs like STRIPS graph states and transitions; this allows efficient bidirectional search; and
- model-based approaches use simulators and models to validate and optimize generated plans.

These algorithms empower mechatronic systems to synthesize efficient, collision-free motion plans, optimized production schedules, and dynamic multi-agent coordination strategies using logical planning representations.

Automated planning provides goal-driven autonomy to mechatronic systems for navigating complexity [10].

1.5.2 Decision-Making Under Uncertainty

Key approaches for decision-making under uncertainty include the following:

- Bayesian networks represent conditional dependencies between variables probabilistically for inference under uncertainty. It facilitates diagnostic reasoning.
- Markov decision processes (MDPs) model sequential decision-making problems as state transitions with probabilistic outcomes. Optimal policies maximize expected utility.
- Partially observable MDPs (POMDPs) handle partial observability and hidden states using belief distributions. It enables acting optimally with incomplete knowledge.
- Reinforcement learning explores the environment, takes actions, and maximizes rewards. There is no need for accurate models.

- Fuzzy logic systems can infer decisions from subjective, linguistic rules, accommodating vagueness.
- Utility theory quantifies preferences over outcomes as utilities for maximizing subjective expected utility. It handles risk attitudes.

These techniques allow mechatronic systems to make optimal, risk-aware decisions under unpredictable, stochastic environments and process variability.

It enables applications like predictive maintenance, adaptive control, diagnostic reasoning, and hazard avoidance.

1.5.3 Applications in Navigation, Manufacturing Automation, Etc.

Key applications that benefit from AI planning and decision-making include the following:

- autonomous robot navigation in dynamic environments via replanning algorithms like D* considering moving obstacles;
- multi-agent coordination for robot swarms accomplishing tasks like search and rescue via distributed constraint optimization;
- optimized scheduling and dispatching of tasks on smart manufacturing assembly lines using timed Petri nets;
- early fault prediction and predictive maintenance by reasoning about equipment failure risks using Bayesian networks;
- natural motion planning for dexterous humanoid robots performing complex manipulation tasks using motion primitives;
- adaptive process control under uncertain operating conditions via reinforcement learning agents;
- intelligent warehouse systems optimizing inventory control and logistics decisions under supply–demand variability;
- smart grids balancing power distribution using probabilistic and learning-based predictive models; and
- self-driving vehicles navigating traffic scenarios of varying risk levels using contextual decision-making.

AI-based reasoning, planning, and decision capabilities will enable the next generation of optimized, self-governing mechatronic systems across automation applications [6].

1.6 Natural Language Interaction

Natural language processing (NLP) enables intuitive human–machine interaction in mechatronic systems through the use of speech recognition and natural language understanding. It facilitates bidirectional communication using natural human language, either spoken or written. This allows users to simply talk to or text mechatronic systems as they would to another person to monitor, instruct, collaborate, and acquire information.

Key elements of NLP include speech recognition to transcribe human speech to text, natural language understanding to analyze text and extract meanings, semantic analysis to represent meaning, dialogue systems to manage conversations, and text generation to formulate responses.

Together, these capabilities allow mechatronic systems ranging from service robots to autonomous vehicles to smart appliances to seamlessly communicate with users, understand commands, clarify ambiguities via dialogue, execute tasks accurately, and respond to queries. This reduces the need for complex programming or control interfaces.

This section covers NLP techniques for mechatronics along with applications in assistive robots, manufacturing assistants, self-driving cars, drone fleet control, and augmented reality systems. By enabling natural, contextual language-based interaction, NLP will transform how humans perceive, instruct, and collaborate with mechatronic systems across industrial and consumer applications in the future [5].

1.6.1 Speech Recognition and Natural Language Processing

Speech recognition transcribes human speech audio into text using machine learning models like deep neural networks. Key elements include acoustic modeling to map audio signals to phonetic units and language modeling using linguistic grammar and probabilities. Beam search decoders enable efficient speech to text conversion.

Natural language processing (NLP) analyzes text to extract meanings and semantics. This involves techniques like part-of-speech tagging, syntactic parsing to model sentence structure, semantic analysis to represent meaning, sentiment analysis to gauge emotional tone, dialogue systems to manage conversations, and text generation to formulate responses.

Together, speech recognition and NLP enable conversational interfaces for mechatronic systems. Users can speak naturally, have their speech recognized correctly, and have the system respond intelligently based on language context and meanings. This makes possible user-friendly, voice-controlled mechatronic systems across applications.

1.6.2 Enabling Intuitive Human–Machine Interaction

Natural language interaction powered by speech recognition and NLP techniques enables more intuitive and seamless communication between humans and mechatronic systems. Key benefits include the following:

- Users can speak naturally and contextually instead of memorizing rigid voice commands.
- Systems comprehend commands, instructions, and queries expressed in natural language.
- Conversational interfaces facilitate clarifying ambiguities via dialogue.
- Users need minimal training compared to programming interfaces.
- Mechatronic systems can respond in natural language providing status, seeking clarification, and answering questions.
- Multimodal interaction combines language with gestures and expressions.

This makes mechatronic systems ranging from consumer robots to industrial assistants much easier to interact with. NLP reduces the complexity of human–machine communication to enable versatile, user-friendly control and collaboration [13].

1.6.3 Use Cases in Service Robots, Intelligent Agents, Human–Robot Collaboration

Key applications of natural language interaction in mechatronics include the following:

- service robots answering user queries, following instructions, and assisting with tasks through conversational interfaces;

- intelligent virtual agents accepting voice commands for appliance/device control and providing natural language responses;
- manufacturing cobots understanding context-based instructions from human coworkers to enable flexible shared workspaces;
- autonomous vehicles interacting with passengers via speech dialogue for navigation assistance and requests;
- multimodal drone fleet control combining speech commands with gesture and vision-based instructions;
- augmented reality systems allowing technicians to query manuals hands-free using voice while working on equipment;
- smart home automation with conversational agents for lighting, climate, and entertainment control.

As NLP continues advancing, natural language capabilities will transform human interaction with mechatronic systems across industrial and consumer domains to be highly intuitive, assisting, and collaborative.

1.7 AI in Mechatronic System Design

The integration of AI techniques is transforming mechatronic system design to be more optimized, adaptable, and intelligent right from the initial design phases. AI facilitates simulation-driven design, intelligent prototyping of mechatronic systems, and automated synthesis of optimal system architectures. This empowers engineers to virtually develop increasingly sophisticated mechatronic systems before physical implementation. Table 1.1 demonstrates the technical capabilities, strengths, and limitations of popular AI methods for mechatronics applications.

Key focus areas for AI in mechatronic design include realistic simulation of machine learning models and AI control strategies to validate performance, virtual prototyping tools to develop and test AI-enabled mechatronic systems, and AI-based automated design optimization approaches. By applying bio-inspired algorithms, engineers can automatically generate optimized mechatronic system designs meeting specified objectives.

This section provides an overview of how AI techniques can enhance modeling, prototyping, simulation, optimization, and verification throughout the mechatronic system design process. With AI, engineers can experiment rapidly with diverse intelligent system designs at a fraction of the cost and time compared to physical build-test iterations. AI-driven tools will transform the

Table 1.1 Technical capabilities, strengths, and limitations of popular AI methods for mechatronics applications.

AI method	Technical capabilities	Strengths	Limitations	Example applications
Supervised Learning	Training predictive models from labeled data; classification and regression tasks	Works well when labels are available; can achieve high accuracy	Requires large amounts of labeled data; may overfit to the training data; not suitable for exploratory analysis	Quality evaluation, fault diagnosis, sensor fusion, trajectory modeling, and process optimization
Unsupervised Learning	Identifying patterns in unlabeled data; clustering, dimensionality reduction, and association rule mining	Can work with unlabeled data; useful for anomaly detection and data compression	Results may be difficult to interpret; sensitive to input data; may not perform well on complex tasks	Anomaly detection, data compression, noise removal, and feature extraction
Reinforcement Learning	Optimizing sequential decision-making via trial-and-error interactions; dynamic programming, Monte Carlo methods, and temporal difference learning	Learns from trial and error; can optimize control policies; capable of handling complex, and dynamic environments	Requires a lot of data and time to train; difficulty in specifying a suitable reward function; may struggle with long-term planning	Mechatronic control optimization, maintenance scheduling, and robot navigation

(Continued)

Table 1.1 Technical capabilities, strengths, and limitations of popular AI methods for mechatronics applications. (*Continued*)

AI method	Technical capabilities	Strengths	Limitations	Example applications
Computer Vision	Perception and understanding of visual data; image classification, object recognition and detection, motion tracking, and 3D mapping	Enables environmental perception; enhances situational awareness; can operate in real-time	Requires large amounts of labeled data for training; sensitive to variations in lighting, perspective, and occlusions; high computational requirements	Visual servo control, intelligent robotic inspection, self-driving vehicles, and augmented reality interfaces
Soft Computing Techniques	Adaptive, flexible information processing; fuzzy logic, artificial neural networks, evolutionary computation, Bayesian networks, and swarm intelligence	Can handle uncertainty and noise; efficient, cost-effective solutions; can model and analyze complex phenomena	May require expert knowledge for designing and tuning; performance can be sensitive to parameter settings	Inverted pendulum balancing, autonomous vehicle steering, subway operation, and design optimization

(*Continued*)

Table 1.1 Technical capabilities, strengths, and limitations of popular AI methods for mechatronics applications. (*Continued*)

AI method	Technical capabilities	Strengths	Limitations	Example applications
AI Planning and Decision-Making	Intelligent planning actions and making rational choices; informed search, probabilistic reasoning, utility maximization	Can optimize sequences of actions; can handle incomplete knowledge and noisy sensor observations; can adapt to dynamic changes	Can be computationally intensive; requires a good model of the environment; may struggle with long-term planning	Autonomous robot navigation, optimized scheduling, predictive maintenance, disaster rescue, and robotic manipulation
Natural Language Interaction	Bidirectional communication using natural human language; speech recognition, natural language understanding, and semantic analysis	Enables intuitive human–machine interaction; reduces the need for complex programming or control interfaces; can provide more natural and efficient communication	Requires handling of ambiguity in language; limited by the quality and diversity of training data; difficulty in understanding complex, nuanced human language	Service robots, Intelligent virtual agents, autonomous vehicles, smart home automation, and augmented reality systems

next generation of mechatronic system design to be highly intelligent, resilient, and efficient right from conceptualization.

1.7.1 Simulation of AI-Based Controllers and Behaviors

Realistic simulation of AI components allows testing mechatronic system performance prior to physical implementation:

- Software platforms like MATLAB and Python enable developing and simulating machine learning models for control policies, computer vision, and predictive analytics.
- Physics engines like MuJoCo facilitate simulating dynamics of robots and mechanisms integrated with learned neural network controllers.
- Co-simulation combines control logic simulation with mechanical CAD models for virtual testing.
- Simulated environments like game engines and Gazebo emulate complex real-world conditions to validate performance.
- Cloud simulation leverages parallel computing for large-scale multi-run testing of AI controllers.
- Hardware-in-the-loop testing integrates physical components with simulated environments and control algorithms.

Thorough simulation-based testing uncovers flaws early, minimizing expensive rework. This accelerates realizing high-performing intelligent mechatronic systems.

1.7.2 Tools for Virtual Prototyping of Intelligent Mechatronics

Integrated environments enable developers to prototype and test mechatronic systems with AI capabilities:

- Multiphysics tools like ANSYS, Comsol, and Adams facilitate coupled simulation of mechanical, electrical, and control systems.
- Platforms like MATLAB/Simulink and LabVIEW integrate with physics engines to simulate dynamics with machine learning components.
- Robot operating system (ROS) and Gazebo provide libraries and 3D virtual worlds to prototype intelligent robot architectures.

- Game engines like Unity and Unreal Engine support simulating AI systems in interactive 3D environments.
- Digital twins create comprehensive virtual models of mechatronic systems mirrored to the real-world counterpart.
- Model-based design using graphical modeling languages like Simulink and SysML.

These tools empower designers to develop and optimize intelligent mechatronic systems virtually, reducing the need for costly physical prototypes.

1.7.3 AI-Driven Design Optimization

AI algorithms can automate optimizing mechatronic system design:

- Genetic algorithms evolve optimal system parameters and architectures over generations to meet design goals.
- Neural architecture search employs reinforcement learning to find optimal network topologies for control policies.
- Bayesian optimization constructs probabilistic models to guide sample selection in design space exploration.
- Swarm intelligence methods like ant colony optimization mimic natural collectives for rapid distributed optimization.
- Machine learning surrogate models enable high-speed prediction of design evaluations to guide optimization.
- Knowledge-based engineering infers design patterns from prior solutions to provide recommendations.
- Explainable AI uncovers influence of design factors on objectives to derive insights.

Such AI-driven optimization automates engineer's tasks like design space exploration, optimizing topologies, parameters, and materials selection. This enables developing high-performance, resilient mechatronic systems orders of magnitude faster.

1.8 Challenges and Future Outlook

While the integration of AI techniques offers tremendous potential for making mechatronic systems more intelligent, autonomous, and versatile, there are important challenges that need to be addressed. Achieving robust, safe AI-enabled mechatronic systems requires overcoming limitations around model accuracy, data availability, transparency, generalization capabilities, and computational constraints. Furthermore, crucial considerations around reliability, security, safety, and ethics arise when deploying mission-critical intelligent systems in the real world.

At the same time, rapid advances on both fronts are unlocking exciting opportunities at the intersection of AI and mechatronics. Areas like embedded intelligence, cloud-enabled capabilities, multi-agent coordination, bio-inspired robotics, and AI chips promise to push boundaries of what is possible through this synergistic fusion. Realizing the full potential would require collaborative efforts leveraging the strengths of both fields to manage risks and accelerate innovations.

This concluding section reflects on current limitations, open problems, safety considerations, as well as emerging trends and opportunities shaping the future of intelligent mechatronics powered by AI. Insights from pioneers driving innovations in this domain provide a valuable perspective on the road ahead for this transformative integration.

1.8.1 Current Limitations in Applying AI to Mechatronics

Key challenges and limitations in integrating AI techniques into mechatronic systems include the following:

- Model accuracy: Machine learning models may fail to generalize well or rely on inaccurate simulated data. Lack of interpretability makes debugging difficult.
- Data scarcity: Insufficient quality training data restricts model performance and robustness for many applications; high costs of data acquisition.
- Computational constraints: Embedded mechatronic systems have limited hardware for real-time inference using complex deep learning models.
- Brittleness: AI systems are vulnerable to adversarial attacks, minor distributional shifts, and edge cases outside training data.
- Explainability: Inability to explain reasoning and decisions hinders trust and verification.
- Hybrid techniques required: No single AI approach solves all aspects. True intelligence requires synergistically combining multiple methods.

Overcoming these limitations through advances in both fields remains an active research priority to realize reliable and versatile AI-enabled mechatronics [7].

1.8.2 Safety, Security, and Robustness Considerations

Deploying mission-critical AI-enabled mechatronic systems poses important challenges around safety, security, and robustness:

- rigorous testing across edge cases required to validate reliability and prevent accidents; redundancy and fail-safes are imperative;
- cybersecurity critical for networked mechatronic systems against threats of adversarial attacks and data breaches;
- techniques like adversarial training, sandboxing, and anomaly detection needed to harden AI systems;
- Regulatory approval for autonomous mechatronic systems in applications like transportation and healthcare;
- accountability and traceability essential for investigating incidents and assigning causality; and
- human oversight and supervision necessary to monitor, interrupt, or override autonomous mechatronic systems when required.

Addressing these considerations through a holistic systems approach will be vital for managing risks and enabling widespread adoption of AI-powered mechatronics.

1.8.3 Emerging Trends and Opportunities

Exciting innovation opportunities lie ahead in integrating AI with mechatronics. Figure 1.9 illustrates annual shipments of industrial robots worldwide:

- Embedded AI will bring machine learning to resource-constrained mechatronic devices for on-device intelligence.
- 5G, edge computing, and cloud robotics will enable real-time coordination of multi-agent robotic systems.
- Generative design powered by AI will automate discovering optimal mechatronic system configurations.

- Bio-inspired soft robotics integrating flexible materials, adaptive morphologies, and neurorobotics for advanced versatility and resilience.
- AI accelerators and neuromorphic chips will overcome latency and power constraints for embedded intelligence.
- Digital twin technologies will facilitate simulation-based design of complex cyber-physical mechatronic systems.
- Ethical AI techniques will enhance safety, security, and reliability of autonomous mechatronics.

With synergistic collaboration, cross-pollination of ideas, and co-design between the AI and mechatronics communities, this multidisciplinary integration can fulfill the enormous promise ahead.

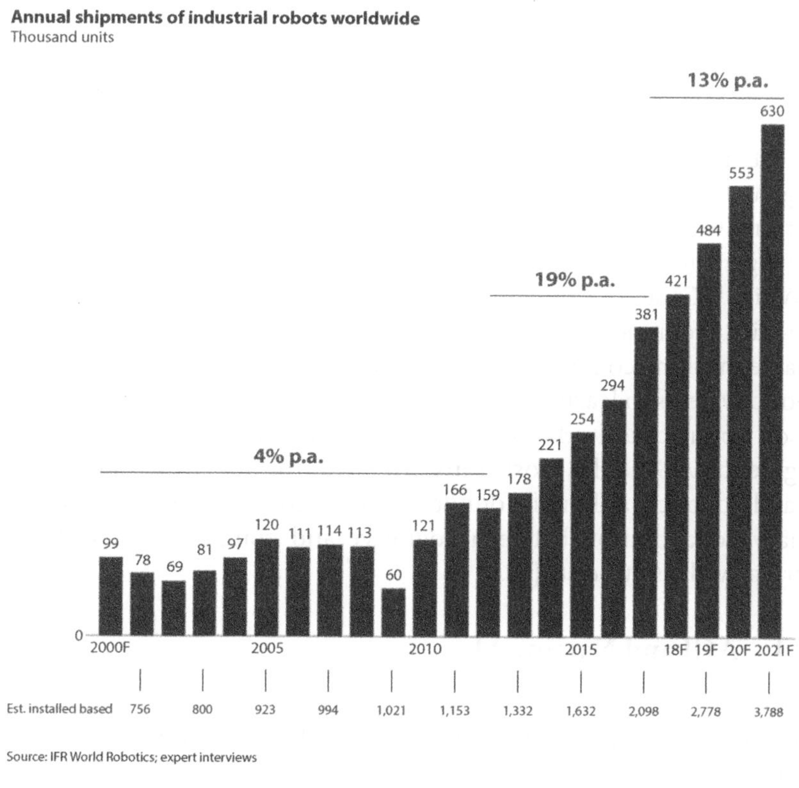

Figure 1.9 Annual shipments of industrial robots worldwide (*thousand units) [16].

1.9 Artificial General Intelligence (AGI)

The realization of artificial general intelligence (AGI) with human-level cognitive versatility and learning capabilities represents the holy grail of AI research today. AGI refers to AI systems that exhibit generalized intelligence and autonomy comparable to humans across a broad range of capabilities. While today's AI demonstrates narrow intelligence limited to specific niche tasks, AGI aims to eventually create AI agents with more open-ended learning aptitude, intuitive reasoning, transfer learning, creative problem solving, and cross-domain competencies.

The achievement of AGI at its fullest envisioned scope represents a transformational grand challenge for the coming decades. However, even incremental progress toward expanding general AI capabilities can profoundly advance the field of mechatronics by introducing increasing autonomy, adaptability, cognitive reasoning, and multi-functionality for intelligent robotic systems. Figure 1.10 demonstrates the timeline graphic showing key milestones in the historic development of artificial general intelligence (AGI).

Advancing AGI requires synergistic integration across capabilities like learning, reasoning, representation, memory, curiosity, social intelligence, embodiment, and marshaling background knowledge. Promising approaches that attempt to combine these competencies in varying degrees include integrated cognitive architectures, developmental robotics, multi-modal neural networks, model-based reinforcement learning, transfer learning, meta-learning, neuro-symbolic integration, and unified theories of intelligence.

While AGI at full human-level versatility remains distant, pragmatic step-wise progress on its key components can already deliver increasingly capable machine cognition. In the context of mechatronics, this can enable robotic systems to learn broader task repertoires, adapt to novel situations, reason about causal relations, interact more naturally, and make ethical judgments. AGI is the most transformative frontier at the intersection of AI and robotics, although foundational progress will require multidisciplinary collaboration between cognitive and computer scientists, engineers, as well as ethicists [8].

1.9.1 AGI and Narrow AI

Narrow AI refers to contemporary AI systems that demonstrate intelligence limited to specific problem domains like computer vision, speech recognition, and game playing. While exhibiting superhuman capabilities in their specialty, narrow AI lacks generalized learning skills, intuitive reasoning, knowledge transfer, and multi-domain adaptability characteristic of human cognition.

In contrast, AGI aims to eventually create AI with more open-ended learning aptitude, abstraction, causal reasoning, creative problem solving, and cross-domain versatility comparable to humans. AGI systems should be able to continuously acquire broad competencies and apply knowledge flexibly [9].

Key differences include the following:

- Narrow AI excels on focused tasks but cannot handle unfamiliar situations beyond training data. AGI aims for adaptable general competence.
- Narrow AI lacks higher-level reasoning, causal models, and explanatory capacity. AGI targets more cognitive capabilities.
- Narrow AI focuses solely on maximizing task performance metrics. AGI would have broader goals, values, and social intelligence.
- Narrow AI needs massive labeled data for training. AGI aims for learning from less data like humans.

Whereas narrow AI has achieved remarkable successes today, AGI remains a grand challenge needing foundational breakthroughs in areas like transfer learning, meta-learning, and robust common sense reasoning.

1.9.2 Historical Development of AGI

Pioneering efforts toward achieving AGI include the following:

- Symbolic reasoning systems like states, operators, and reasoning system (SOAR) and adaptive control of thought-rational (ACT-R) aim to represent knowledge and perform inference and planning through high-level abstractions.
- Integrated cognitive architectures like CLARION combine symbolic and neural techniques for unified cognition.
- Developmental robotics focuses on lifelong accumulative learning of skills grounded in embodiment. Table 1.2 demonstrates the capabilities and approaches of AI vs. AGI.
- Integrated agent systems synthesize capabilities like vision, language, and motor control for general competence.
- Computational models of brain function take inspiration from neural mechanisms for learning and memory.

Table 1.2 AI vs. artificial general intelligence (AGI) capabilities and approaches.

	Narrow AI	Artificial General Intelligence (AGI)
Definition	AI systems that demonstrate intelligence limited to specific problem domains like computer vision, speech recognition, and game playing.	AI with open-ended learning aptitude, abstraction, causal reasoning, creative problem solving, and cross-domain versatility comparable to humans.
Capabilities	Excels on focused tasks but cannot handle unfamiliar situations beyond training data. Lacks higher-level reasoning, causal models, and explanatory capacity; focuses solely on maximizing task performance metrics.	Aims for adaptable general competence; targets more cognitive capabilities; would have broader goals, values, and social intelligence.
Approach	Needs massive labeled data for training.	Aims for learning from less data like humans.
Current Status	Has achieved remarkable successes today.	Remains a grand challenge needing foundational breakthroughs in areas like transfer learning, meta-learning, and robust common sense reasoning.
Understanding of Context	Struggles to understand and adapt to context changes.	Expected to understand and adapt to context changes like humans.
Creativity	Limited creativity, mostly operates within the boundaries of its training data.	Expected to exhibit creativity and problem-solving abilities similar to humans.

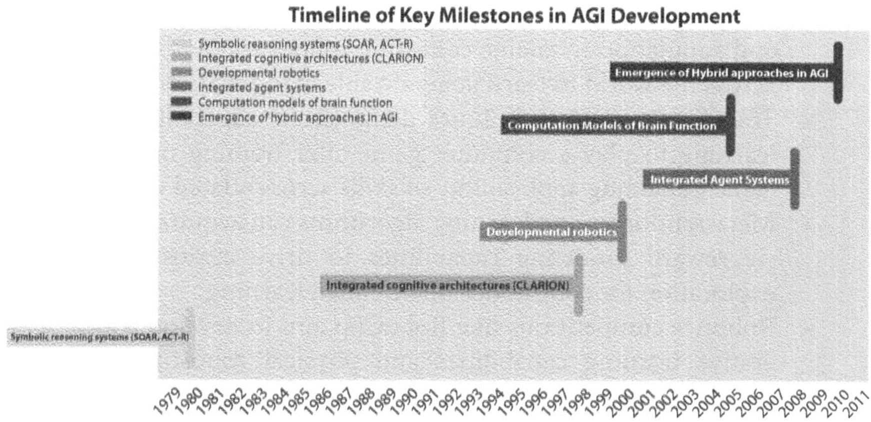

Figure 1.10 Timeline graphic showing key milestones in the historic development of artificial general intelligence (AGI) [20].

- Hybrid approaches leverage synergies between symbolic AI, machine learning, cognitive psychology, and neuroscience.

However, replicating the breadth of human intelligence in machines remains an extremely challenging open problem. Advancing AGI calls for long-term interdisciplinary collaboration spanning across fields like cognitive science, neuroscience, computer science, psychology, linguistics, and philosophy. While major breakthroughs are needed, incremental progress continues driven by advances in algorithms, data, and computing power.

1.9.3 State of AGI in Mechatronics Today

While true full-fledged AGI remains a distant prospect, some promising work in recent years demonstrates early progress toward generalist AI agents and architectures in the context of mechatronics and robotics:

- Virtual test environments like DeepMind Lab and AI Habitat provide customizable 3D platforms to train embodied agents on simulation-based tasks aimed at developing general competencies around navigation, object interactions, and language grounding.
- Multimodal deep neural architectures combining computer vision, NLP, and robotic motor control have been developed that take first steps toward integrating perceptual, linguistic, and motor learning.

- Cloud-based robot learning approaches leverage big data and simulation to enable collecting robot experiences from across the world to learn generalizable manipulation skills.
- Transfer learning techniques demonstrate capabilities like learning to play a computer game after training on other games, enabling applicability of skills across related tasks.
- Meta-reinforcement learning algorithms can acquire intrinsic reward modeling capabilities to drive curiosity and exploration for continuous open-ended learning.
- Robotics competitions like RoboCup aim to develop cooperative teaming capabilities and physical robustness for general-purpose soccer playing humanoid robots.

While these efforts remain limited and narrow compared to human-level AGI, they represent promising directions toward expanding task versatility, adaptability, learning potential, and cognition in mechatronic and robotic systems through a foundational, incremental approach.

1.9.4 Future Possibilities

As foundational AGI capabilities slowly mature in the coming decades through steady research progress, some exciting possibilities could emerge in applying AGI for transformative mechatronics. Figure 1.11 exhibits the future possibilities of robotic chefs cooking food:

- intelligent robotic assistants capable of versatile perception, cognition, and task-learning that can actively help out humans at home, workplaces, and public spaces with a range of cognitive functions;
- adaptive modular self-reconfigurable robotics that can autonomously assemble, morph, and optimize their structure and functionality to suit dynamic task requirements;
- decentralized multi-agent robotic ecosystems that can perceive environments, reason about optimal collaborations, share learning, and coordinate flexibly like ant colonies to accomplish collective goals;
- immersive human–robot collaboration enabled by intuitive bidirectional communication, explanation and transparency of reasoning between AGI agents and people working seamlessly alongside in shared workspaces;

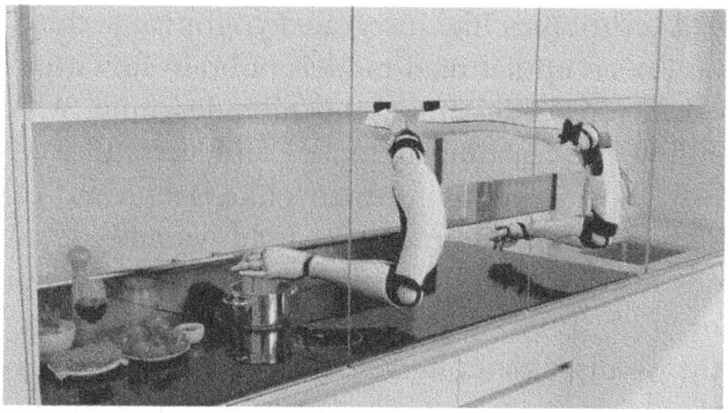

Figure 1.11 Robotic chef cooking food [17].

- transformable fabless production facilities that leverage generalist AGIs for rapid adaptation and reconfiguration to different products, small batches, and changing constraints; and
- open-ended robotic discovery in domains like space exploration, where robots driven by curiosity continuously explore, learn, and innovate beyond their original capabilities over long timeframes.

Realizing such futuristic possibilities will likely require extensive foundational progress in AGI over multiple decades. However, focused incremental efforts today to expand task versatility and learning potential will already unlock increasingly capable machine cognition for the mechatronic systems of tomorrow.

1.10 Conclusion

This concludes our extensive tour exploring the transformative potential of AGI for advancing mechatronics and robotics. We have covered significant ground across key topics in this emerging frontier. Let us now recap the highlights, reflect on insights gained, discuss learnings for students and researchers, and conclude with a motivating message.

Through this chapter, we aimed to provide a comprehensive overview of AGI and its landscape—past, present, and future possibilities. We distinguished narrow AI that excels in specialized tasks, from the more ambitious vision of AGI targeting generalized human-level intelligence. Promising approaches toward AGI integrate synergies across capabilities like learning, reasoning, knowledge representation, memory, curiosity, social intelligence, and embodiment.

We surveyed techniques like integrated cognitive architectures, developmental robotics, computational models of brain function, multi-modal neural networks, transfer learning, and meta-learning, which take steps toward expanding the learning potential and versatility of AI systems, even if narrowly at present. Environments like DeepMind Lab that enable training embodied agents in simulation hold promise. Exciting headline demonstrations also hint at the future potential, even as breakthroughs remain distant.

Importantly, we discussed ongoing AGI research and developments in the context of mechatronics and robotics. While true autonomous cognition remains a grand challenge, noteworthy examples today, like versatile robot learning in simulation, cloud-based shared robot learning, model-based transfer learning, and virtual testing environments, demonstrate initial building blocks. These spark optimism by showing that pragmatic step-wise progress is underway.

For students and researchers, multidisciplinary education spanning AI, cognitive science, neuroscience, ethics, and mechatronic engineering will be invaluable. Hands-on tinkering with real-world robot systems provides important pragmatic grounding. Developing a mindset embracing patience, persistence, and open-mindedness to ideas is vital to drive incremental long-term progress.

In conclusion, the realization of AGI integrated with mechatronics and robotics systems remains a distant yet exciting vision. The path ahead is filled with uncertainties and major challenges necessitating extensive foundational progress. But, the limitless possibilities make it a journey worth undertaking with optimism. We hope this chapter has provided you with informative starting points to begin your own explorations in this landscape. Bon voyage!

1.10.1 Insights Into AGI and Mechatronics Education

Throughout this chapter, let us reflect on key insights regarding education for students and researchers aiming to drive advances in AGI and mechatronics:

- Cultivating strong foundational knowledge in topics like machine learning, neural networks, fuzzy logic, optimization algorithms, control theory, calculus, linear algebra, data structures, and algorithms will provide you with core competencies to build upon.

- Complementing your core engineering foundations with coursework and projects spanning cognitive science, psychology, neuroscience, and linguistics will enable you to learn from complementary perspectives.
- Hands-on experience tinkering with real-world mechatronic systems like robots, drones, or automated vehicles via student projects, internships, or makerspace workshops can provide invaluable pragmatic insights into translating theory into practice.
- Leveraging virtual simulators and sandbox environments for safely testing unconventional AGI ideas for mechatronic systems will help accelerate your rapid prototyping skills.
- Developing prowess across relevant tools like MATLAB, Python, embedded programming, electronics, control systems, Linux, ROS, TensorFlow, Unity, and simulated design tools will equip you with diverse technical skills.
- Participating in collaborative special interest groups, competitions, hackathons, and multidisciplinary team projects around topics like developmental robotics, generative design, and lifelong learning will enable collective thinking.
- Imbibing soft skills like creative thinking, intellectual curiosity, persistence in problem-solving, adaptability, and effective communication are essential to thrive in this emerging frontier.
- The long-term nature of ambitious AGI research necessitates cultivating patience, diligence, and willingness to incrementally build on the works of others who came before you.

We hope that these insights can help guide passionate students and researchers aspiring to advance this landscape through their future endeavors.

1.10.2 Motivating Message

As we reach the concluding pages of this extensive chapter exploring the frontier of AGI and its possibilities for transforming mechatronics, I hope it has ignited excitement and optimism within you about the profound potential impact of this work.

The realization of AGI integrated with robotic systems can herald an era of capable machine assistants providing immense value across industrial, healthcare, domestic, and educational contexts. It can enable resilient adaptive systems and robots that continue learning, innovating, and collaborating with humans to enrich society. We may look back decades from now and see today's incremental advancements as seminal beginnings of that revolution.

The challenges along this journey are undoubtedly enormous. It will require extensive foundational progress spanning decades before we get glimpses of AGI's full transformative power coming to fruition. There will be setbacks, criticisms, and unknowns along the way, requiring perseverance. But, the limitless possibilities make it a trailblazing effort worth undertaking for the benefit of society.

My advice to you is to stay open-minded, intellectually curious, and cognizant of developments across fields but maintain grounded pragmatism. Build your expertise and contribute advancements step by step through whatever approach resonates with you most. Collaborate and share ideas freely. And allow your passion for this topic to drive you through the inevitable ups and downs in the years ahead. You already have the spark within you that brought you to learn this far. Keep that flame alive!

The future awaits those who dare to dream big and tirelessly pursue goals that may take their entire lifetimes or more to fully realize. With dedication and wisdom, we will get there. I wish you the very best!

References

1. Liagkou, V., Stylios, C., Pappa, L., Petunin, A., Challenges and Opportunities in Industry 4.0 for Mechatronics, Artificial Intelligence and Cybernetics. *Electronics*, 10, 16, 2001, 2021. https://doi.org/10.3390/electronics10162001.
2. Harashima, F., Tomizuka, M., Fukuda, T., Mechatronics - "What Is It, Why, and How?" An editorial. *IEEE/ASME Trans. Mechatron.*, 1, 1, 1–4, March 1996, doi: 10.1109/TMECH.1996.7827930.
3. Martinez-de Dios, J.R., Serna, C., Ollero, A., Computer Vision and Robotics Techniques in Fish Farms. *Robotica*, 21, 3, 233–43, 2003, doi: 10.1017/S0263574702004733.
4. Dogmus, Z., Erdem, E., Patoglu, V., ReAct!: An Interactive Educational Tool for AI Planning for Robotics. *IEEE Trans. Educ.*, 58, 1, 15–24, Feb. 2015, doi: 10.1109/TE.2014.2318678.
5. Ellen, F.M., Natural language generation for social robotics: opportunities and challenges. *Phil. Trans. R. Soc.*, 1, 2–5, 2019. B3742018002720180027, http://doi.org/10.1098/rstb.2018.0027.
6. Ott, C. *et al.*, A Humanoid Two-Arm System for Dexterous Manipulation. *2006 6th IEEE-RAS International Conference on Humanoid Robots*, Genova, Italy, pp. 276–283, 2006, doi: 10.1109/ICHR.2006.321397.

7. Ollero, A., Boverie, S., Goodall, R., Sasiadek, J., Erbe, H., Zuehlke, D., Mechatronics, Robotics and Components for Automation and Control. *Annu. Rev. Control*, Elsevier BV, 1, 42–52, January 2006, https://doi.org/10.1016/j.arcontrol.2006.02.002.
8. Pei, J., Deng, L., Song, S. *et al.*, Towards artificial general intelligence with hybrid Tianjic chip architecture. *Nature*, 572, 106–111, 2019, https://doi.org/10.1038/s41586-019-1424-8.
9. Page, J., Bain, M., Mukhlish, F., The Risks of Low Level Narrow Artificial Intelligence. *2018 IEEE International Conference on Intelligence and Safety for Robotics (ISR)*, Shenyang, China, pp. 1–6, 2018, doi: 10.1109/IISR.2018.8535903.
10. Kuru, K. and Yetgin, H., Transformation to Advanced Mechatronics Systems Within New Industrial Revolution: A Novel Framework in Automation of Everything (AoE). *IEEE Access*, 7, 41395–41415, 2019, doi: 10.1109/ACCESS.2019.2907809.
11. Ibrahim, D., An Overview of Soft Computing. *Proc. Comput. Sci.*, Elsevier BV, 1, 35–37, 2016. https://doi.org/10.1016/j.procs.2016.09.366.
12. Berman, S., Halász, Á., Kumar, V., Pratt, S., Algorithms for the Analysis and Synthesis of a Bio-inspired Swarm Robotic System, in: *Swarm Robotics*. SR 2006. Lecture Notes in Computer Science, vol. 4433, E. Şahin, W.M. Spears, A.F.T. Winfield (Eds.), Springer, Berlin, Heidelberg, 2007, https://doi.org/10.1007/978-3-540-71541-2_5.
13. Hoc, J.-M., From human – machine interaction to human – machine cooperation. *Ergonomics*, 43, 7, 833–843, 2000, DOI: 10.1080/001401300409044.
14. Kober, J., Bagnell, J.A., Peters, J., Reinforcement learning in robotics: A survey. *Int. J. Rob. Res.*, 32, 11, 1238–1274, 2013, doi: 10.1177/0278364913495721.
15. Seneff, S. *et al.*, Integration of speech recognition and natural language processing in the MIT VOYAGER system, in: *Acoustics, Speech, and Signal Processing, IEEE International Conference*, Toronto, Ont., Canada, pp. 713–716, 1991, doi: 10.1109/ICASSP.1991.150439.
16. Growth Dynamics in Industrial Robotics, *Industrial robotics Insights into the sector's future growth dynamics*, Technical Report, 1, 13, 2019, [Link: McKinsey; https://shorturl.at/sHMS0].
17. Robotic Chef Will Fill Your Stomach and Haunt Your Dreams, *The World's First Home Robotic Chef Can Cook Over 100 Meals*, 1, 2016 Source [Link: Forbes; https://shorturl.at/koMR7].
18. Elon Musk Defies Lockdown Orders, Reopens Tesla Factory, *Elon Musk Defies Lockdown Orders and Reopens Tesla's Factory*, 1, 2020 Source, [Link: Wired; https://shorturl.at/pAHPV].
19. Atlas - Boston Dynamics, *Boston Dynamics' Atlas Robot: Revolutionizing The Future of Robotics*, 1, 2023, Source, Link: [Boston Dynamics; https://shorturl.at/jlV69].
20. History of Artificial Intelligence, 1, 2006, Fig - 10 – Source, [Link: Wikipedia; https://shorturl.at/qvIO9].

2

Thermodynamics for Mechatronics

Yadav Krishnakumar Rajnath[1]*, Shrikant Tiwari[2] and Virendra Kumar[3]

[1]Department of Industrial Production Engineering Institute of Engineering and Rural Technology (IERT), Prayagraj (Allahabad), Uttar Pradesh, India
[2]Department of Computer Science and Engineering Galgotias University, Greater Noida, Uttar Pradesh, India
[3]Department of Industrial Production Engineering Institute of Engineering and Rural Technology (IERT), Prayagraj (Allahabad), Uttar Pradesh, India

Abstract

Mechatronics stands as a cross-disciplinary domain amalgamating mechanical engineering, electronics, computer science, and control engineering to conceive intelligent systems. Recently, a surge of interest has emerged in fusing thermodynamics into mechatronics to bolster mechatronic systems' efficiency, dependability, and comprehensive functionality. This union holds substantial potential across diverse sectors including robotics, renewable energy, and automotive engineering. Exploration delves into how classical thermodynamics and heat transfer principles can optimize the blueprint, performance, and regulation of mechatronic apparatus. This amalgamation introduces a fresh facet to crafting energy-efficient systems—an imperative in a world grappling with sustainability imperatives. The integration prompts the creation of adaptable systems adept at maneuvering varying thermal states and energy conservation. This entails sophisticated control algorithms mindful of thermal influences, heat dissipation techniques, and energy restoration procedures. Furthermore, the chapter underscores the virtues of integrating phase change materials, thermoelectric generators, and thermodynamic constituents to elevate mechatronic systems via reutilizing wasted heat and augmenting overall efficiency. In tandem, these disciplines open novel avenues for research, pushing past conventional mechatronics boundaries. As technology progresses, the seamless fusion of thermodynamics and mechatronics emerges as a linchpin, charting the trajectory toward intelligent, energy-frugal systems spanning diverse industries.

**Corresponding author*: krishna007.yadav@gmail.com

Kolla Bhanu Prakash, Satish Kumar Peddapelli, Ivan C. K. Tam, Wai Lok Woo and Vishal Jain (eds.) Computational Intelligent Techniques in Mechatronics, (41–82) © 2024 Scrivener Publishing LLC

***Keywords*:** Mechatronics, thermodynamics, efficiency, heat transfer, energy-efficient systems, sustainability, thermoelectric generators, energy-frugal systems

2.1 Introduction

The emergence of mechatronics, an interdisciplinary field blending mechanical engineering, electronics, computer science, and control engineering, has been a result of the evolution of engineering disciplines [16]. This convergence has given rise to intelligent systems spanning consumer electronics to industrial automation. Simultaneously, integrating thermodynamics fundamental to energy conversion and heat transfer into mechatronics has emerged as a compelling frontier.

Thermodynamics, governing energy transformations, is intertwined with mechatronics to enhance the efficiency, reliability, and performance of mechatronic systems. These fusion holds promise in areas like robotics, renewable energy, and automotive engineering [25]. With technology advancing rapidly, the synergy between thermodynamics and mechatronics offers unprecedented opportunities for innovation.

This exploration probes the symbiotic relationship between thermodynamics and mechatronics, exploring how classical thermodynamic concepts optimize mechatronic device design, operation, and control. Adapting thermodynamic principles to mechatronics introduces a new dimension of energy-efficient solutions an imperative in a sustainability-focused world. The chapter dissects the implications of this integration, charting the path to adaptive systems that navigate thermal fluctuations and advance energy optimization. Moreover, it investigates potential benefits by integrating phase change materials, thermoelectric generators, and other thermodynamic elements to enhance mechatronics, accentuating their roles in waste heat utilization and boosting performance [24].

The goal is to highlight the transformative power of merging thermodynamics with mechatronics. This integration not only broadens conventional mechatronic horizons but also forges pathways to innovative solutions that are responsive, efficient, and sustainable. As we delve into this exploration, we embark on a journey through the intricate interaction of thermodynamics and mechatronics, unveiling their multifaceted collaboration's impact on the engineering and technological landscape [23].

2.2 Defining Mechatronics and Its Interdisciplinary Nature

Mechatronics is a cutting-edge field that encompasses the seamless integration of various engineering disciplines, giving rise to intelligent systems that embody the synergy of mechanical engineering, electronics, computer science, and control engineering as shown in Figure 2.1 [16, 26]. At its core, mechatronics transcends the boundaries of individual disciplines, instead forming a cohesive framework where these once-separate fields harmoniously converge. This interdisciplinary nature allows engineers and researchers to design and create innovative solutions that leverage the strengths of each component while addressing complex real-world challenges [7].

In mechatronics, mechanical elements are infused with electronic components and computational intelligence, enabling systems to interact with their environment and adapt to changing conditions. This multidisciplinary approach empowers the creation of highly functional and versatile systems, ranging from advanced manufacturing robots to precision medical devices [27]. By seamlessly integrating mechanics, electronics, and computational expertise, mechatronics yields solutions that exceed the capabilities of each isolated discipline, offering a holistic perspective to tackle the intricate engineering problems of today's technological landscape [15].

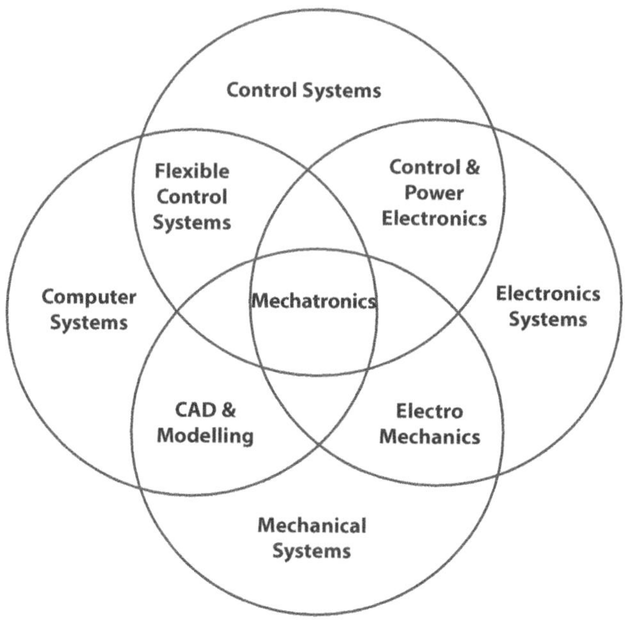

Figure 2.1 Key components of mechatronic systems [26].

2.2.1 The Role of Thermodynamics in Engineering Innovation

Thermodynamic principles form the foundation of energy systems driving human progress and innovation. Their elegance and universal applicability, as acknowledged by C. P. Snow, have inspired minds like Albert Einstein and shaped engineering over centuries. In today's context, these principles serve as a scientific basis for identifying process enhancements. Concepts such as "enthalpy" quantify energy consumption, whereas "exergy" gauges its quality [5, 11, 40]. However, comprehending the second law "entropy" remains a challenge, originating from Clausius' analysis of the Carnot cycle. Beyond engineering, entropy has intrigued various disciplines, including economics, specifically environmental economics. Here, thermodynamic concepts have been harnessed to rethink sustainability, drawing parallels to energy processes. Notably, Nicholas Georgescu-Roegen pioneered this with the "Entropy Law," using entropy to measure economic scarcity [3]. His work fuels the discourse on natural resource scarcity, even though contrasting perspectives arise about the applicability of thermodynamics in understanding these intricate matters.

Practitioners across diverse disciplines, spanning ecology, economics, and engineering, have tapped into the reservoir of thermodynamic concepts, as highlighted in Figure 2.2. A shared concern about resource scarcity and environmental deterioration unites these fields, each driven by

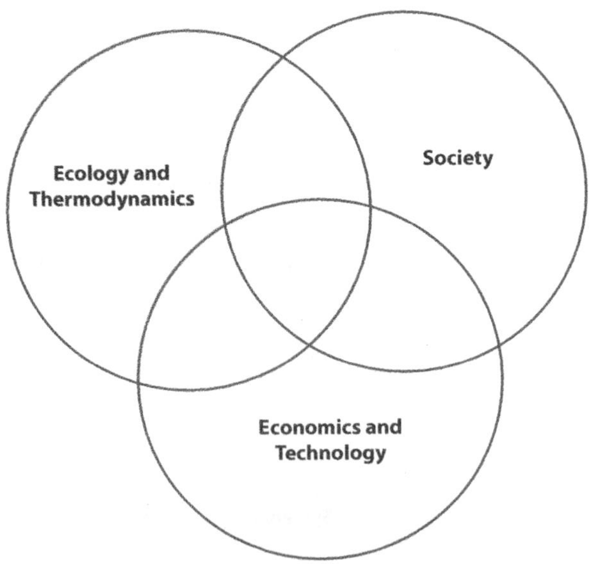

Figure 2.2 Venn diagram representation of "The Three Pillars" of sustainability [32].

a common goal of understanding and mitigating the impacts of human development [32]. Notably, the adverse outcomes of such progress often resonate with the core ideas and analytical methods of thermodynamics, a connection evident in the early work of Mueller at the US Goddard Space Flight Center. Mueller adeptly draws a parallel between resource flows in economic systems and energy, and, implicitly, exergy flows in thermodynamics. This alignment has led to intriguing yet debated notions, such as the analogy between the "technology of man" and heat engines.

These concepts have served as inspiration for thinkers like Parkin and others who perceive thermodynamic principles and laws as potential guiding lights for engineers navigating the labyrinth of environmental sustainability [8]. The exploration of such parallels has sparked discussions about whether these principles, originally rooted in energy transformations, can be harnessed to steer engineering practices toward a more sustainable trajectory. As the call for sustainability echoes across disciplines, the fusion of thermodynamics and environmental concerns unveils novel avenues for research, solutions, and a shared commitment to a more balanced coexistence with our environment [33]. In this way, thermodynamics serves as a driving force behind engineering innovation, shaping the landscape of progress and fostering a greener and more efficient future.

2.2.2 Significance of Integrating Thermodynamics in Mechatronics

The integration of thermodynamics within the realm of mechatronics carries profound significance, reshaping the landscape of engineering innovation. By fusing the principles of thermodynamics with the multifaceted domain of mechatronics, a transformative synergy emerges [34]. This union empowers engineers to design intelligent systems that not only excel in functionality but also optimize energy utilization and promote sustainability.

Mechatronics inherently demands seamless coordination of mechanical, electronic, and computational components. The infusion of thermodynamics adds an extra layer of complexity, enabling engineers to incorporate energy-efficient strategies, manage heat dissipation, and leverage waste heat [2]. This integration holds particular relevance in today's era of heightened environmental consciousness and resource constraints, as it offers a means to curtail energy waste and minimize the ecological footprint of mechatronic systems.

Furthermore, the integration of thermodynamics fosters adaptability in mechatronic devices, enabling them to respond intelligently to varying thermal conditions. This adaptability translates into improved reliability, enhanced performance, and extended life span, all of which are crucial benchmarks for modern engineering solutions. By unifying these domains, engineers are equipped to transcend traditional design boundaries, creating innovative systems that harmonize technological advancement with sustainable practices a testament to the enduring significance of integrating thermodynamics into mechatronics [31].

2.3 Fundamentals of Thermodynamics for Mechatronics

The fundamentals of thermodynamics constitute a cornerstone in the dynamic realm of mechatronics, underpinning the efficiency, performance, and sustainability of intelligent systems [35]. At its core, thermodynamics deals with the intricate interplay of energy, heat transfer, and matter, offering a comprehensive framework for understanding and optimizing engineering processes.

Within mechatronics, the laws of thermodynamics take on an essential role. The first law, often referred to as the law of conservation of energy, governs the conversion of energy from one form to another a principle pivotal in designing energy-efficient mechatronic devices. The second law introduces the concept of entropy, shedding light on the direction of heat flow and the limitations imposed by the inevitable increase in disorder within systems [37].

Heat transfer mechanisms, a focal point in thermodynamics, resonate deeply in mechatronics. Concepts of conduction, convection, and radiation find practical application in managing thermal loads, enabling engineers to devise sophisticated thermal management strategies for enhanced efficiency and durability [6].

Furthermore, thermodynamics introduces the notion of exergy, reflecting the quality of energy within a system. This concept becomes instrumental in gauging the potential for useful work, guiding engineers in optimizing processes to maximize the conversion of energy into desired outputs [4].

As mechatronics inherently involves dynamic systems with intricate energy exchanges, a comprehensive grasp of thermodynamics empowers engineers to unlock innovative solutions. These fundamental principles enable the creation of energy-frugal systems that harmonize functionality, adaptability, and sustainability key objectives in the evolution of modern mechatronics.

2.3.1 Laws of Thermodynamics: Concepts and Implications

The laws of thermodynamics are fundamental principles that govern the behavior of energy and matter in the universe. These laws provide a framework for understanding the physical processes that occur in various systems and have far-reaching implications across various scientific disciplines and everyday life [9]. The four laws are as follows:

- Zeroth Law of Thermodynamics: If two systems are each in thermal equilibrium with a third system, then they are in thermal equilibrium with each other. This law lays the foundation for temperature measurement and the concept of thermal equilibrium.
- First Law of Thermodynamics (Law of Conservation of Energy): Energy cannot be created or destroyed, only converted from one form to another. This law emphasizes the principle of energy conservation, demonstrating that the total energy in a closed system remains constant.
- Second Law of Thermodynamics: The total entropy of an isolated system can only increase over time. Entropy is a measure of the disorder or randomness within a system. This law introduces the concept of irreversibility and the tendency for natural processes to move toward a state of greater disorder.
- Third Law of Thermodynamics: As the temperature of a system approaches absolute zero, the entropy of the system approaches a minimum value. This law highlights the behavior of systems at extremely low temperatures and is related to the unattainability of absolute zero.

Implications of these laws are as follows:

- Energy Conservation: The first law underscores that energy is a finite resource and must be used judiciously to ensure sustainability.
- Efficiency Limits: The second law implies the existence of limits to the efficiency of energy conversion processes, such as heat engines, and the necessity for waste heat generation.

- Entropy and Disorder: The second law focuses on entropy reveals why natural processes tend to move from ordered states to disordered states, underpinning concepts [36] in fields like chemistry and biology.
- Thermodynamic Equilibrium: These laws help explain the concept of equilibrium and the conditions necessary for a system to be in a stable state.
- Cryogenics and Absolute Zero: The third law is critical in understanding the behavior of matter at ultra-low temperatures, with applications in fields like superconductivity and superfluidity [14].
- Ecosystems and Earth Systems: Thermodynamics contributes to understanding the flow of energy and matter in ecosystems and the Earth's climate system, highlighting the interconnectedness of natural processes [10].
- Technological Innovation: The laws guide the design of engines, refrigeration systems, and other technologies that impact daily life, promoting efficiency, and sustainability.

In conclusion, the laws of thermodynamics provide a framework that governs energy and matter in the universe. Their concepts and implications span a wide range of scientific, technological, and philosophical domains, shaping our understanding of the physical world and influencing advancements in various fields.

2.3.2 Heat Transfer Mechanisms and Applications in Mechatronics

Heat transfer mechanisms play a pivotal role in mechatronics, a multidisciplinary field that integrates mechanics, electronics, and computer science to design and create intelligent systems. Understanding and effectively managing heat transfer is crucial for maintaining the performance, reliability, and efficiency of mechatronic devices and systems. There are three main heat transfer mechanisms: conduction, convection, and radiation, each with its own applications in mechatronics as revealed in Figure 2.3 [20].

Conduction:

Conduction is the transfer of heat through direct contact between solids or particles within a substance. In mechatronics, conduction is relevant

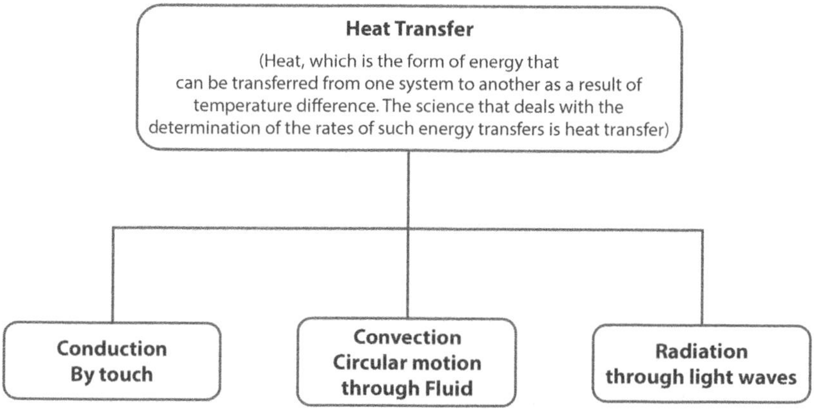

Figure 2.3 Mechanism of heat transfer [20].

in scenarios where components are physically connected or in direct contact with one another. For instance, heat sinks are used to dissipate excess heat generated by electronic components such as microprocessors, ensuring they operate within safe temperature ranges [18]. Thermal conductive materials are employed to efficiently transfer heat between interconnected components, enhancing the overall performance and life span of the system.

Convection:

Convection involves the transfer of heat through the movement of fluids (liquids or gases). In mechatronics, convection is harnessed to manage heat in various ways. Cooling fans, often integrated into electronic systems, help enhance heat dissipation by forcing air over heat sinks or other heat-generating components. Fluid-based cooling systems, like liquid cooling, are employed in high-performance mechatronic systems to efficiently absorb and transport heat away from critical components. This method ensures stable operation even under heavy loads.

Radiation:

Radiation is the transfer of heat through electromagnetic waves, such as infrared radiation. While radiation might be less commonly applied directly in mechatronics, it plays a role in thermal management [18]. Enclosures and housings for mechatronic devices are designed to reflect or absorb radiation, preventing excessive heat buildup and maintaining optimal operating temperatures. Additionally, radiative heat transfer is considered when designing energy-efficient systems that utilize solar panels or thermal imaging sensors.

Applications of heat transfer mechanisms in mechatronics:

- Robotics: In robotic systems, heat management is crucial to prevent overheating of motors, sensors, and control electronics. Proper heat dissipation techniques ensure reliable and accurate robot operation in various environments [21].
- Automotive Systems: Mechatronics is prevalent in modern vehicles, and effective heat transfer mechanisms are used to manage the heat generated by engines, exhaust systems, and electronic components. Cooling systems, heat shields, and thermal insulators contribute to optimal vehicle performance and passenger comfort.
- Consumer Electronics: Heat transfer mechanisms are critical in devices like smartphones, laptops, and gaming consoles. Effective heat dissipation helps prevent thermal throttling and extends the life span of components.
- Industrial Automation: Mechatronic systems used in industrial settings, such as manufacturing and assembly lines, require efficient heat management to ensure reliable operation and minimize downtime.
- Renewable Energy Systems: Heat transfer plays a role in mechatronic systems designed for harnessing renewable energy, such as solar panels and wind turbines. Proper heat management improves energy conversion efficiency and overall system performance.

Heat transfer mechanisms are essential for the design, operation, and longevity of mechatronic systems. By effectively managing heat generated by electronic components and mechanical systems, mechatronics engineers can create more reliable, efficient, and innovative devices that serve a wide range of applications.

2.3.3 Energy Conversion Principles and Efficiency Considerations

Energy conversion principles form the backbone of mechatronics, a multidisciplinary field integrating mechanics, electronics, and computer science.

The application of thermodynamics concepts within mechatronics is pivotal to optimize energy usage and ensure efficient system performance [41].

Mechatronic systems often involve the conversion of one form of energy into another, such as mechanical to electrical or vice versa. These conversions are governed by the laws of thermodynamics, which dictate fundamental principles of energy conservation and efficiency. The first law states that energy cannot be created or destroyed, only converted from one form to another. In mechatronics, this law underscores the need to design systems that utilize energy effectively, minimizing losses [28].

Efficiency considerations are paramount in mechatronic designs to achieve optimal performance. The second law of thermodynamics introduces the concept of entropy and the tendency of systems to move toward increased disorder [30]. Applied to mechatronics, this law highlights the importance of mitigating losses due to heat and friction. To maximize efficiency, engineers employ heat sinks, lubricants, and precision engineering techniques to minimize energy dissipation.

Moreover, the third law of thermodynamics influences mechatronic designs involving extreme temperatures. Cryogenic applications, for instance, leverage the behavior of matter at ultra-low temperatures, where superconductivity and superfluidity emerge. Understanding the relationship between energy conversion and temperature enables the creation of cutting-edge technologies like superconducting magnets in magnetic resonance imaging [12].

In mechatronics, energy conversion occurs within devices such as electric motors, generators, and sensors. These components employ principles of electromagnetism and thermal management to achieve desired outcomes. Optimizing energy conversion efficiency in electric motors, for instance, involves minimizing electrical losses, maximizing magnetic coupling, and utilizing advanced control algorithms.

Efficiency considerations extend to the integration of renewable energy sources within mechatronic systems. Solar panels and wind turbines harness clean energy, but their effectiveness relies on efficient energy conversion and storage mechanisms. Battery technology, an integral part of mechatronics, plays a role in maintaining energy stability and ensuring power availability [38].

In conclusion, energy conversion principles rooted in thermodynamics are paramount in the realm of mechatronics. By aligning designs with thermodynamic laws and focusing on efficiency, mechatronics engineers create systems that not only perform optimally but also contribute to sustainable energy usage. This synergy between energy conversion, thermodynamics, and mechatronics fosters innovation in various industries and shapes the technological landscape of the future [39].

2.4 Enhancing Efficiency in Mechatronics Through Thermodynamics

Enhancing efficiency in mechatronics through the application of thermodynamics is a critical endeavor that drives innovation, sustainability, and optimal performance in complex integrated systems. By harnessing the principles of energy conservation and heat transfer, mechatronics engineers can create more efficient and reliable devices with far-reaching implications.

Thermodynamics principles dictate that energy cannot be created or destroyed, only transformed from one form to another [17]. In mechatronics, this principle guides engineers to design systems that maximize energy utilization, minimizing waste and losses. By optimizing the energy conversion processes within components like motors, actuators, and sensors, engineers can achieve higher overall system efficiency.

Heat transfer mechanisms play a pivotal role in mechatronic designs. Efficient thermal management is essential to prevent overheating, which can lead to component degradation and reduced life span. Incorporating heat sinks, cooling fans, and conductive materials ensures that critical components operate within safe temperature ranges, avoiding efficiency-reducing thermal throttling.

The concept of entropy, as outlined by the second law of thermodynamics, underscores the importance of minimizing losses due to friction and heat dissipation. Precision engineering and friction-reducing coatings enable smoother mechanical operation, enhancing efficiency in mechanisms such as robotic joints and linear actuators.

Thermodynamics also guides the integration of renewable energy sources into mechatronic systems. Solar panels and energy harvesting devices capture ambient energy, but their efficiency depends on how well energy conversion processes align with thermodynamic principles. Advanced control algorithms and energy storage mechanisms ensure continuous and stable power availability.

Efforts to enhance efficiency extend to mechatronic systems employed in industrial automation and manufacturing. Through predictive maintenance techniques based on thermodynamic principles, engineers can identify and address potential efficiency-reducing issues before they escalate.

This minimizes downtime, optimizes energy usage, and enhances overall system reliability.

Innovations like regenerative braking in electric vehicles exemplify the synergy between thermodynamics and mechatronics. By converting kinetic energy back into electrical energy during braking, these systems enhance efficiency and extend battery life. Similarly, the development of micro- and nano-scale mechatronic systems benefits from thermodynamic insights to overcome challenges posed by scale and heat dissipation.

In conclusion, enhancing efficiency in mechatronics through thermodynamics is an integral aspect of modern engineering. By aligning designs with the laws of energy conservation, heat transfer, and entropy, engineers can create mechatronic systems that perform optimally, contribute to sustainability, and drive technological advancements across various industries. This interdisciplinary approach ensures that mechatronics remains at the forefront of innovation, shaping the future of intelligent systems.

2.4.1 Thermodynamics-Driven Design Optimization for Mechatronic Systems

The fusion of mechanical, electronic, and control components in mechatronic systems has revolutionized various industries, from manufacturing to healthcare. However, this integration brings about complex energy interactions that can significantly impact performance, efficiency, and overall system reliability. In response, a novel approach known as "Thermodynamics-Driven Design Optimization" is gaining traction, aiming to leverage thermodynamic principles to enhance the design and operation of mechatronic systems.

Traditionally, mechatronic design has primarily focused on achieving functional objectives, often neglecting the intricate energy exchanges occurring within the system. Thermodynamics offers a fresh perspective by analyzing energy transformations and their inherent limitations [29]. By applying fundamental concepts like the laws of thermodynamics and entropy, engineers can unveil hidden inefficiencies and energy loss mechanisms.

One of the fundamental considerations in this approach is entropy generation. The second law of thermodynamics asserts that all energy transformations result in some degree of entropy increase. Integrating this understanding into mechatronic design prompts engineers to seek pathways that minimize this entropy production, leading to more efficient and sustainable systems.

Central to thermodynamics-driven optimization is the selection of materials and components. Thermodynamic properties of materials, such as thermal conductivity and electrical resistivity, become pivotal in making informed choices. These properties impact heat dissipation, energy transmission, and overall system thermal equilibrium. By selecting materials that align with thermodynamic objectives, engineers can mitigate energy losses and improve system performance [13].

Moreover, thermodynamics aids in devising intelligent energy management strategies. By modeling energy flows and accounting for thermodynamic constraints, sophisticated control algorithms can be developed. These algorithms ensure that mechatronic systems operate in modes that conserve energy, adapt to varying loads, and minimize waste.

Exergy analysis, another thermodynamic concept, provides a metric to assess the quality of energy within a system. This enables engineers to identify energy streams with high potential for performing useful work [1]. By capitalizing on these high-quality energy sources, mechatronic systems can be optimized for maximum efficiency and reduced environmental impact.

However, integrating thermodynamics into mechatronic design is not without challenges. It demands interdisciplinary collaboration among mechanical engineers, electrical engineers, and control theorists. Additionally, the complexity of thermodynamic analysis and the need to account for real-world uncertainties can be daunting. Yet, the benefits are compelling. Thermodynamics-driven design optimization leads to energy-efficient systems, prolonged component lifetimes, and enhanced system robustness.

In conclusion, the marriage of thermodynamics and mechatronics through design optimization signifies a leap forward in engineering innovation. By delving into the energy intricacies of mechatronic systems, engineers can create solutions that not only meet functional requirements but also operate in harmony with the laws of thermodynamics. This approach aligns with the global pursuit of sustainable technologies and positions thermodynamics-driven design optimization as a cornerstone of future mechatronic advancements.

2.4.2 Thermal Management Strategies: Heat Dissipation and Regulation

In the world of electronics and engineering, the efficient management of heat has become a critical consideration. As devices become smaller, more powerful, and densely packed, the generation of heat has intensified, posing significant challenges to their performance, reliability, and longevity. Thermal management strategies, specifically focused on heat dissipation and regulation, have emerged as essential components of modern engineering to ensure optimal operation and prevent overheating-induced failures.

Heat Dissipation Strategies:

- Heat Sinks: Heat sinks are widely used passive cooling solutions that enhance heat dissipation by increasing the surface area exposed to the surrounding air. These metallic structures absorb heat from hot components and transfer it to the air through convection.
- Fans and Blowers: Active cooling solutions, such as fans and blowers, facilitate convective heat transfer by increasing airflow over heat sinks or heat exchangers. They are effective for dissipating heat from confined spaces and preventing localized hotspots.
- Liquid Cooling: Liquid cooling systems employ fluids to carry away heat from components. These systems can offer higher efficiency compared to air cooling, as liquids have better heat-carrying capacity. Liquid cooling is often used in high-performance computing and industrial applications.
- Phase Change Materials: Phase change materials absorb and release heat during phase transitions. They can store and dissipate heat efficiently, stabilizing component temperatures by absorbing excess heat during high-load periods and releasing it during low-load periods.

Heat Regulation Strategies:

- Thermal Interface Materials (TIMs): TIMs enhance the thermal conductivity between heat-generating components and heat sinks, ensuring efficient heat transfer. High-quality TIMs minimize thermal resistance and reduce the risk of overheating.
- Thermal Design and Layout: Proper component placement and arrangement within a system can help distribute heat more evenly and prevent localized hotspots. Adequate spacing and airflow channels are critical considerations in this strategy.

- Thermal Sensors and Feedback Control: Embedded temperature sensors continuously monitor component temperatures. Feedback control systems adjust cooling mechanisms, such as fan speed or voltage, based on real-time temperature data, maintaining safe operating temperatures.
- Dynamic Voltage and Frequency Scaling (DVFS): DVFS adjusts the operating voltage and frequency of processors based on workload. This strategy reduces heat generation during light loads, conserving energy and minimizing heat buildup.
- Adaptive Cooling: Intelligent systems can dynamically adjust cooling solutions based on the system's current thermal state. This ensures that cooling mechanisms are utilized only when necessary, optimizing energy usage.
- Predictive Analytics: Machine learning and data-driven approaches can predict future thermal behavior based on historical data. This enables proactive adjustments to cooling strategies to prevent overheating scenarios.

Effective thermal management strategies are paramount to maintaining device performance, reliability, and life span. Engineers must carefully balance considerations like size, cost, efficiency, and environmental impact when selecting and implementing these strategies. In a technology landscape driven by continuous miniaturization and increasing computational demands, mastering the art of heat dissipation and regulation remains a crucial aspect of engineering design and innovation.

2.4.3 Energy Efficiency Techniques and Heat Recovery in Mechatronics

In the realm of mechatronics, where mechanical and electronic systems converge, optimizing energy usage is paramount for achieving sustainable and cost-effective solutions. Energy efficiency techniques, coupled with innovative heat recovery strategies, play a pivotal role in ensuring optimal performance, reducing waste, and contributing to a greener technological landscape.

- Energy Efficiency Techniques
- Component Selection: Opting for energy-efficient components, such as low-power processors and motors, can

significantly reduce overall energy consumption while maintaining desired functionality.

- Power Management: Implementing intelligent power management schemes, such as sleep modes and power gating, conserves energy by dynamically adjusting component power states based on workload requirements.
- Efficient Actuators: Choosing actuators with high energy conversion efficiency, like brushless DC motors or piezoelectric actuators, minimizes energy losses and improves system responsiveness.
- Regenerative Braking: In mechatronic systems involving motion, regenerative braking converts kinetic energy into electrical energy during deceleration, which can be stored or reused, thus saving energy.
- Variable Speed Drives: Variable frequency drives allow motors to operate at different speeds based on load requirements, preventing unnecessary energy consumption during light loads.
- Energy-Aware Control Algorithms: Developing control algorithms that optimize energy usage without compromising performance is crucial. These algorithms can adaptively adjust system parameters to minimize energy consumption.
- Efficient Sensor Networks: Utilizing energy-efficient sensors and sensor networks that activate only when needed reduces constant power consumption and extends battery life in wireless applications.

Heat Recovery Strategies:

- Thermoelectric Generators: Thermoelectric materials can convert waste heat directly into electrical energy, making them suitable for recovering energy from hot components in mechatronic systems.
- Heat Exchangers: Heat exchangers capture and transfer heat from components to a heat sink or fluid, which can then be used for other purposes, such as space heating or preheating fluids.
- Cogeneration: In industrial mechatronic systems, cogeneration captures waste heat from processes and uses it to produce electricity, simultaneously meeting multiple energy needs.

- Heat Pipes: Heat pipes are efficient heat transfer devices that can transport heat from hot to cold regions within a system. They are particularly useful for dissipating heat from compact, high-power components.
- Energy Storage: Utilizing phase change materials or thermal storage systems can temporarily capture excess heat and release it when needed, reducing the load on active cooling mechanisms.
- Combined Heat and Power (CHP) Systems: In larger mechatronic applications, CHP systems recover waste heat from power generation and utilize it for heating or other industrial processes [22].

Efficient energy usage and heat recovery not only lower operational costs but also contribute to environmental conservation by reducing the overall demand for energy resources. Mechatronic systems that embrace these techniques align with the global push toward sustainable technologies and embody the principles of responsible engineering and innovation.

2.5 Sustainability and Thermodynamics in Mechatronics

The integration of sustainability principles and thermodynamics concepts within the field of mechatronics presents a powerful synergy that can reshape the way engineering solutions are designed, developed, and operated. Mechatronics, which combines mechanical, electronic, and control systems, has the potential to address global sustainability challenges through energy-efficient design, reduced resource consumption, and minimized environmental impact.

Energy Efficiency as a Sustainability Pillar:

Sustainability in mechatronics begins with optimizing energy consumption. Thermodynamics, as the study of energy transformations, provides a framework to analyze and improve the efficiency of energy utilization within mechatronic systems. By applying thermodynamic principles, engineers can identify opportunities to minimize energy losses, enhance energy conversion, and reduce waste heat generation.

Thermodynamic Principles for Sustainability:

- Second Law Efficiency: The second law of thermodynamics highlights the limitations of energy conversion processes. By striving to increase second law efficiency, engineers can create mechatronic systems that make better use of available energy, thereby reducing the overall environmental impact.
- Entropy Reduction: Minimizing entropy generation, a measure of energy degradation, aligns with sustainability goals. Engineers can focus on designing systems that produce less waste heat and operate closer to thermodynamic reversibility.
- Exergy Analysis: Exergy, or available work potential, provides a means to quantify the quality of energy within a system. Integrating exergy analysis allows engineers to identify energy streams that can be more effectively harnessed, leading to increased efficiency.

Sustainable Design Principles:

- Life Cycle Assessment (LCA): Incorporating LCA into mechatronic design enables engineers to evaluate the environmental impact of a system throughout its entire life cycle, from raw material extraction to disposal. This holistic approach helps identify opportunities for reducing resource use and emissions.
- Materials Selection: Thermodynamics principles can guide the selection of materials with favorable properties, such as high thermal conductivity or low energy consumption during fabrication, contributing to more sustainable design choices.
- Renewable Energy Integration: Mechatronic systems can be designed to incorporate renewable energy sources, such as solar panels or kinetic energy harvesters, leveraging thermodynamics to maximize energy capture and utilization.
- Design for Energy Recovery: Integrating heat recovery mechanisms, inspired by thermodynamics, enables the extraction of useful energy from waste heat, reducing the need for additional energy inputs.

Advanced Control and Monitoring:

Thermodynamics-informed control strategies can enhance sustainability by optimizing energy consumption in real time. Feedback control systems that adjust parameters based on thermodynamic principles can minimize energy waste and improve overall system efficiency.

Challenges and Opportunities:

The application of thermodynamics in mechatronics for sustainability is not without challenges. Complexities arise from the multidisciplinary nature of mechatronics and the need to balance competing design objectives. However, these challenges present opportunities for innovation, collaboration, and the development of new methodologies that unite thermodynamics and sustainability in mechatronic systems.

In conclusion, the integration of sustainability principles and thermodynamics in mechatronics has the potential to foster a new era of responsible engineering. By focusing on energy efficiency, waste reduction, and system optimization, engineers can create mechatronic solutions that address global sustainability challenges and contribute to a more environmentally conscious and efficient technological landscape.

2.5.1 Mechatronics as a Catalyst for Sustainable Engineering

In today's rapidly evolving world, the imperative for sustainable engineering solutions has taken center stage. Mechatronics, the interdisciplinary fusion of mechanical, electronic, and control engineering, is emerging as a powerful catalyst for advancing sustainability across various industries. By integrating intelligent design, energy-efficient components, and innovative control strategies, mechatronics has the potential to drive transformative changes that align with the principles of environmental responsibility and resource conservation.

- **Integration of Functionalities:** Mechatronic systems excel at integrating multiple functionalities into a single, unified system. This convergence not only enhances efficiency but also reduces the need for separate components, thereby minimizing resource consumption, manufacturing complexity, and waste generation.
- **Energy Efficiency at the Core:** At the heart of mechatronics lies the pursuit of energy efficiency. By combining

mechanical components with intelligent control systems, mechatronics optimizes energy usage across diverse applications. This results in lower energy consumption, reduced greenhouse gas emissions, and extended operational life spans for products.

- **Compact Design and Footprint:** Mechatronic systems often enable compact and space-efficient designs. This attribute is essential for resource conservation and sustainable urban development, where minimizing the physical footprint of devices, machinery, and equipment is paramount.
- **Smart and Adaptive Control:** The integration of advanced control algorithms within mechatronic systems enables real-time monitoring, adaptive adjustments, and predictive responses. This intelligence leads to optimized energy consumption, reduced wear and tear, and more efficient resource utilization.
- **Renewable Energy Integration:** Mechatronics seamlessly incorporates renewable energy sources like solar panels, wind turbines, and kinetic energy harvesters. By intelligently managing the interplay between these sources and the overall system, mechatronics maximizes energy harvesting and reduces reliance on non-renewable resources.
- **Waste Reduction and Recycling:** Mechatronics supports the development of systems with modular and interchangeable components. This modularity enhances repairability and facilitates the recycling of individual parts, reducing electronic waste and promoting circular economy principles.
- **Precision Manufacturing and Assembly:** Mechatronic systems often demand precision manufacturing and assembly processes. This focus on accuracy minimizes material waste and contributes to the creation of high-quality, longer-lasting products.
- **Innovative Mobility Solutions:** Mechatronics is instrumental in the advancement of sustainable transportation solutions, including electric vehicles, smart traffic management, and autonomous mobility. These innovations contribute to reduced emissions, congestion, and energy consumption.
- **Healthcare and Biomedical Applications:** Mechatronics plays a crucial role in medical devices and equipment. It leads to better patient care through more accurate diagnostics, minimally invasive procedures, and enhanced treatment

efficacy, while also contributing to sustainable healthcare practices.

- **Challenges and Collaboration:** While mechatronics holds immense potential for sustainable engineering, it also presents challenges such as complex integration, potential for electronic waste, and energy-intensive manufacturing processes. Addressing these challenges requires interdisciplinary collaboration, stringent design-for-sustainability principles, and ongoing research and development.

In conclusion, mechatronics is poised to revolutionize sustainable engineering by leveraging its interdisciplinary nature, energy efficiency focus, and intelligent control systems. Through its innovative applications and continuous advancement, mechatronics can drive positive change across industries, facilitating the transition toward a more sustainable and responsible future.

2.5.2 Environmental Benefits of Energy-Efficient Mechatronics

Energy-efficient mechatronics, which combine mechanical engineering, electronics, and control systems to optimize energy consumption, bring forth a multitude of environmental benefits. As society increasingly emphasizes sustainability and conservation, energy-efficient mechatronics play a crucial role in mitigating environmental impacts across various industries. Here are some significant environmental benefits:

- **Reduced Greenhouse Gas Emissions:** Energy-efficient mechatronics lower energy consumption, leading to a direct reduction in greenhouse gas emissions. By optimizing energy usage and minimizing waste, these systems contribute to combating climate change and decreasing the carbon footprint associated with energy production.
- **Resource Conservation:** Mechatronic systems that operate more efficiently require fewer resources to accomplish the same tasks. This translates into reduced demand for raw materials, less manufacturing input, and lowered waste generation. The conservation of resources helps preserve ecosystems, minimize habitat destruction, and maintain biodiversity.
- **Lower Energy Demand:** Energy-efficient mechatronics lessen the overall demand for energy, which eases the strain

on power generation systems. By reducing the need for fossil fuel-based energy sources, these systems contribute to a more sustainable energy mix and alleviate the pressure on non-renewable resources.

- **Extended Product Life spans:** Energy-efficient mechatronics are designed to operate optimally over extended periods. This longevity reduces the frequency of replacements and disposals, minimizing the waste stream and the environmental impact associated with producing, transporting, and disposing of products.
- **Efficient Manufacturing Processes:** Energy-efficient mechatronics often require precision manufacturing, which typically uses fewer resources and generates less waste compared to less refined production methods. This aligns with sustainable manufacturing practices and contributes to reduced industrial environmental impact.
- **Renewable Energy Integration:** Energy-efficient mechatronics are well-suited to integrate renewable energy sources like solar, wind, and hydropower. By effectively utilizing clean energy, these systems promote the adoption of sustainable energy solutions and decrease reliance on fossil fuels.
- **Responsible Resource Management:** Energy-efficient mechatronics encourage responsible resource management throughout a product's life cycle. From design and manufacturing to operation and end-of-life disposal, these systems prioritize efficient energy use and minimize waste, thereby reducing environmental harm.
- **Support for Sustainable Infrastructure:** Energy-efficient mechatronics are a fundamental component of sustainable infrastructure. From smart buildings to efficient transportation systems, they contribute to reduced energy consumption in urban environments, fostering cleaner and greener cities.
- **Global Environmental Responsibility:** As energy-efficient mechatronics become the norm, industries around the world contribute collectively to global environmental responsibility. These systems promote a shift toward cleaner technologies and more sustainable practices on a larger scale.

In conclusion, the environmental benefits of energy-efficient mechatronics are manifold and crucial for creating a more sustainable future.

By optimizing energy consumption, conserving resources, and minimizing waste, these systems play a pivotal role in reducing the environmental impact of human activities while advancing technological progress.

2.5.3 Utilizing Thermodynamics for Sustainable Resource Management

Harnessing the principles of thermodynamics offers a compelling avenue for achieving sustainable resource management on both local and global scales. By applying the laws of energy conservation, heat transfer, and entropy, industries can optimize resource utilization, reduce waste generation, and mitigate environmental impacts.

- Energy-Efficient Processes: Thermodynamics aids in designing processes that maximize energy conversion efficiency, reducing energy consumption and dependence on finite resources.
- Waste Heat Utilization: Thermodynamics-driven waste heat recovery systems capture and repurpose industrial waste heat, minimizing energy waste and enhancing overall efficiency.
- Exergy Analysis: Exergy, a measure of energy quality, guides industries in identifying areas for energy optimization, ensuring more effective resource utilization.
- Renewable Energy Integration: Thermodynamic insights inform the integration of renewable energy sources, promoting sustainable power generation while decreasing reliance on non-renewables.
- Efficient Cooling and Heating: Thermodynamics principles optimize heating, ventilation, and cooling systems, reducing energy consumption and greenhouse gas emissions.
- Closed-Loop Systems: Thermodynamics-based closed-loop systems enable resource recycling, minimizing extraction, waste, and pollution.
- Materials Efficiency: By considering thermodynamic properties, industries can select materials that optimize resource utilization and minimize environmental impact.
- Circular Economy Alignment: Thermodynamics aligns with circular economy principles, guiding the creation of regenerative systems that promote sustainable resource reuse.

- Smart Grid Management: Thermodynamics aids in optimizing energy distribution, load management, and integration of distributed energy sources in smart grids.
- Energy-Efficient Building Design: Thermodynamic principles drive the design of energy-efficient buildings, reducing energy consumption and contributing to sustainable urban development.
- Sustainable Manufacturing: Applying thermodynamics optimizes manufacturing processes, reducing energy use, emissions, and material waste.
- Decentralized Energy Systems: Thermodynamics informs the design of decentralized energy systems, enhancing energy resilience and minimizing transmission losses.
- Life Cycle Assessment (LCA): Thermodynamics enhances LCA accuracy, guiding decisions for more sustainable product design and manufacturing.
- Urban Planning for Efficiency: Thermodynamic insights guide urban planners in designing energy-efficient cities with optimized infrastructure.
- Optimized Resource Allocation: By analyzing energy flows, industries can optimize resource allocation, minimizing waste and ensuring sustainable resource use.
- Eco-Friendly Transportation: Thermodynamics informs vehicle design for improved fuel efficiency and reduced emissions in transportation systems.
- Smart Appliance Development: Incorporating thermodynamics into appliance design results in energy-efficient products that lower energy bills and overall resource demand.
- Water Management Efficiency: Thermodynamics guides the design of energy-efficient water treatment and desalination systems, addressing water scarcity sustainably.
- Environmental Policy Support: Utilizing thermodynamics-based data aids in crafting effective environmental policies that prioritize sustainable resource management.
- Advancing Global Sustainability: Integrating thermodynamics into resource management strategies contributes to a more sustainable and resilient future, aligning with global sustainability goals.

Incorporating thermodynamics principles into resource management strategies empowers industries to make informed decisions that support environmental stewardship and long-term sustainability objectives. By optimizing energy usage, reducing waste, and promoting responsible resource utilization, organizations play a vital role in shaping a more sustainable world.

2.6 Innovative Applications and Future Trends

In the dynamic landscape of technological innovation, mechatronics continues to play a pivotal role in reshaping industries and enhancing our daily lives. This interdisciplinary field, merging mechanical engineering, electronics, and control systems, has given rise to a multitude of innovative applications and holds the promise of exciting future trends that will transform how we interact with the world.

Innovative Applications:

- Assistive Robotics: Mechatronics powers advanced assistive devices that empower individuals with disabilities, enabling them to regain mobility, communicate, and perform everyday tasks independently.
- Advanced Manufacturing: Precision mechatronic systems are revolutionizing manufacturing, from 3D printing with intricate designs to automated assembly lines that optimize efficiency and quality [27].
- Smart Healthcare: Medical robots and wearable devices, driven by mechatronics, are advancing diagnostics, surgeries, and patient care, enhancing precision and minimizing invasiveness.
- Precision Agriculture: Mechatronics is transforming agriculture with autonomous drones, robotic harvesters, and data-driven tools that optimize crop management and resource utilization.
- Autonomous Transportation: Self-driving cars, drones, and delivery robots are showcasing the potential of mechatronics in revolutionizing transportation for efficiency and safety.
- Renewable Energy Systems: Mechatronic solutions, including solar tracking systems and wind turbine control, enhance the efficiency of renewable energy generation and storage.

- Human-Robot Interaction: Collaborative robots, or cobots, work alongside humans in manufacturing, healthcare, and logistics, streamlining operations and ensuring safety.

Future Trends:

- AI and Machine Learning Integration: The fusion of mechatronics with artificial intelligence and machine learning will enable autonomous decision-making, learning, and adaptability.
- Soft Robotics: Flexible, adaptable robots inspired by natural organisms will find applications in delicate tasks and human interaction, minimizing safety concerns.
- Neuromorphic Engineering: Developing systems that mimic the neural architecture of the brain will lead to highly efficient computation and more human-like interactions.
- Bionic Integration: Mechatronics will play a pivotal role in seamlessly integrating bionic limbs and organs into the human body, restoring functionality to those in need.
- Biohybrid Systems: The merger of biological and mechatronic components will lead to innovative solutions that harness the best of both worlds for enhanced performance.
- Energy Harvesting Innovations: Mechatronics will further refine energy harvesting techniques, allowing devices to collect energy from diverse sources to power electronics and sensors.
- Miniaturization and Nanotechnology: Advances in micro and nanomechatronics will open doors to new applications, from medical implants to environmental sensors.
- Sustainable Solutions: Mechatronics will drive the creation of sustainable technologies, from energy-efficient buildings to waste management systems, promoting a greener future.

As we look ahead, mechatronics holds the key to addressing complex challenges, enhancing our quality of life, and shaping a more technologically advanced, interconnected, and sustainable world. The synergy between engineering disciplines, coupled with the relentless pursuit of innovation, will continue to drive mechatronics toward transformative applications and pioneering trends.

2.6.1 Harnessing Waste Heat: Thermoelectric Generators in Mechatronics

In the pursuit of energy efficiency, mechatronics is increasingly turning to innovative solutions like thermoelectric generators (TEGs) to capture and convert waste heat into usable energy. TEGs utilize the Seebeck effect to generate electrical power from temperature differences across a material. In mechatronics, where heat is often a byproduct of various processes, TEGs offer a promising way to transform waste heat into a valuable resource. From automotive systems to industrial machinery, TEGs can enhance energy sustainability by reducing reliance on external power sources and decreasing overall energy consumption.

2.6.2 Embracing Energy-Frugal Systems: Future Trends and Challenges

As technology continues to advance, the demand for energy-efficient and environmentally friendly systems is growing. The concept of energy-frugal systems encompasses the development of technologies that not only perform tasks efficiently but also minimize energy usage and environmental impact. The integration of renewable energy sources, intelligent control algorithms, and advanced materials is shaping the landscape of energy-frugal mechatronics. However, challenges remain, including optimizing the balance between energy efficiency and system complexity, as well as addressing potential trade-offs between efficiency and performance. Energy-frugal systems prioritize efficiency, reduced resource consumption, and minimal environmental impact. As we look to the future, several trends and challenges emerge in the endeavor to embrace energy-frugal systems:

Future Trends:

- Intelligent Automation and AI: AI-driven technologies enable real-time monitoring, predictive maintenance, and optimized control in mechatronic systems. Smart algorithms can adapt system parameters to minimize energy usage while maintaining performance.
- Decentralized Energy Generation: The integration of renewable energy sources like solar panels, wind turbines, and energy storage systems contributes to localized energy production, reducing reliance on traditional power grids.

- Energy Harvesting: Advancements in energy harvesting technologies enable the capture and utilization of ambient energy, such as vibrations, heat differentials, and light, to power low-energy devices and sensors.
- Materials Innovation: Development of energy-efficient materials with enhanced thermal and mechanical properties can contribute to reduced energy losses in mechatronic systems, improving overall efficiency.
- Circular Economy Practices: Implementing circular economy principles, such as recycling, refurbishment, and remanufacturing, can significantly extend the life span of mechatronic devices and reduce the need for new resource extraction.
- Edge Computing: Processing data locally at the edge of the network minimizes data transmission and processing requirements, leading to lower energy consumption in mechatronic systems.

Challenges:

- Complexity vs. Efficiency: Balancing the complexity of energy-frugal systems with their energy efficiency is a challenge. Adding more intelligent components can increase complexity, which might counteract energy savings.
- Integration of Renewable Sources: Incorporating renewable energy sources into mechatronic systems requires careful integration and control strategies to ensure consistent performance and reliability.
- Interdisciplinary Collaboration: Achieving energy-frugality necessitates collaboration among mechanical engineers, electrical engineers, software developers, and sustainability experts, posing challenges in communication and knowledge sharing.
- Data Privacy and Security: Gathering and analyzing real-time data for optimization can raise concerns about data privacy and security. Ensuring the protection of sensitive information is crucial.
- Initial Costs and Investment: Implementing energy-efficient technologies often requires higher initial investments. Overcoming financial barriers and demonstrating long-term cost savings is essential.

- Education and Training: Developing a workforce skilled in designing, operating, and maintaining energy-frugal systems is vital. Education and training programs need to adapt to incorporate these evolving trends.
- Behavioral Change: Encouraging behavioral change among consumers, operators, and engineers to prioritize energy efficiency and conservation can be challenging due to established habits and perceptions.
- Regulatory Frameworks: Developing regulatory frameworks that incentivize energy efficiency and support the adoption of energy-frugal systems is necessary to accelerate the transition toward sustainability.

Embracing energy-frugal systems requires a concerted effort from various stakeholders, including researchers, engineers, policymakers, and consumers. Addressing the challenges and leveraging the trends will lead to the development of innovative solutions that not only minimize energy consumption but also contribute to a more sustainable and resilient future.

2.6.3 Challenges in Implementing Future Trends

While the potential for innovation in mechatronics is immense, several challenges must be addressed. Complex interdisciplinary collaborations are necessary to bridge the gap between mechanical, electronic, and control engineering. The trade-offs between energy efficiency, performance, and cost must also be carefully managed. Moreover, ensuring the security and reliability of interconnected mechatronic systems, particularly in critical applications, presents a significant challenge. Here are some challenges that educators and practitioners might encounter when implementing future trends in thermodynamics for mechatronics:

- Interdisciplinary Knowledge Gap: Mechatronics requires a deep understanding of various fields. Bridging the gap between traditional thermodynamics and modern mechatronics concepts can be challenging. Educators need to design curricula that seamlessly integrate these disciplines, enabling students to grasp thermodynamic principles while applying them to complex mechatronic systems.

- Complex System Dynamics: Mechatronic systems often involve intricate interactions between mechanical components, electronics, and software. Traditional thermodynamics might need to be extended to address dynamic behaviors and transient responses in such systems. Incorporating these complexities while maintaining the clarity of fundamental thermodynamic concepts can be demanding.
- Miniaturization and Microscale Effects: Many mechatronic devices are becoming increasingly miniaturized, leading to microscale thermodynamic phenomena. Teaching and understanding how classical thermodynamic laws apply at such small scales can be a challenge. Exploring nanoscale effects and thermodynamics in emerging fields like nanomechatronics requires novel approaches [19].
- Energy Efficiency and Sustainability: Future trends in mechatronics emphasize energy-efficient designs and sustainable technologies. Integrating these concepts with thermodynamics involves not only understanding energy transfer and conversion but also considering environmental impacts. Educators must incorporate discussions about renewable energy sources, waste heat recovery, and life cycle analysis.
- Multidisciplinary Collaboration: Mechatronics projects often involve collaboration between engineers with diverse backgrounds. Effective communication between mechanical, electrical, and software engineers is crucial for successful implementation. Teaching students how to effectively communicate and collaborate across disciplines while considering thermodynamic principles is a challenge.
- Rapid Technological Advancements: The field of mechatronics evolves rapidly due to technological advancements. Educators must stay updated with the latest developments in both mechatronics and thermodynamics to provide relevant and current instruction. Incorporating emerging technologies such as AI, machine learning, and Industry 4.0 concepts into thermodynamics education adds complexity.
- Laboratory Infrastructure: Hands-on experiences are vital for understanding thermodynamics in mechatronics. However, setting up advanced laboratories with state-of-the-art

equipment for both thermodynamics and mechatronics can be resource-intensive. Overcoming financial and logistical challenges to provide practical experiences is a hurdle.

- Ethical and Societal Considerations: As mechatronics technologies continue to shape society, ethical considerations related to automation, robotics, and AI become more pronounced. Teaching students to integrate ethical perspectives with thermodynamic principles in mechatronics design adds a layer of complexity to the curriculum.

Addressing these challenges requires a balanced approach that combines foundational thermodynamics education with innovative mechatronics concepts. Educators must encourage critical thinking, interdisciplinary collaboration, and adaptability in students to prepare them for the evolving landscape of mechatronics and its intersection with thermodynamics.

2.7 Educational and Professional Implications

The integration of thermodynamics principles in the field of mechatronics has profound implications for both education and professional practice. This synergy between disciplines offers unique opportunities and challenges that impact how engineers are educated and how they contribute to the advancement of technology and sustainability.

Educational Implications:

- Interdisciplinary Learning: Educators need to emphasize the interdisciplinary nature of mechatronics, fostering collaboration between mechanical, electrical, and software engineering. Incorporating thermodynamics concepts helps students appreciate the interconnectedness of these fields.
- Holistic Problem Solving: Integrating thermodynamics into mechatronics curricula encourages students to approach problems holistically. They learn to analyze not only the mechanical and electronic aspects but also the energy implications of their designs, fostering well-rounded engineers.
- Real-World Relevance: Thermodynamics provides a tangible link to real-world applications, reinforcing the practicality of mechatronic systems. Incorporating case studies and

practical projects helps students grasp how thermodynamic principles impact the performance and efficiency of their designs.

- Sustainability Awareness: Including thermodynamics in mechatronics education instills an early understanding of energy efficiency and sustainability. Students learn to prioritize eco-friendly designs, aligning with global efforts toward environmentally conscious engineering.
- Critical Thinking: Thermodynamics introduces complex concepts that encourage critical thinking and problem-solving skills. Students learn to optimize energy usage while maintaining functionality, driving innovation in their designs.

Professional Implications:

- Design Optimization: Engineers equipped with thermodynamics knowledge can optimize mechatronic systems for maximum energy efficiency. This translates to products with longer battery life, reduced operational costs, and minimized environmental impact.
- Innovative Solutions: Professionals who can blend mechatronics and thermodynamics are well-positioned to develop innovative solutions in areas like renewable energy integration, smart manufacturing, and autonomous systems.
- Sustainability Champions: Engineers with a strong foundation in thermodynamics become advocates for sustainable practices in their industries. They drive the adoption of energy-frugal technologies, positively influencing business strategies and societal well-being.
- Systems Thinking: Thermodynamics encourages engineers to view mechatronic systems as holistic entities with interconnected energy flows. This systems thinking approach prevents suboptimal optimizations by considering the broader impact of design choices.
- Cross-Disciplinary Collaboration: Professionals who understand both mechatronics and thermodynamics are effective collaborators. They can bridge communication gaps between different engineering departments, enhancing project efficiency and outcomes.

- Ethical Considerations: The inclusion of thermodynamics highlights the ethical responsibility of engineers to design energy-efficient solutions. Professionals are compelled to consider the environmental and societal implications of their work.
- Continual Learning: Rapid technological advancements require professionals to stay updated. Thermodynamics-educated engineers are more adaptable to changes, enabling them to remain at the forefront of mechatronics innovations.

The integration of thermodynamics principles into mechatronics education and practice transforms engineers into versatile problem solvers who prioritize efficiency, sustainability, and innovation. This integration not only enhances their ability to design cutting-edge systems but also empowers them to contribute positively to global challenges in energy consumption and environmental conservation.

2.7.1 Emphasizing the Importance of Incorporating Thermodynamics Education in Mechatronics Programs

The integration of thermodynamics education within mechatronics programs is a strategic imperative that equips students with a comprehensive skill set and a deeper understanding of energy interactions in complex systems. The marriage of these two disciplines fosters engineers who are not only adept at designing functional mechatronic systems but also mindful of energy efficiency and sustainability. Here is why incorporating thermodynamics education is crucial in mechatronics programs:

- Holistic Understanding: Mechatronics involves a synergistic blend of mechanical, electronic, and software engineering. By incorporating thermodynamics, students gain a holistic understanding of energy transfer, conversion, and utilization, completing the multidisciplinary knowledge required for successful mechatronic design.
- Real-World Relevance: Thermodynamics provides practical relevance by bridging theoretical concepts with real-world applications. Students can grasp how energy efficiency influences the functionality and performance of mechatronic systems, preparing them for industry demands.

- Energy-Frugal Design: Educating students about thermodynamics instills a culture of energy consciousness. Graduates will prioritize designing energy-frugal mechatronic systems, contributing to reduced operational costs and minimized environmental impact.
- Sustainability: Incorporating thermodynamics education aligns with global sustainability goals. Engineers equipped with energy-efficient design skills play a pivotal role in addressing energy consumption challenges and advancing sustainable engineering practices.
- Innovation Potential: A solid understanding of thermodynamics empowers students to think creatively. They can identify innovative solutions that harness energy interactions for novel applications, pushing the boundaries of mechatronics.
- System-Level Thinking: Thermodynamics encourages students to consider mechatronic systems as interconnected networks of energy flows. This system-level thinking helps them optimize the overall energy performance of complex systems.
- Competitive Advantage: Graduates with thermodynamics knowledge possess a unique advantage in the job market. They can offer a broader skill set, making them valuable assets for companies seeking engineers who can tackle multifaceted challenges.
- Global Challenges: As the world grapples with energy crises and environmental concerns, engineers armed with thermodynamics education become agents of change. They are equipped to address pressing global challenges related to energy consumption and climate change.
- Interdisciplinary Collaboration: Engineers well-versed in both mechatronics and thermodynamics can facilitate effective communication between different engineering domains. This collaborative ability enhances teamwork and problem-solving in multidisciplinary projects.
- Ethical Responsibility: Incorporating thermodynamics highlights the ethical duty of engineers to consider the energy impact of their designs. Graduates are equipped to make ethically sound decisions that prioritize sustainability and societal well-being.

Incorporating thermodynamics education in mechatronics programs is an investment in producing engineers who are not only technically proficient but also environmentally conscious and adaptable to the evolving needs of society. By imparting a strong foundation in thermodynamics, educational institutions empower students to be innovative, responsible, and transformative contributors to the engineering landscape.

2.7.2 Encouraging Interdisciplinary Collaboration Among Engineers to Optimize Energy-Frugal Mechatronic Systems

Optimizing energy-frugal mechatronic systems requires a concerted effort from engineers with diverse expertise. Encouraging interdisciplinary collaboration is essential to harness the full potential of various disciplines and achieve innovative solutions that prioritize energy efficiency and sustainability. Here is how to promote such collaboration:

- Integrated Curriculum: Educational institutions should design curricula that emphasize interdisciplinary projects. Collaborative assignments that require mechanical, electrical, and software engineering students to work together on energy-efficient mechatronic designs foster cross-disciplinary understanding and teamwork.
- Joint Workshops and Seminars: Organize workshops and seminars that bring together experts from different fields to discuss the intersection of mechatronics and energy efficiency. These events create a platform for knowledge exchange and spark collaborative ideas.
- Cross-Functional Teams: In industry and research settings, form cross-functional teams comprising engineers from diverse backgrounds. Encourage them to collectively address energy efficiency challenges in mechatronics, pooling their knowledge and skills.
- Problem-Based Learning: Present interdisciplinary challenges related to energy-frugal mechatronic systems to students. Require them to collaborate in solving these problems, integrating their expertise and learning from one another.
- Interdepartmental Projects: Collaborate with other engineering departments to undertake projects that require a

multidisciplinary approach. For instance, combine mechanical engineering with materials science to design energy-efficient components.

- Research Collaborations: Establish research partnerships between different engineering departments or even with non-engineering fields like environmental science. These collaborations can lead to breakthroughs in sustainable mechatronic solutions.
- Joint Research Centers: Create research centers that focus on interdisciplinary research at the intersection of mechatronics and energy efficiency. Such centers can serve as hubs for collaboration, innovation, and knowledge dissemination.
- Cross-Disciplinary Courses: Develop courses that span multiple engineering disciplines and focus on energy-efficient mechatronics. Encourage students to enroll in these courses to broaden their skill set and perspective.
- Shared Resources: Provide access to shared resources like laboratories, software, and equipment. This facilitates interaction among engineers from different backgrounds and encourages them to work together on energy-frugal projects.
- Networking Events: Organize networking events that facilitate interactions between engineers, researchers, and students from various disciplines. These events can spark new collaborative initiatives and foster a sense of community.
- Communication Platforms: Create digital platforms or spaces where engineers can share their expertise, ask questions, and seek collaboration. Online forums, discussion boards, or social media groups can facilitate virtual interdisciplinary interactions.
- Recognition and Rewards: Recognize and reward interdisciplinary collaboration through awards, certificates, or acknowledgments. This motivates engineers to actively engage in cross-disciplinary projects.

By promoting interdisciplinary collaboration, engineers can collectively tackle the multifaceted challenges of designing energy-frugal mechatronic systems. This collaborative approach not only leads to more comprehensive and innovative solutions but also nurtures a culture of teamwork and continuous learning among professionals from various engineering disciplines.

2.7.3 Conclusion: Leveraging Thermodynamics for Energy-Efficient Mechatronic Designs

In the journey toward sustainable engineering solutions, the marriage of thermodynamics and mechatronics emerges as a formidable force. This chapter has explored the intricate relationship between these disciplines and highlighted the significance of incorporating thermodynamics principles in mechatronic systems. As we conclude, let us recap the key takeaways and reinforce the importance of utilizing thermodynamics for energy-efficient and sustainable mechatronic designs.

Key Takeaways:

- Foundation of Efficiency: Thermodynamics forms the bedrock of energy interactions, guiding the transfer, conversion, and dissipation of energy in mechatronic systems. It provides engineers with the tools to analyze, optimize, and design systems that minimize energy losses and maximize performance.
- Holistic Approach: The integration of thermodynamics widens the lens through which mechatronic systems are viewed. Engineers gain a comprehensive understanding of how mechanical, electronic, and software components interact, considering both functional and energy-related aspects.
- Sustainability Imperative: In an era marked by environmental concerns, incorporating thermodynamics is not just an option but a responsibility. Engineers equipped with thermodynamics knowledge contribute to the creation of energy-frugal designs that reduce resource consumption and environmental impact.
- Innovation and Adaptation: Thermodynamics fosters innovation by encouraging engineers to explore unconventional methods of energy conversion, harvesting, and utilization. This creativity drives the development of novel mechatronic solutions that address emerging challenges.
- Interdisciplinary Synergy: The integration of thermodynamics into mechatronics bridges gaps between different engineering domains. It paves the way for interdisciplinary collaboration, resulting in systems that seamlessly balance mechanical, electronic, and energy considerations.

Reinforcing the Significance:

The significance of utilizing thermodynamics for energy-efficient and sustainable mechatronic designs cannot be overstated. By embracing thermodynamics principles, engineers become architects of change in an energy-conscious world:

- They design mechatronic systems that excel not only in functionality but also in energy performance, prolonging battery life, and reducing operational costs.
- They forge the path toward a greener future by optimizing renewable energy integration and minimizing waste heat.
- They embody ethical engineering practices, prioritizing sustainability and acting as stewards of the environment.
- They respond effectively to global energy challenges, aligning their skills with the needs of society and industry.

In closing, the integration of thermodynamics in mechatronics is not merely an enhancement—it is an evolution. It is a shift toward a more conscientious and responsible approach to engineering. As the journey continues, let us remember that every line of code, every mechanical component, and every electronic circuit can be optimized to make a significant difference. Utilizing thermodynamics, we pave the way for energy-efficient mechatronic designs that illuminate a brighter, more sustainable tomorrow.

References

1. Achinas, S. and Euverink, G.J.W., Rambling facets of manure-based biogas production in Europe: A briefing. *Renew. Sustain. Energy Rev.*, 119, 109566, 2020.
2. Andrady, A.L. (Ed.), *Plastics and the Environment*, John Wiley & Sons, United States, 2003.
3. Beinhocker, E.D., *The origin of wealth: Evolution, complexity, and the radical remaking of economics*, Harvard Business Press, United States, 2006.
4. Bejan, A., *Advanced engineering thermodynamics*, John Wiley & Sons, United States, 2016.
5. Bronowski, J., *Science and human values*, Faber & Faber, United Kingdom, 2011.
6. Bronzino, J.D., *Biomedical engineering handbook*, vol. 2, CRC Press, United States, 1999.

7. Brundiers, K. and Wiek, A., Do we teach what we preach? An international comparison of problem-and project-based learning courses in sustainability. *Sustainability*, 5, 4, 1725–1746, 2013.
8. Buckland, P.D., The unsuspected teachers: Environmental identity and sustainability education in the Anthropocene, The Pennsylvania State University, United States, 2015.
9. Daly, H.E. and Farley, J., *Ecological economics: principles and applications*, Island Press, United States, 2011.
10. De Groot, R.S., Wilson, M.A., Boumans, R.M., A typology for the classification, description and valuation of ecosystem functions, goods and services. *Ecol. Econ.*, 41, 3, 393–408, 2002.
11. Ferguson, E.S., *Engineering and the Mind's Eye*, MIT Press, United States, 1994.
12. Filippini, J.P., A search for wimp dark matter using the first five-tower run of the cryogenic dark matter search, University of California, Berkeley, 2008.
13. Gao, M., Zhu, L., Peh, C.K., Ho, G.W., Solar absorber material and system designs for photothermal water vaporization towards clean water and energy production. *Energy Environ. Sci.*, 12, 3, 841–864, 2019.
14. Glampedakis, K., Andersson, N., Samuelsson, L., Magnetohydrodynamics of superfluid and superconducting neutron star cores. *Mon. Not. R. Astron. Soc.*, 410, 2, 805–829, 2011.
15. Gunes, V., Peter, S., Givargis, T., Vahid, F., A survey on concepts, applications, and challenges in cyber-physical systems. *KSII Trans. Internet Inf. Syst.*, 8, 12, 4242–4268, 2014.
16. Habib, M.K., Mechatronics-A unifying interdisciplinary and intelligent engineering science paradigm. *IEEE Ind. Electron. Mag.*, 1, 2, 12–24, 2007.
17. Horowitz, J.M. and Esposito, M., Thermodynamics with continuous information flow. *Phys. Rev. X*, 4, 3, 031015, 2014.
18. Howell, J.R., Mengüç, M.P., Daun, K., Siegel, R., *Thermal radiation heat transfer*, CRC press, United States, 2020.
19. Hsu, T.R., *MEMS and microsystems: design, manufacture, and nanoscale engineering*, John Wiley & Sons, United States, 2008.
20. Jani, J.M., Leary, M., Subic, A., Gibson, M.A., A review of shape memory alloy research, applications and opportunities. *Mater. Des. (1980-2015)*, 56, 1078–1113, 2014.
21. Jouhara, H., Khordehgah, N., Almahmoud, S., Delpech, B., Chauhan, A., Tassou, S.A., Waste heat recovery technologies and applications. *Therm. Sci. Eng. Prog.*, 6, 268–289, 2018.
22. Kolarevic, B. and Parlac, V. (Eds.), *Building dynamics: exploring architecture of change*, Routledge, United Kingdom, 2015.
23. Komolafe, A., Zaghari, B., Torah, R., Weddell, A.S., Khanbareh, H., Tsikriteas, Z.M., Beeby, S., E-textile technology review–from materials to application. *IEEE Access*, 9, 97152–97179, 2021.

24. Lamnabhi-Lagarrigue, F., Annaswamy, A., Engell, S., Isaksson, A., Khargonekar, P., Murray, R.M., Van den Hof, P., Systems & control for the future of humanity, research agenda: Current and future roles, impact and grand challenges. *Annu. Rev. Control*, 43, 1–64, 2017.
25. Marculescu, D., Marculescu, R., Zamora, N.H., Stanley-Marbell, P., Khosla, P.K., Park, S., Nakad, Z., Electronic textiles: A platform for pervasive computing. *Proc. IEEE*, 91, 12, 1995–2018, 2003.
26. Nof, S.Y., Morel, G., Monostori, L., Molina, A., Filip, F., From plant and logistics control to multi-enterprise collaboration. *Annu. Rev. Control*, 30, 1, 55–68, 2006.
27. Roslan, M.F., Hannan, M.A., Ker, P.J., Uddin, M.N., Microgrid control methods toward achieving sustainable energy management. *Appl. Energy*, 240, 583–607, 2019.
28. Ruiz-Mirazo, K., Briones, C., de la Escosura, A., Prebiotic systems chemistry: new perspectives for the origins of life. *Chem. Rev.*, 114, 1, 285–366, 2014.
29. Schneider, E.D. and Kay, J.J., Life as a manifestation of the second law of thermodynamics. *Math. Comput. Modell.*, 19, 6-8, 25–48, 1994.
30. Scott, J.C., *Seeing like a state: How certain schemes to improve the human condition have failed*, yale university Press, United States, 2020.
31. Sine, W.D. and David, R.J., Environmental jolts, institutional change, and the creation of entrepreneurial opportunity in the US electric power industry. *Res. Policy*, 32, 2, 185–207, 2003.
32. Smith, B.H. and Weintraub, E.R., *Complexities: social studies of knowledge practices*, Duke University Press, United States, 2002.
33. Taha, W.M., Taha, A.E.M., Thunberg, J., *Cyber-Physical Systems: A Model-Based Approach*, p. 187, Springer Nature, Germany, 2021.
34. Trianni, V., *Evolutionary swarm robotics: evolving self-organisingbehaviours in groups of autonomous robots*, vol. 108, Springer, Germany, 2008.
35. Vedral, V., *Decoding reality: the universe as quantum information*, Oxford University Press, Germany, 2018.
36. Von Bertalanffy, L., The meaning of general system theory, in: *General system theory: Foundations, development, applications*, vol. 30, p. 53, 1973.
37. Wang, H., Jasim, A., Chen, X., Energy harvesting technologies in roadway and bridge for different applications–A comprehensive review. *Appl. Energy*, 212, 1083–1094, 2018.
38. Weaver, P., Jansen, L., Van Grootveld, G., Van Spiegel, E., Vergragt, P., *Sustainable technology development*, Routledge, United Kingdom, 2017.
39. Wilson, E.O., *Consilience: The unity of knowledge*, vol. 31, Vintage, United States, 1999.
40. Zhou, L.Y., Fu, J., He, Y., A review of 3D printing technologies for soft polymer materials. *Adv. Funct. Mater.*, 30, 28, 2000187, 2020.

3

Role of Data Acquisition, Sensors, and Actuators in Mechatronics Industry

Harpreet Kaur Channi

Department of Electrical Engineering, Chandigarh University, Mohali, Punjab, India

Abstract
Mechatronics creates intelligent and automated systems using mechanical, electrical, computer, and control engineering. Data, sensors, and actuators make mechatronic systems smart and functional. Data collection starts with environment or system data. It digitalizes analog signals like temperature, pressure, and position. These data guide control algorithms. Mechatronics relies on sensors to detect environmental changes. These signals provide the system with real-time feedback, enabling it to adjust. Cameras, proximity sensors, accelerometers, and encoders make mechatronic plans more flexible. Actuators change the system based on processing unit control signals. These devices convert electrical impulses into mechanical motion or specialized energy. Actuators power robotics, industrial automation, and precision control. Data acquisition, sensors, and actuators often collaborate. Autonomous automobiles perceive their environment through LiDAR, cameras, and radar. Data control motors and brakes to drive the automobile safely. Industrial mechatronic systems monitor temperature, pressure, and vibration. Actuators control process variables to improve production efficiency and quality. Mechatronics needs sensors, actuators, and data. Their seamless integration allows intelligent decision-making, precise control, and automation, creating complex systems that alter industries and improve lives. Technology will advance mechatronics using these components. Industry 4.0 and mechatronics may change sustainability. Mechatronics provides cyber-physical systems and intelligent automation. Sensors, data, and artificial intelligence–driven algorithms optimize industrial operations and save resources. Robotics and actuators save time and energy. This combination improves logistics, transportation, healthcare,

Email: harpreetchanni@yahoo.in

Kolla Bhanu Prakash, Satish Kumar Peddapelli, Ivan C. K. Tam, Wai Lok Woo and Vishal Jain (eds.) Computational Intelligent Techniques in Mechatronics, (83–108) © 2024 Scrivener Publishing LLC

energy management, and manufacturing sustainability. Smart manufacturing in Industry 4.0 mechatronics aids global sustainability.

***Keywords*:** Mechatronics, data, sensors, actuators, Industry 4.0, sustainability, robotics

3.1 Introduction

The mechatronics sector combines mechanical engineering with electronics, computer science, and control engineering to develop innovative goods and infrastructure. At the center of this discipline lie data collection, sensors, and actuators, which play critical roles in these complex systems' flawless working and management. Accurate analysis and management are made possible using data acquisition systems, which gather and digitize raw data. In contrast, sensors act as the mechatronic equivalent of the human senses, picking up on and recording data about things like temperature, pressure, and motion. They are essential inputs that let the system react to and adapt to its surroundings. A mechatronic system interacting with its environment relies on actuators, critical effectors that translate electrical impulses into physical actions. Together, these parts represent the backbone of the mechatronics industry, paving the way for creating intelligent and adaptive systems in fields as diverse as advanced manufacturing, robotics, the automobile industry, and healthcare [1]. To design, build, and run intelligent and automated systems, mechatronics is an interdisciplinary discipline that draws on mechanical engineering, electronics, computer science, and control engineering, among others. In other words, the term "mechatronics" is a combination of the words "mechanics" and "electronics." Mechatronics aims to improve mechanical systems' efficiency, accuracy, and automation by integrating them with electronic sensors, actuators, and control algorithms. Robotics, factory automation, home appliances, car systems, healthcare equipment, and other fields use mechatronic methods. Mechatronics is the study and use of scenarios in which mechanical components provide the necessary physical structure and mechanical motion. Data and information are gathered from the system's surroundings or internal operations through electronics and sensors [2]. A precise and regulated action is carried out using this information via actuators, which convert electrical impulses into mechanical motion or other types of energy. The "brain" of a mechatronic system is often a microcontroller or microprocessor, which executes complex control algorithms and makes intelligent decisions to enhance the system's operation. Mechatronic systems can adapt, react, and interact with their environment in real time due to mechanical, electrical, and computational aspects. As a whole, mechatronics is a dynamic, forward-thinking discipline that is reshaping several sectors and helping to progress automation and intelligent systems [3].

Mechatronics relies heavily on data gathering, sensors, and actuators to create automation and control systems. Analog-to-digital converters (ADCs) in data collection systems sample and measure a wide range of physical qualities, providing a pathway for converting physical phenomena into digital signals for further processing and analysis by the control systems. To precisely monitor and alter system parameters in real time, the control unit relies on data from sensors, which may range from a primary temperature sensor to more complex vision and motion sensors. Electric motors, hydraulic cylinders, and pneumatic valves are all actuators that may carry out predetermined actions and interact with their surroundings in response to input signals [4]. The mechatronics industry has revolutionized the creation of intelligent systems through the holistic integration of these components, resulting in state-of-the-art robotics, automated manufacturing processes, and cutting-edge technologies across various fields. Mechatronic systems rely heavily on data gathering, sensors, and actuators to perform their functions and exhibit intelligence. The data acquisition process is the backbone of any information processing system because it takes analog information from its surroundings or internal operations and turns it into digital data. Sensing physical changes and environmental variables in real time is the sensors' job. This input is crucial for making educated decisions and exercising precise control, allowing mechatronic systems to dynamically adapt and react to changing circumstances [5]. However, actuators carry out operations in response to the processed data by translating electrical signals into mechanical motion or other types of energy. Applications ranging from robots and industrial automation to medical equipment and consumer electronics may all benefit from this ability's capacity to automate, precisely regulate, and efficiently operate. Mechatronic systems are crucial to developing cutting-edge technologies because they enable the execution of complicated tasks, the optimization of processes, and the revolution of numerous sectors via the coordinated use of data collection, sensors, and actuators, as shown in Figure 3.1 [6].

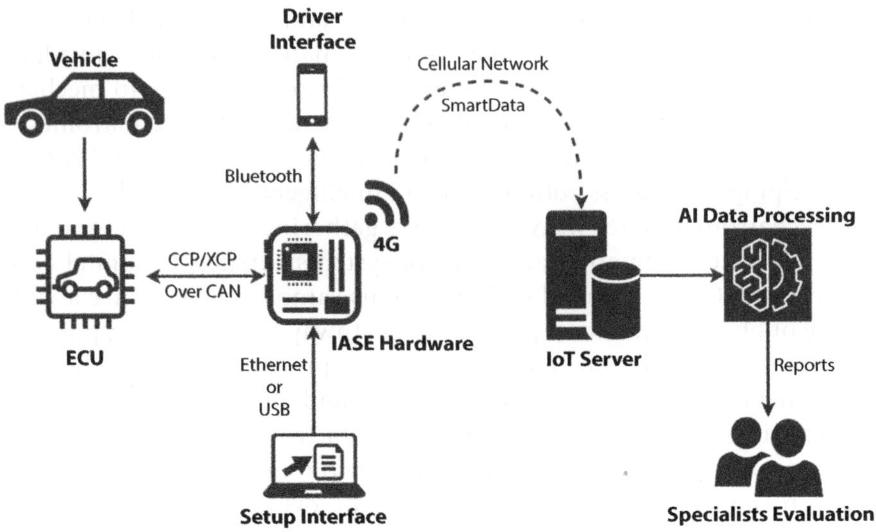

Figure 3.1 Data flow overview in mechatronic industry [6] (CC BY. 4.0).

3.2 Literature Survey

Aleksic (2019) discussed that, in the Internet of Things (IoT), a massive number of sensors, actuators, and other equipment for data acquisition and processing will be interconnected, utilizing an omnipresent communication network able to support heterogeneous transmission technologies and applications efficiently, as the data acquisition and manipulation capabilities are becoming the main competencies in the industries [7]. Turan *et al.* (2019) discussed using visible light communications within the industrial IoT paradigm. Such a promising role of data analytics has been so far without a proper, quantitative inspection of the cost-improvement trade-off nor has the process of acquiring sensors and extracting valuable information from their captured data formalized in a series of methodological steps [8]. Para *et al.* (2019) introduced the Analyze, Sense, Preprocess, Predict, Implement, and Deploy methodology. This iterative decision workflow spans from the acquisition of sensing equipment to the quantitative assessment of the contribution of their captured data to enhance the production step under focus. The seamless integration of industrial robotic arms with server computers, sensors, and actuators can revolutionize how automated non-destructive testing (NDT) is performed and conceived [9]. Mineo *et al.* (2020) presented a toolbox with highly efficient algorithms and software functions developed to be used through high-level programming language platforms (for example, MATLAB, LabVIEW, and Python) and integrated within low-level language (for example, C# and C++) applications [10]. Franklin (2020) described the overall design to build a cost-effective cyber-physical system (CPS) with rich communication options. Integrated data acquisition services are used in real-time applications and autonomous decision-making processes [11]. Fitriani *et al.* (2021) studied internet-based remote settings and data acquisition for fuel cells. A remote setting and DAQ for a fuel cell stack are presented. Google Apps Script was used to manage data communication with Arduino Yun, and the data were stored on Google Drive. The electro-mechanical actuators (EMAs) play an essential role in the new-generation aircraft, making the fault diagnosis of EMA a hot topic in the industry [12]. To tackle the challenge, Zhang *et al.* (2021) presented an innovative algorithm that combines a hybrid-spatial and temporal attention–based gated recurrent unit with seasonal-trend decomposition procedures based on loess to predict multiple time-series data for more failure information. Industrial automation traditionally relies on supervisory control and data acquisition, and remote I/O is an essential element of this infrastructure [13]. Pontarolli *et al.* (2022) described the development of an open-source and low-cost distributed input/output as a service for industrial automation [14].

3.3 Fundamentals of Data Acquisition

Data acquisition in mechatronics is the systematic and planned gathering of information from various sources, such as sensors and instruments. Mechatronic systems rely on this technique, which entails digitizing analog signals representing physical quantities like temperature, pressure, location, or voltage [15]. Mechatronics relies on data capture to offer real-time details about the status of a system, its surroundings, or a process under observation. The mechatronic system can react appropriately and carry out its intended activities with the help of these data, which acts as vital input for control algorithms and decision-making processes. The system can learn to adjust to new conditions and improve its performance because it always gathers new data about the world [16]. The primary goals and advantages of data collecting in mechatronics are shown in Figure 3.2. Fundamentals of data acquisition comprise the critical processes involved in acquiring and transforming real-world events into digital data for analysis and control in mechatronic systems. Using sensors and transducers that convert physical factors like temperature, pressure, and motion into electrical impulses is crucial to this procedure [17]. These signals are

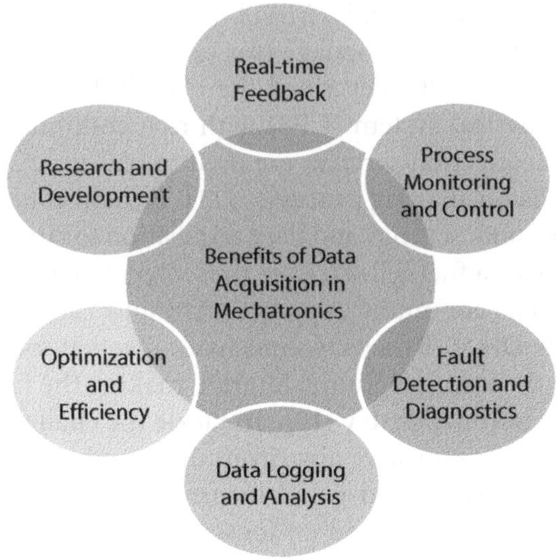

Figure 3.2 Benefits of data acquisition in mechatronics.

first subjected to signal conditioning to ensure the highest precision and dependability during the analog-to-digital conversion (ADC). Accurate timing and synchronization are essential for protecting the quality of the data gathered, and sound data processing and storage procedures allow for the efficient use of the obtained data in system analysis and control. A firm understanding of these basic concepts is required to build and execute the reliable data-gathering systems that are the backbone of modern mechatronic applications [18]. The first stage is choosing and installing sensors and transducers that can detect and convert different physical factors into electrical signals. Signal conditioning methods are subsequently performed to enhance these analog signals, including amplification, filtering, and isolation, assuring their compatibility with the following digital conversion process. To facilitate processing and analysis by the computer system, ADCs play a crucial role by converting continuous analog signals into discrete digital values.

Furthermore, accurate timing and synchronization procedures are necessary to ensure the correctness and dependability of the gathered data by ensuring coherence between the data collection process and the overall system functioning. To effectively handle and use the collected data for real-time monitoring, control, and decision-making within the mechatronics framework, specialized software and hardware generally support effective data processing and storage techniques. The successful integration and operation of complex mechatronic applications across various industrial sectors rely on engineers and practitioners having a firm grasp of these fundamental characteristics [19].

3.3.1 Types of Data Acquisition Systems

In the field of mechatronics, a few different kinds of data acquisition systems are used to gather and analyze data from other types of sensors and instruments. These systems are not all the same; they range in capability, complexity, and applications. The following is a list of typical data collection system mechatronics used [20].

- **PC-Based Data Acquisition Systems:** A PC controls and manages data gathering in these systems. They commonly use USB, Ethernet, or PCI interfaces to link external data-gathering gear to the PC. Through specialized software, PC-based systems enable flexibility, integration, and significant data processing.
- **Standalone Data Loggers:** Standalone data loggers capture and store data without a computer. Remote or portable applications that need continuous data recording employ them. These loggers may include built-in memory or storage media to store and analyze data.
- **Programmable Logic Controller (PLC)–Based Data Acquisition Systems:** PLCs are frequently utilized in industrial automation. Some PLCs may directly communicate with sensors and actuators through data-collecting modules. Manufacturing and process sectors may use these systems for real-time control and monitoring.
- **Data Acquisition Modules and Cards:** Compact data-collecting modules and cards may be incorporated into larger systems or embedded into electrical devices. They enable ADC, digital input/output, and data collection through USB, PCI, or Ethernet interfaces.
- **Distributed Data Acquisition Systems:** Distributed data acquisition systems link data collection modules to a central controller or network. This method is employed in large-scale applications with several sensors and actuators scattered over a considerable region.
- **Wireless Data Acquisition Systems:** Wireless data acquisition systems provide data between sensors, actuators, and

the CPU via Wi-Fi, Bluetooth, or Zigbee. These systems are flexible, easy to install, and ideal for situations where wired connections are impossible.

- **Data Acquisition Software:** Data acquisition systems need specialized software for data management, configuration, and analysis in addition to the hardware components. Users can configure data-collecting settings, see data in real time, and process and analyze the collected information with the help of this program [21, 22].

3.3.2 Analog-to-Digital Conversion

Different data-collecting systems are used depending on the needs of the mechanical system. The number of sensors, the necessary data rate, the required degree of data processing, and the ambient circumstances in which the system works are all important considerations. The Conversion of analog signals to digital values, known as analog-to-digital conversion (ADC), is a crucial step in the field of mechatronics. Temperature, pressure, position, and voltage sensors are just a few examples of mechatronic system sensors that provide analog signals that reflect physical values [23]. Numerous electrical gadgets, such as microcontrollers and digital processors, rely on digital signals, making the ADC process crucial. Thanks to ADC, the system can process, analyze, and store data

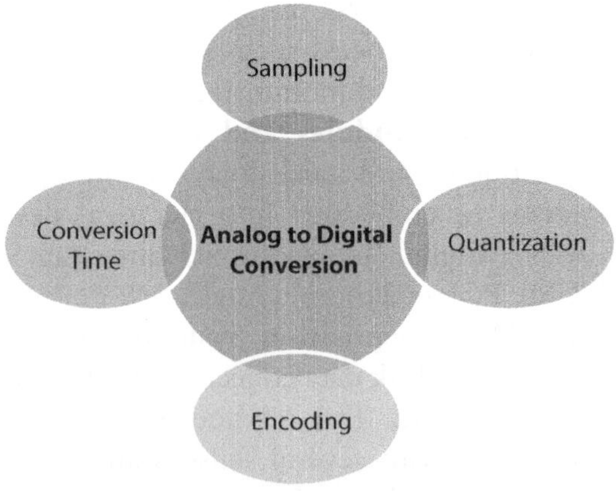

Figure 3.3 Analog-to-digital conversion.

using digital electronics and algorithms. Figure 3.3 shows a high-level diagram of the ADC procedure used in mechatronics. Because sensors often provide continuous analog signals, ADCs find widespread use in mechatronic systems. Once the signals are digital, they may be used in control algorithms, data analysis, decision-making, and inter-electronic device connection by microcontrollers or digital processors. The reliability and performance of mechatronic systems that depend on accurate measurement and control of physical quantities are directly tied to the ADC's precision and resolution [24].

3.3.3 Sampling and Signal Conditioning

Mechatronics relies heavily on sampling and signal conditioning to handle and analyze analog sensor data. Accurate and dependable data collection and control are impossible without these steps in mechatronic systems. Sampling is the practice of taking measurements of analog signals at regular intervals. Voltage, temperature, and pressure are continuous analog signals with unlimited resolution and timing precision [25]. However, for digital systems to handle and evaluate these signals, they must first be transformed into discrete values at predetermined intervals. Snapshots of the continuous signal are taken at regular intervals determined by the sampling frequency or sample rate. The sample rate is carefully considered to prevent aliasing and provide accurate signal reconstruction. The Nyquist-Shannon sampling theorem states that, to avoid data loss, the sampling frequency must be double the maximum frequency contained in the analog signal. To prepare the raw analog signal for further processing, digitalization, and control, signal conditioning is performed. This process includes methods for cleaning up the signal, boosting its quality, and making it compatible with the data collection system's input specifications [26]. The following are examples of standard signal conditioning methods used in mechatronics. The precision, detail, and dependability of data collection, control, and decision-making are all directly impacted by the quality of the sampling and signal conditioning used in mechatronics. Mechatronic systems, used in fields as diverse as industrial automation, robotics, and medical devices, benefit from accurate control thanks to carefully sampled and conditioned signals [27].

3.3.4 Sensors in Mechatronics

In mechatronics, sensors play a crucial role as the bridge between the natural environment and the computer systems used for controlling and making decisions. Mechatronic systems rely on these sensors to detect and measure a wide range of physical properties and then translate those readings into electrical signals for further processing [28]. Sensors are crucial in giving real-time feedback to achieve adaptive and intelligent behavior in mechatronic applications. Some typical sensor types used in mechatronics are shown in Figure 3.4.

These are just a handful of the many different types of sensors used in mechatronics. There are many different kinds of sensors, and they all help mechatronic systems perform better and become more versatile and intelligent [23]. Sensors are crucial to the success of current technological advances because they provide real-time data on the status of the system and the environment, allowing mechatronic systems to carry out complicated tasks, automate processes, and intelligently adjust to changing circumstances.

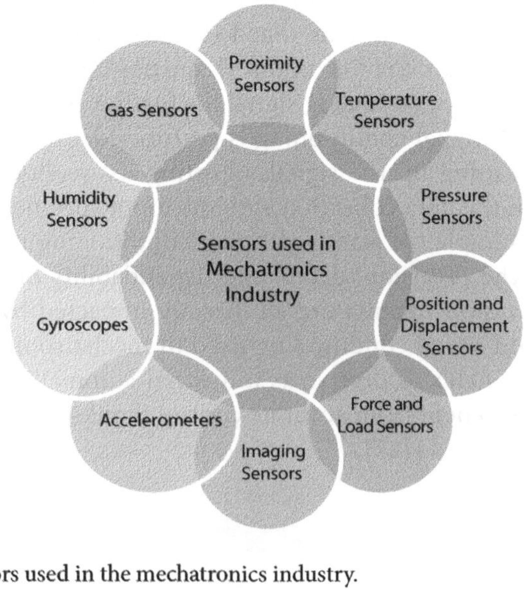

Figure 3.4 Sensors used in the mechatronics industry.

3.3.5 Actuators in Mechatronics

Mechatronic systems need actuators to turn electrical information into physical motion or activity and interact with their surroundings. These devices help execute system control signals to position, move, grasp, or apply force. Electric motors, hydraulic cylinders, pneumatic actuators, and others are used in varied applications based on power, speed, accuracy, and environment. DC, stepper, and servo motors are utilized in mechatronic systems for their precision, efficiency, and adaptability in industrial applications. Hydraulic actuators use hydraulic fluids' incompressibility to create force for heavy-duty applications that need high power. Compressed air drives linear or rotary actuators, making them quick and cost-effective for jobs requiring moderate strength and reaction [24]. Shape memory alloys and piezoelectric actuators are excellent for niche applications due to their accurate small-scale control and shape-changing capabilities. Integrating actuators into mechatronic systems requires careful consideration of power, speed, accuracy, and control mechanisms to maximize performance and efficiency. The actuator type chosen relies on the application's motion, load capacity, ambient circumstances, and system complexity. To build and execute reliable mechatronic systems for various industrial and commercial applications, one must understand actuator types and their capabilities [25]. For a mechatronic system to interact with its surroundings, it requires actuators, the parts of the system responsible for translating electrical impulses into mechanical motion or action. Necessary for positioning, moving, grasping, and delivering force, these devices operate in response to control signals from the system. Electric motors, hydraulic cylinders, pneumatic actuators, and other actuators are available, each with its own advantages and disadvantages in terms of power, speed, accuracy, and environmental conditions. Electric motors, such as DC, stepper, and servo motors, are extensively employed in mechatronic systems for their precise control, high efficiency, and adaptability in diverse industrial applications. Because hydraulic fluids are incompressible, actuators may create enormous force, making them useful for heavy-duty applications requiring high power output [26]. Compressed air provides linear or rotational motion in pneumatic actuators, making them a quick and inexpensive option for jobs requiring a modest amount of force and a short reaction time. Various specialized actuators, including shape memory alloy and piezoelectric actuators, are well suited to technical niche applications because of their unique characteristics, such as precise control at tiny scales and the ability to change form [27].

3.4 Coordination and Synchronization in Mechatronic Systems

Coordination and synchronization are critical features in the functioning of mechatronic systems, ensuring smooth integration and interaction between diverse components to provide accurate and efficient performance. Because mechatronic methods often include mechanical, electrical, and computational elements, these disparate subsystems must be able to communicate and work together effectively. Accurate operation and control of a system rely on the timely interchange of information and instructions, which can only be achieved via proper coordination between data collection, sensors, actuators, and control systems. In a mechatronic architecture, synchronization techniques are essential for preserving temporal coherence and alignment among various processes and subsystems [28]. This is important in applications like robotics, automated manufacturing, and control systems, which depend heavily on accurate time. By coordinating the activities of many system parts, synchronization helps keep things running smoothly, with as few interruptions as possible. Additionally, by using sophisticated communication protocols and feedback loops, real-time data may be continuously exchanged, allowing for rapid modifications and alterations in response to changing environmental circumstances or user inputs. Mechatronic systems rely on precise and efficient coordination and synchronization to carry out complicated operations with high accuracy and consistency in various industrial and commercial settings [29].

3.4.1 Interplay Between Data Acquisition, Sensors, and Actuators

Mechatronic systems that are both effective and quick to react are built on the foundation of data collecting, sensors, and actuators. The first step is data acquisition, which involves gathering and digitizing information from the physical world for use in the following phases of analysis and management. Sensors that detect and measure biological factors offer these data, which is essential for the system to perceive and react to its surroundings. Sensors play a critical role in monitoring and recording vital information to ensure that the system can make educated judgements and modifications based on sensory input [30]. In turn, actuators let the system interact with its environment and carry out predetermined tasks by translating electrical impulses into physical actions. A feedback loop is formed between sensors, data acquisition, and actuators, whereby sensor data are used to guide the activities of the actuators, and the resultant system behavior is continually monitored and altered via data collection. The mechatronic system's ability to react to inputs, carry out complicated tasks, and adjust to new conditions is a direct result of the high degree of coordination and synchronization between its many parts. This complex interaction is vital to the smooth functioning of many mechatronic applications, such as robotics, automated production, and control systems, allowing prompt and accurate answers to ever-changing industrial needs. A complete knowledge of this interdependent connection is crucial for building and implementing resilient and adaptable mechatronic procedures across varied industrial sectors [31].

3.5 Industrial Automation and Robotics

Industrial automation and robots have changed industry demographics by providing cutting-edge methods for increasing productivity and quality. Robotic industrial systems and PLCs are two examples of control systems used in industrial automation. This comprises material handling, assembly, packing, and quality control. Industrial automation allows for the development of smart factories that can function independently and adjust to fluctuating production needs by integrating various mechatronic components [32]. Industrial automation relies heavily on robotics because of its role in the efficient and accurate completion of complex and repetitive activities. Industrial robots with specialized end-effectors and sensors can perform various operations with great precision and consistency, including welding, painting, and transporting materials. Alongside human operators, these robots boost productivity, protect workers from harm, and refine production. Reduced downtime, higher output rates, and better product quality are only some of the benefits of the synergy between industrial automation and robots, which has led to the creation of flexible manufacturing systems. Moreover, the integration of advanced technologies such as artificial intelligence (AI), machine learning, and the IoT has further augmented the capabilities of industrial automation and robotics, enabling predictive maintenance, real-time monitoring, and data-driven decision-making in manufacturing processes. This has led to the widespread use of automation and robots in industry, which has boosted productivity, profitability, and competitiveness in the manufacturing sector [33].

3.5.1 Automotive and Transportation Systems

Vehicle performance, safety, and the driving experience have all been drastically improved because of the incorporation of mechatronic technology into today's automotive and transit systems. The use of sensors, actuators, and control systems to improve vehicle safety and help drivers deal with changing road conditions has been made possible thanks to advancements in mechatronics. Adaptive cruise control, lane departure warning, collision avoidance, and parking assistance are just some of the functions included in these systems, all of which work together to make driving safer and more convenient [34]. Electric and hybrid cars, which use cutting-edge propulsion systems and energy management technologies to cut pollution and fuel costs drastically, are another product of mechatronic advancements. Some examples of mechatronic components that have helped expand eco-friendly transportation options include electric power steering systems, regenerative braking, and cutting-edge engine control units. Automated assembly lines, robotic welding, and precise machining are just some of the ways that mechatronics has changed the automobile industry for the better in terms of productivity, quality, and cost. Intelligent traffic control, computerized ticketing, and vehicle monitoring are just a few examples of intelligent transportation systems made possible by mechatronic components that have improved the effectiveness and security of public transit. Future mobility solutions that are safer, more efficient, and more sustainable are on the horizon, thanks to the ongoing advancement and integration of mechatronic technology in automotive and transportation systems [35].

3.5.2 Healthcare and Biomedical Applications

Patient care, diagnostic processes, and treatment methods have all been revolutionized using mechatronics in healthcare and biomedical applications. Robotic surgical systems, automated medication administration systems, and state-of-the-art diagnostic imaging tools are just a few examples of the cutting-edge medical technology made possible by mechatronics. These tools allow for minimally invasive surgeries with pinpoint accuracy, improving surgical results, shortening patient recoveries, and reducing the likelihood of complications.

Sensors and actuators are only two examples of the mechatronic components crucial in creating assistive technologies, including prosthetic limbs, exoskeletons, and wheelchairs. These gadgets are meant to emulate natural human motions, offering patients increased comfort, control, and usefulness in their everyday lives. Mechatronics has helped progress drug discovery and biotechnology research by enabling the creation of automated laboratory equipment, high-precision microscopes, and robotic systems [36]. With the help of these tools, scientists may analyze biological material and carry out laborious processes with pinpoint precision and ease. Intelligent healthcare systems, such as remote patient monitoring devices, wearable health trackers, and telemedicine platforms, have been developed thanks to mechatronics integration, allowing healthcare professionals to remotely monitor patient health, provide timely interventions, and deliver individualized care to people in a variety of locations. Overall, the continued incorporation of mechatronics in healthcare and biomedical applications promises to provide more effective, personalized, and accessible healthcare solutions, thereby enhancing patient outcomes and contributing to advances in medical research and treatment methodologies [37].

3.6 Technical Challenges in Integration and Compatibility

There are several technological issues ensuring component compatibility, interoperability, and system integration in the mechatronics sector. Maintaining interoperability and communication between hardware and software from multiple vendors is a significant obstacle. Employing varied communication protocols and interfaces across different devices typically complicates their seamless integration, resulting in interoperability challenges and system inefficiencies. In addition, mechatronic systems' complicated design makes it difficult to achieve perfect synchronization and coordination between the many mechanical, electrical, and software components. Sensors, actuators, and control systems must be precisely timed, synchronized, and controlled for the system to function at its highest efficiency and dependability. While this synchronization is desirable, it is not always easy to achieve and typically requires complex calibration and fine-tuning [38]. In addition, data administration, processing, and security become more complicated when cutting-edge technologies like AI, ML, and IoT are included in mechatronic systems. Robust and sophisticated data management and security methods are required to handle the vast amounts of real-time data produced by sensors and actuators, guarantee data accuracy, and safeguard data integrity from possible cybersecurity attacks.

Furthermore, difficulties arise regarding system upgradability, maintenance, and lifecycle management due to the fast advancement of mechatronic technologies and the ongoing introduction of new components and functionality. To prevent system obsolescence and guarantee long-term sustainability, it is crucial to ensure components' backward and forward compatibility and provide proper support and maintenance for legacy systems. Overcoming these technological issues needs a multidisciplinary approach, requiring cooperation among engineers, academics, and industry experts to build standardized communication protocols, robust synchronization mechanisms, and efficient data management and security solutions. The mechatronics industry must overcome these obstacles to create more powerful and cutting-edge mechatronic systems for use in a wide range of industrial contexts [39].

3.6.1 Innovations Driving Mechatronics Advancements

As a result of mechatronics breakthroughs, several sectors are experiencing a sea change and new, cutting-edge solutions with improved functionality and performance are being developed. One of the primary drivers of mechatronics innovation is the increasing integration of AI and machine learning, allowing mechatronic systems to analyze complicated data, make autonomous choices, and adapt to dynamic settings in real time. Optimizing production processes, improving diagnostic capabilities, and allowing autonomous navigation and control are examples of how AI-powered mechatronic applications disrupt industries as varied as manufacturing, healthcare, and transportation [40]. Furthermore, the spread of the IoT has revolutionized mechatronic systems by providing seamless communication and data sharing across networked devices and components. Internet of Things (IoT)–enabled mechatronics applications to use sensor data, cloud computing, and data analytics to improve operational efficiency and productivity in various industrial domains via remote monitoring, predictive maintenance, and efficient resource management. The creation of lightweight, long-lasting, and multifunctional components with improved performance and capacities has been made possible by innovative materials, including smart materials and nanomaterials, which have played a crucial part in driving mechatronics breakthroughs. These cutting-edge materials have broadened the scope of mechatronic system design, enabling the development of smaller, more efficient, and more adaptable devices for a wide range of uses [41].

In addition, the ongoing innovations in robotics, such as the creation of collaborative robots (cobots) and bio-inspired robots, have broadened the capabilities and applications of mechatronic systems, making for more secure human–robot interactions, more excellent agility and greater flexibility in challenging settings. The cumulative effect of these mechatronics advancements is nothing short of revolutionary, opening the stage for realizing a more technologically sophisticated and interconnected future by supporting the creation of more intelligent, connected, and adaptable systems [42].

3.6.2 Mechatronics Industry and Industry 4.0

Industry 4.0, the fourth industrial revolution, relies heavily on the mechatronics sector for its implementation and progress. Manufacturing and industrial processes are defined by incorporating CPSs, IoT, cloud computing, and AI in what is known as Industry 4.0. With its emphasis on combining mechanical, electrical, and software elements, mechatronics is at the forefront of this change. In the context of Industry 4.0, mechatronic systems allow for efficient machine-to-machine connection, data sharing, and intelligent decision-making. Manufacturing processes may have their characteristics constantly monitored and improved thanks to data-collecting systems and sensors [43]. Autonomous and accurate management of equipment and robots is made possible by actuators and control systems, leading to increased productivity and adaptability in industrial settings.

Artificial intelligence (AI) and machine learning (ML) enable mechatronic systems to evaluate massive amounts of data, forecast when necessary maintenance, reduce unplanned downtime, and maximize efficiency. Robots and humans working side by side, with the help of mechatronic devices, have been shown to improve production processes regarding both safety and productivity. In addition to the manufacturing sector, other areas such as logistics, healthcare, agriculture, and transportation may benefit from mechatronics in Industry 4.0. The employment of intelligent mechatronic systems in logistics paves the way for real-time monitoring, route optimization, and robotic storage facilities [44]. Mechanized medical tools improve surgical operations, rehabilitation, and telemedicine, benefiting patients. The mechatronics industry's widespread use of Industry 4.0 has sparked new developments, raised productivity, and altered several fields. The industry as we know it is being revolutionized by the rise of intelligent mechatronic systems, which are making formerly manual tasks more accessible, reducing environmental impact, and preparing us for a more sophisticated future [45–47].

The convergence of mechatronics and Industry 4.0 has promising implications for sustainability initiatives. Increased data insights and AI-driven optimization help improve industrial operations, reducing waste and optimizing the use of resources. Overproduction and surplus inventories may be decreased because of intelligent systems' ability to change production in response to demand dynamically [48]. Reduced carbon emissions align with environmental sustainability goals; energy-efficient mechatronic systems facilitate this. Mechatronics' use in Industry 4.0, furthermore, goes well beyond the realm of conventional manufacturing to include such areas as intelligent logistics, transportation, and infrastructure. As a result, sustainable techniques may be implemented across a broader range of industries, including supply chain management, transportation, and energy. Mechatronics and Industry 4.0 help businesses become more sustainable regarding their use of resources, environmental effects, and long-term viability [49]. Organizations may help reach global sustainability objectives and safeguard current and future generations' economic, social, and ecological well-being by adopting "smart manufacturing" methods [50].

3.7 Future Trends and Implications

Many significant developments and their ramifications will shape the future of the mechatronics industry during the next several years. Artificial intelligence (AI) and machine learning are being increasingly integrated into mechatronic systems, which is one of the most notable developments. More intelligent and responsive mechatronic applications may be created via this integration's facilitation of better decision-making, autonomous operation, and adaptive features [51]. The increasing focus on the IoT will also encourage the widespread use of mechatronic devices and systems that can communicate with one another, share data, and be remotely monitored. Increased operational efficiency and the ability to make preemptive decisions in industrial and commercial settings are two outcomes that stand to benefit from this trend. The deployment of human–robot cooperation and the subsequent growth of collaborative robotics (cobots) will have far-reaching effects on the mechatronics sector. Increased productivity, enhanced workplace safety, and new prospects for automating jobs requiring human dexterity and decision-making are all possible outcomes of these breakthroughs in human–robot interaction [52]. Greener and more environmentally friendly mechatronic solutions will be developed in response to the growing concern for sustainability and energy efficiency. This movement will encourage using renewable energy sources, energy-efficient components, and sustainable materials in developing mechatronic systems, leading to a more eco-friendly and resource-conserving industrial landscape [53]. Overall, these trends and their ramifications signal a move toward more intelligent, connected, and sustainable mechatronic systems, which should encourage innovation, efficiency, and flexibility across various industries and propel the sector forward into a more sustainable and cutting-edge era [54].

3.7.1 Advancements in Sensor Technology

Many significant developments and their ramifications will shape the future of the mechatronics industry during the next several years. Artificial intelligence (AI) and machine learning are being increasingly integrated into mechatronic systems, which is one of the most notable developments. More intelligent and responsive mechatronic applications may be created via this integration's facilitation of better decision-making, autonomous operation, and adaptive features. The increasing focus on the IoT will also encourage the widespread use of mechatronic devices and systems that can communicate with one another, share data, and be remotely monitored. Increased operational efficiency and the ability to make preemptive decisions in industrial and commercial settings are two outcomes that stand to benefit from this trend [55].

The deployment of human–robot cooperation and the subsequent growth of collaborative robotics (cobots) will have far-reaching effects on the mechatronics sector. Increased productivity, enhanced workplace safety, and new prospects for automating jobs requiring human dexterity and decision-making are all possible outcomes of these breakthroughs in human–robot interaction. Greener and more environmentally friendly mechatronic solutions will be developed in response to the growing concern for sustainability and energy efficiency [56]. This movement will encourage using renewable energy sources, energy-efficient components, and sustainable materials in developing mechatronic systems, leading to a more eco-friendly and resource-conserving industrial landscape. Overall, these trends and their ramifications signal a move toward more intelligent, connected, and sustainable mechatronic systems, which should encourage innovation, efficiency, and flexibility across various industries and propel the sector forward into a more sustainable and cutting-edge era [57].

3.7.2 Integration of AI and IoT in Mechatronic Systems

A new age of connected, intelligent, and adaptable industrial applications has begun with the advent of AI and IoT integration in mechatronic systems. Mechatronic systems' intelligence and autonomy may be significantly improved with AI technologies such as machine learning and deep learning algorithms, which allow them to analyze massive volumes of data, recognize patterns, and make data-driven choices in real time. In addition, mechatronic systems that use the IoT connection may easily share information and coordinate the actions of their many sensors and actuators [58]. The ability to gather data in real-time from various sources is a boon to remote monitoring, predictive maintenance, and effective resource management in manufacturing. Improved efficiency, productivity, and decision-making are just some of the outcomes of combining AI and IoT in mechatronics, which promotes the creation of intelligent, networked systems that can adapt to changing environmental circumstances and operating needs. Proactive maintenance and the early identification of potential difficulties or breakdowns in crucial components are made possible by integrating AI and IoT in mechatronic systems via predictive analytics and condition monitoring technologies. Mechatronic systems that are maintained in a preventative manner have fewer breakdowns over time, spend less money on repairs, and last longer because of it [59].

Further, the incorporation of AI and IoT technologies in mechatronics has promoted the growth of autonomous and collaborative robotic systems, leading to improved human–robot communication, more intelligent robot decision-making, and the smooth incorporation of robotics into a wide range of manufacturing procedures. Manufacturing systems, healthcare options, and automation capacities across various industries have all benefited from these developments. When taken as a whole, the convergence of AI and IoT in mechatronic systems represents a dramatic shift in the industrial landscape, paving the way for the creation of more intelligent, more interconnected, and more adaptive systems that can improve performance, efficiency, and creativity in a wide range of sectors [60].

3.8 Conclusion

In conclusion, the convergence of mechatronics with Industry 4.0 offers a game-changing chance to advance sustainability initiatives in a wide range of industries. The development of CPSs and intelligent automation is greatly aided by mechatronics, which is concerned with the integration of mechanical, electrical, and software elements. Greater resource efficiency, less waste, and less environmental effect are all fostered by the implementation of Industry 4.0 concepts in the mechatronics sector. More data-centric and efficient manufacturing processes are made possible using intelligent sensors, data-collecting systems, and AI-driven algorithms. This helps preserve the environment by decreasing energy needs, lowering emissions, and streamlining manufacturing. Actuators and robots that work together seamlessly improve output, accuracy, and adaptability, resulting in less waste and more efficient use of resources. In addition to manufacturing, other sectors, including logistics, transportation, healthcare, and energy management, benefit from incorporating mechatronics into Industry 4.0. Thanks to this adaptability, sustainable practices are encouraged across various sectors, which boosts supply chain effectiveness, optimizes transportation routes, and allows for more effective use of available energy. The importance of environmental considerations grows as the mechatronics sector advances in the age of Industry 4.0. Organizations may help achieve global sustainability objectives by adopting ingenious manufacturing methods and data-driven decision-making, leading to a greener, more resource-efficient, and more socially responsible future. The Mechatronics industry in Industry 4.0 catalyzes good change, building a greener and more sustainable environment for future generations via creative solutions, multidisciplinary cooperation, and a dedication to sustainability.

References

1. Pontarolli, R.P., Bigheti, J.A., Domingues, F.O., de Sá, L.B., Godoy, E.P., Distributed I/O as a service: A data acquisition solution to Industry 4.0. *HardwareX*, *12*, e00355, 2022.
2. Sharma, A., Bello, O., Teodoriu, C., Karami, H., Design and Implementation of a Laboratory Sucker Rod Pumping Unit Using Industry 4.0 Concepts. *J. Energy Power Technol.*, *3*, 2, 1–14, 2021.
3. Abrishambaf, R., Structural Modeling and Implementation of Smart Sensor and Actuator Networks using IEEE 1451, in: *2020 IEEE International Instrumentation and Measurement Technology Conference (I2MTC)*, pp. 1–5, IEEE, 2020, May.
4. Faheem, M., Butt, R.A., Ali, R., Raza, B., Ngadi, M.A., Gungor, V.C., CBI4. 0: A cross-layer approach for big data gathering for active monitoring and maintenance in the manufacturing industry 4.0. *J. Ind. Inf. Integr.*, *24*, 100236, 2021.
5. Barriault, M., Alexander, I., Tasnim, N., O'Brien, A., Najjaran, H., Hoorfar, M., Classification and regression of binary hydrocarbon mixtures using single metal oxide semiconductor sensor with application to natural gas detection. *Sens. Actuators, B*, *326*, 129012, 2021.
6. Bedretchuk, J.P., Arribas García, S., Nogiri Igarashi, T., Canal, R., Wedderhoff Spengler, A., Gracioli, G., Low-Cost Data Acquisition System for Automotive Electronic Control Units. *Sensors*, *23*, 4, 2319, 2023.

7. Aleksic, S., A survey on optical technologies for IoT, smart industry, and smart infrastructures. *J. Sens. Actuator Netw.*, *8*, 3, 47, 2019.
8. Turan, B., Demir, K.A., Soner, B., Ergen, S.C., Visible light communications in industrial internet of things (IIOT), in: *The Internet of Things in the Industrial Sector: Security and Device Connectivity, Smart Environments, and Industry 4.0*, pp. 163–191, Cham: Springer International Publishing, U.K., London, 2019.
9. Para, J., Del Ser, J., Nebro, A.J., Zurutuza, U., Herrera, F., Analyze, sense, preprocess, predict, implement, and deploy (ASPPID): An incremental methodology based on data analytics for cost-efficiently monitoring the industry 4.0. *Eng. Appl. Artif. Intell.*, *82*, 30–43, 2019.
10. Mineo, C., Vasilev, M., Cowan, B., MacLeod, C.N., Pierce, S.G., Wong, C., Cross, E.J., Enabling robotic adaptive behaviour capabilities for new industry 4.0 automated quality inspection paradigms. *Insight: Non-Destr. Test. Cond. Monit.*, *62*, 6, 338–344, 2020.
11. Franklin, R.J., Industry 4.0: Real Time Embedded System with Integrated Data Acquisition, in: *2020 Third International Conference on Smart Systems and Inventive Technology (ICSSIT)*, pp. 637–642, IEEE, 2020, August.
12. Fitriani, R.Z., Kuncoro, C., Kuan, Y.D., Internet-based Remote Setting and Data Acquisition for Fuel Cell. *Sens. Mater.*, 33, 3903–3915, 2021.
13. Zhang, X., Tang, L., Chen, J., Fault diagnosis for electro-mechanical actuators based on STL-HSTA-GRU and SM. *IEEE Trans. Instrum. Meas.*, *70*, 1–16, 2021.
14. Pontarolli, R.P., Bigheti, J.A., Domingues, F.O., de Sá, L.B., Godoy, E.P., Distributed I/O as a service: A data acquisition solution to Industry 4.0. *HardwareX*, *12*, e00355, 2022.
15. Gouriveau, R., Medjaher, K., Zerhouni, N., From prognostics and health systems management to predictive maintenance 1: *Monitoring and prognostics*, John Wiley & Sons, Hoboken, New Jersey, 2016.
16. Saad, S.M., Bahadori, R., Jafarnejad, H., The smart SME technology readiness assessment methodology in the context of industry 4.0. *J. Manuf. Technol. Manage.*, *32*, 5, 1037–1065, 2021.
17. Abdelsattar, A., Park, E.J., Marzouk, A., An OPC UA Client/Gateway-Based Digital Twin Architecture of a SCADA System with Embedded System Connections, in: *2022 IEEE/ASME International Conference on Advanced Intelligent Mechatronics (AIM)*, pp. 798–803, IEEE, 2022, July.
18. Abrishambaf, R., Da Rocha, H., Espírito-Santo, A., IEC 61499 and IEEE 1451 for distributed control and measurement systems, in: *IECON 2021–47th Annual Conference of the IEEE Industrial Electronics Society*, pp. 1-6, IEEE, 2021, October.
19. Ala-Laurinaho, R., Autiosalo, J., Tammi, K., Open sensor manager for IIoT. *J. Sens. Actuator Netw.*, *9*, 2, 30, 2020.
20. Urbikain, G. and de Lacalle, L.L., MoniThor: A complete monitoring tool for machining data acquisition based on FPGA programming. *SoftwareX*, *11*, 100387, 2020.

21. Phuyal, S., Bista, D., Izykowski, J., Bista, R., Design and implementation of cost efficient SCADA system for industrial automation. *Int. J. Eng. Manuf.*, *10*, 2, 15, 2020.
22. Li, D., Wang, Y., Wang, J., Wang, C., Duan, Y., Recent advances in sensor fault diagnosis: A review. *Sens. Actuators, A*, *309*, 111990, 2020.
23. Yakimov, P., II, Asparuhova, K.K., Grigorova, T.G., Shehova, D.A., Industry 4.0 and the Challenges Faced by STEM Education, in: *2020 XXIX International Scientific Conference Electronics (ET)*, pp. 1–4, IEEE, 2020, September.
24. Pivoto, D.G., de Almeida, L.F., da Rosa Righi, R., Rodrigues, J.J., Lugli, A.B., Alberti, A.M., Cyber-physical systems architectures for industrial internet of things applications in Industry 4.0: A literature review. *J. Manuf. Syst.*, *58*, 176–192, 2021.
25. Dafflon, B., Moalla, N., Ouzrout, Y., The challenges, approaches, and used techniques of CPS for manufacturing in Industry 4.0: A literature review. *Int. J. Adv. Manuf. Technol.*, *113*, 2395–2412, 2021.
26. Tarjan, L., Šenk, I., Erić Obućina, J., Stankovski, S., Ostojić, G., Extending legacy industrial machines by a low-cost easy-to-use iot module for data acquisition. *Symmetry*, *12*, 9, 1486, 2020.
27. Akanmu, A.A., Anumba, C.J., Ogunseiju, O.O., Towards next generation cyber-physical systems and digital twins for construction. *J. Inf. Technol. Construct.*, *26*, 505–525, 2021.
28. Moideen, S., Deepthi, H., Dorothy, J., Augmenting Cyber Physical Systems through Data Collection and Machine Learning: A Perspective. *Int. J. Digit. Technol.*, *2*, 1, 90–95, 2023.
29. Jerman, A., Pejić Bach, M., Aleksić, A., Transformation towards smart factory system: Examining new job profiles and competencies. *Syst. Res. Behav. Sci.*, *37*, 2, 388–402, 2020.
30. Hossain, M.T., Badsha, S., Shen, H., Porch: A novel consensus mechanism for blockchain-enabled future scada systems in smart grids and industry 4.0, in: *2020 IEEE International IOT, Electronics and Mechatronics Conference (IEMTRONICS)*, pp. 1–7, IEEE, 2020, September.
31. Santos, R.C. and Martinho, J.L., An Industry 4.0 maturity model proposal. *J. Manuf. Technol. Manage.*, *31*, 5, 1023–1043, 2020.
32. Çeven, S. and Albayrak, A., Design and implementation of modular test equipment for process measurements in mechatronics education. *Comput. Appl. Eng. Educ.*, *28*, 2, 324–337, 2020.
33. Salah, M. and Saleem, A., Hysteresis compensation-based robust output feedback control for long-stroke piezoelectric actuators at high frequency. *Sens. Actuators, A*, *319*, 112542, 2021.
34. Bersani, C., Ruggiero, C., Sacile, R., Soussi, A., Zero, E., Internet of things approaches for monitoring and control of smart greenhouses in industry 4.0. *Energies*, *15*, 10, 3834, 2022.

35. Saad, S.M., Bahadori, R., Jafarnejad, H., Putra, M.F., Smart production planning and control: technology readiness assessment. *Procedia Comput. Sci.*, *180*, 618–627, 2021.
36. Khan, I.H. and Javaid, M., Role of Internet of Things (IoT) in adoption of Industry 4.0. *J. Ind. Integr. Manage.*, *7*, 04, 515–533, 2022.
37. Singh, A., Madaan, G., Hr, S., Kumar, A., Smart manufacturing systems: a futuristics roadmap towards application of industry 4.0 technologies. *Int. J. Comput. Integr. Manuf.*, *36*, 3, 411–428, 2023.
38. Jazdi, N., Talkhestani, B.A., Maschler, B., Weyrich, M., Realization of AI-enhanced industrial automation systems using intelligent Digital Twins. *Procedia CIRP*, *97*, 396–400, 2021.
39. Dall'Ora, N., Alamin, K., Fraccaroli, E., Poncino, M., Quaglia, D., Vinco, S., Digital transformation of a production line: Network design, online data collection and energy monitoring. *IEEE Trans. Emerging Top. Comput.*, *10*, 1, 46–59, 2021.
40. Rathee, G., Kerrache, C.A., Ferrag, M.A., A blockchain-based intrusion detection system using viterbi algorithm and indirect trust for iiot systems. *J. Sens. Actuator Netw.*, *11*, 4, 71, 2022.
41. Ajith, J.B., Manimegalai, R., Ilayaraja, V., An IoT based smart water quality monitoring system using cloud, in: *2020 International conference on emerging trends in information technology and engineering (ic-ETITE)*, pp. 1–7, IEEE, 2020, February.
42. Ruvinov, I., Zamani, N., Zhou, N.Y., Khan, M., II, Novel laser processed shape memory alloy actuator design with an embedded strain gauge sensor using dual resistance measurements. Part II: Recurrent neural network-based position and force estimation. *Sens. Actuators, A*, *313*, 112188, 2020.
43. Herarsi, N.E.H., Mkedder, M.E.A., Hassam, A., Meliani, S.M., Development of a supervisory and control Interface for the distribution station of the FESTO MPS 500 system using WinCC software, in: *2021 1st International Conference On Cyber Management And Engineering (CyMaEn)*, pp. 1–4, IEEE, 2021, May.
44. Priyanka, E.B., Maheswari, C., Thangavel, S., Proactive decision making based IoT framework for an oil pipeline transportation system, in: *Proceeding of the International Conference on Computer Networks, Big Data and IoT (ICCBI-2018)*, Springer International Publishing, pp. 108–119, 2020.
45. Channi, H.K. and Kumar, R., The role of smart sensors in smart city, in: *Smart Sensor Networks: Analytics, Sharing and Control*, pp. 27–48, Cham: Springer International Publishing, U.K., London, 2021.
46. Sanjiwani, Y., Risonata, V.Y., Afif, I., Subarkah, S., Effect of SCADA Implementation to Productivity in Ammunition Industry: A Review. *Int. J. Mech. Eng. Technol. Appl.*, *1*, 2, 37–42, 2020.
47. Pan, J., Qu, L., Peng, K., Sensor and actuator fault diagnosis for robot joint based on deep CNN. *Entropy*, *23*, 6, 751, 2021.

48. Pajak, G., Krutz, P., Patalas-Maliszewska, J., Rehm, M., Pajak, I., Dix, M., An approach to sport activities recognition based on an inertial sensor and deep learning. *Sens. Actuators, A, 345,* 113773, 2022.
49. Ramirez, R., Chang, C.K., Liang, S.H., PLC cyber-security challenges in Industrial Networks, in: *2022 18th IEEE/ASME International Conference on Mechatronic and Embedded Systems and Applications (MESA),* pp. 1–6, IEEE, 2022, November.
50. Kumar, V.S. and Krishnamoorthi, C., Development of electrical transduction based wearable tactile sensors for human vital signs monitor: Fundamentals, methodologies and applications. *Sens. Actuators, A, 321,* 112582, 2021.
51. Aguero, M., Maharjan, D., Rodriguez, M.D.P., Mascarenas, D.D.L., Moreu, F., Design and implementation of a connection between augmented reality and sensors. *Robotics, 9,* 1, 3, 2020.
52. Bevilacqua, M., Bottani, E., Ciarapica, F.E., Costantino, F., Di Donato, L., Ferraro, A., Vignali, G., Digital twin reference model development to prevent operators' risk in process plants. *Sustainability, 12,* 3, 1088, 2020.
53. Dikhanbayeva, D., Shaikholla, S., Suleiman, Z., Turkyilmaz, A., Assessment of industry 4.0 maturity models by design principles. *Sustainability, 12,* 23, 9927, 2020.
54. Yucesan, Y.A., Dourado, A., Viana, F.A., A survey of modeling for prognosis and health management of industrial equipment. *Adv. Eng. Inf., 50,* 101404, 2021.
55. Abid, A., Khan, M.T., Iqbal, J., A review on fault detection and diagnosis techniques: basics and beyond. *Artif. Intell. Rev., 54,* 3639–3664, 2021.
56. Wang, T., Guan, Y., Pang, J., Li, N., Nie, J., Xie, J., An accurate dew point sensor based on MEMS piezoelectric resonator and piecewise fitting method. *Sens. Actuators, B, 370,* 132411, 2022.
57. Trunzer, E., Prata, P., Vieira, S., Vogel-Heuser, B., Concept and evaluation of a technology-independent data collection architecture for industrial automation, in: *IECON 2019-45th Annual Conference of the IEEE Industrial Electronics Society,* vol. 1, pp. 2830–2836, IEEE, 2019, October.
58. Velásquez, D., Sánchez, A., Sarmiento, S., Velásquez, C., Toro, M., Montoya, E., Sierra, B., A Cyber-Physical Data Collection System Integrating Remote Sensing and Wireless Sensor Networks for Coffee Leaf Rust Diagnosis. *Sensors, 21,* 16, 5474, 2021.
59. Coito, T., Martins, M.S., Viegas, J.L., Firme, B., Figueiredo, J., Vieira, S.M., Sousa, J.M., A middleware platform for intelligent automation: An industrial prototype implementation. *Comput. Ind., 123,* 103329, 2020.
60. Qian, P., Pu, C., Liu, L., Lv, P., Páez, L.M.R., A novel pneumatic actuator based on high-frequency longitudinal vibration friction reduction. *Sens. Actuators, A, 344,* 113731, 2022.

4

Optimization Techniques for Mechatronics: A Comprehensive Review and Future Directions

Ikvinderpal Singh[1*] and Sapandeep Kaur Dhillon[2†]

[1]PG Department of Computer Science and Applications, Trai Shatabdi GGS Khalsa College, Amritsar, Punjab, India
[2]Department of Computer Science, Guru Nanak Dev University, Amritsar, Punjab, India

Abstract
Mechatronics, the integration of mechanical, electrical, and computer engineering disciplines, plays a pivotal role in designing and developing advanced systems in various industries. As mechatronic systems grow in complexity, the need for efficient and effective optimization techniques becomes increasingly significant. This research paper provides a comprehensive review of optimization techniques applied in the field of mechatronics. We explore various optimization methodologies, including mathematical programming, evolutionary algorithms, machine learning, and metaheuristic techniques, highlighting their applications in mechatronics design, control, and manufacturing processes. Furthermore, we discuss the challenges and opportunities in employing these techniques and propose future research directions to optimize mechatronic systems.

Keywords: Mechatronics, optimization techniques, mathematical programming, evolutionary algorithms, metaheuristics, machine learning, real-time optimization, multi-objective optimization

**Corresponding author*: ikvinderpal@gmail.com
†*Corresponding author*: sapandhillon@gmail.com

Kolla Bhanu Prakash, Satish Kumar Peddapelli, Ivan C. K. Tam, Wai Lok Woo and Vishal Jain (eds.) Computational Intelligent Techniques in Mechatronics, (109–134) © 2024 Scrivener Publishing LLC

4.1 Introduction

Mechatronics is an interdisciplinary field that integrates mechanical engineering, electronics, computer science, and control engineering to design and develop smart systems and products [1]. The term "mechatronics" is a combination of "mechanics" and "electronics," emphasizing the seamless integration of these disciplines to create advanced and intelligent systems. Mechatronic systems are characterized by their ability to sense, process information, and respond accordingly to their environment, often in real time [2, 3]. The fundamental principle behind mechatronics is the synergistic combination of mechanical components, sensors, actuators, and microcontrollers, all working in harmony to achieve specific functions with precision and efficiency. The integration of mechanical components with sensors, actuators, and embedded control systems has led to the creation of intelligent and adaptive mechatronic systems capable of performing complex tasks with precision and efficiency. However, as mechatronic systems become increasingly sophisticated and intricate, the need for optimal solutions to address complex engineering challenges becomes paramount [4]. Optimization techniques play a fundamental role in the design, analysis, and control of mechatronic systems. These techniques are essential for maximizing performance, reducing costs, improving reliability, and ensuring energy efficiency in mechatronics applications. The aim of this research paper is to provide a comprehensive review of optimization methodologies used in mechatronics and to explore their diverse applications across different facets of the field. The insights presented in this paper will be valuable for researchers, engineers, and practitioners seeking to optimize mechatronic systems and unlock their full potential in the modern technological landscape.

4.1.1 Key Components of Mechatronics

The following are the key components of mechatronics:

- *Mechanical Systems* [5]: The mechanical aspect of mechatronics involves the design of physical components, such as gears, motors, linkages, and structures that perform mechanical tasks. These components form the foundation upon which the entire mechatronic system operates.
- *Electronics*: Electronics plays a crucial role in mechatronics by providing the means to control and interact with the

mechanical components. This includes microcontrollers, sensors, data acquisition systems, and other electronic circuits responsible for signal processing and actuation.

- *Computer Science*: Software and computer programming are integral to mechatronics, enabling the implementation of control algorithms, data analysis, and system communication. Computer science aspects also encompass human–machine interfaces, allowing users to interact with and monitor the mechatronic systems.
- *Control Engineering*: Mechatronic systems rely on control engineering principles to regulate and optimize their behavior. Control algorithms are used to maintain stability, achieve desired performance, and adapt to changing conditions.

4.2 Related Work

The optimization of mechatronic systems has been a subject of extensive research, leading to various related works that have contributed to the advancement of optimization techniques in mechatronics. This section presents a summary of some notable related works in the field, highlighting their key findings and contributions.

A Comparative Study of Optimization Techniques for Mechatronic Design

- Researchers conducted a comprehensive comparative study of optimization techniques applied in mechatronic design. They evaluated the performance of traditional mathematical programming methods, evolutionary algorithms (EAs), and metaheuristic techniques in optimizing the design of a robotic manipulator. The study revealed that EAs, such as genetic algorithms and particle swarm optimization, outperformed other methods in finding optimal design parameters, offering a more efficient and effective approach for mechatronic design optimization.

Optimal Control of Autonomous Vehicles Using Model Predictive Control

- This work focused on the optimal control of autonomous vehicles using model predictive control (MPC). The researchers developed a real-time MPC-based optimization algorithm to enable precise trajectory tracking and obstacle avoidance in autonomous vehicles. The proposed approach demonstrated improved vehicle performance and safety, showcasing the potential of MPC-based optimization in enhancing the control of mechatronic systems operating in dynamic environments.

Multi-Objective Optimization for Sustainable Manufacturing in Mechatronics

- In this study, researchers explored the application of multi-objective optimization (MOO) in sustainable manufacturing processes within mechatronics. They developed an optimization framework that considered multiple objectives, including energy efficiency, material waste reduction, and production cost minimization. The results indicated that MOO allowed manufacturers to strike a balance between economic, environmental, and social considerations, leading to more sustainable manufacturing practices in mechatronics.

Real-Time Optimization of Robotic Assembly Lines using Reinforcement Learning

- This work investigated the use of reinforcement learning for real-time optimization of robotic assembly lines. Researchers designed a reinforcement learning algorithm to dynamically adjust robot configurations and task allocations to minimize cycle times and maximize throughput. The implementation demonstrated improved assembly line efficiency, providing insights into the potential of reinforcement learning–based optimization in mechatronics manufacturing.

Hybrid Optimization Techniques for Mechatronic System Control

- Hybrid optimization techniques, combining genetic algorithms with gradient-based methods, were proposed for mechatronic system control in this research. The hybrid approach aimed to enhance convergence speed and solution accuracy in mechatronic control tasks. Simulation results showcased improved control performance, making a case for the adoption of hybrid optimization methods in mechatronics applications.

Machine Learning–Based Multi-Objective Optimization in Mechatronics

- This study explored the integration of machine learning with MOO techniques in mechatronics. Researchers developed a machine learning model to predict the performance of mechatronic systems under different operating conditions. This predictive model was then integrated into a MOO framework, enabling the identification of optimal solutions based on the predicted performance. The research demonstrated the potential of machine learning in assisting MOO tasks in mechatronics.

These related works contribute significantly to the advancement of optimization techniques in mechatronics. From exploring the efficiency of different optimization methodologies to investigating real-time control and sustainable manufacturing applications, these studies highlight the diverse applications and potential of optimization in enhancing mechatronic system design, control, and manufacturing processes. As the field continues to evolve, these research findings provide valuable insights and pave the way for further advancements in optimization techniques for mechatronics.

4.3 Optimization in Mechatronics Design

Mechatronics design involves the integration of mechanical, electrical, and computer engineering disciplines to create intelligent and adaptive systems [6]. Optimization techniques play a critical role in enhancing the design of mechatronic components and systems, enabling engineers to find optimal solutions that meet performance requirements, minimize costs, and improve energy efficiency. This section explores the application of optimization techniques in mechatronics design, highlighting their impact on various aspects of the design process.

Structural Optimization

- Structural optimization is essential in designing lightweight and robust mechatronic components. Optimization techniques, such as topology optimization, shape optimization, and size optimization, are employed to determine the optimal geometry and material distribution of components like robotic arms, sensor housings, and actuators. These methods help reduce weight while ensuring structural integrity, leading to improved overall system performance and energy efficiency.

Material Selection

- Optimization methodologies assist in selecting appropriate materials for mechatronic components based on their mechanical, electrical, and thermal properties. Multi-objective optimization is often used to consider factors such as material cost, weight, and performance characteristics to identify materials that best meet the design requirements. The optimal material selection contributes to the durability, reliability, and cost-effectiveness of mechatronic systems.

Kinematic and Dynamic Analysis

- Mechatronic systems often involve complex kinematic and dynamic interactions between various components. Optimization techniques are utilized to analyze and optimize the motion and force distribution in systems like robotic manipulators and motion platforms. Through dynamic optimization, engineers can optimize control strategies to achieve precise and smooth motion, reducing wear and tear on mechanical components.

Energy Efficiency Optimization

- Energy efficiency is a key consideration in mechatronics design, especially for battery-powered or autonomous systems. Optimization techniques are employed to minimize energy consumption while maintaining performance levels. This involves optimizing control strategies, sensor usage, and actuation to achieve the desired functionality with minimal energy consumption, thereby prolonging the system's operating time and reducing overall energy costs.

Multi-Domain Optimization

- Mechatronic systems often involve multiple domains, such as mechanical, electrical, and thermal domains, with interdependencies between them. Multi-domain optimization approaches are utilized to optimize the design parameters across these domains simultaneously, considering their mutual effects. This integrated optimization results in mechatronic systems that are not only energy-efficient but also highly coordinated and responsive.

Sensitivity Analysis

- Sensitivity analysis is an integral part of mechatronics design optimization. Engineers use sensitivity analysis to identify how changes in design parameters affect system performance and optimize the most influential parameters accordingly. This approach helps in understanding the trade-offs between design variables, guiding the design process toward optimal solutions.

Reliability and Robustness Optimization

- Reliability and robustness are crucial aspects of mechatronic systems, particularly in safety-critical applications. Optimization techniques are employed to enhance the system's reliability by identifying optimal redundancy, fault tolerance, and reliability-based design. Robust optimization ensures that the mechatronic system performs consistently under uncertainties and disturbances, making it more adaptable and reliable in real-world scenarios.

Manufacturing-Oriented Design Optimization

- In addition to performance optimization, mechatronics design must consider manufacturability and cost-effectiveness. Manufacturing-oriented design optimization aims to find design solutions that are easily manufacturable and minimize production costs. Techniques like manufacturability analysis and design for assembly are integrated with optimization to ensure the feasibility of the design in the production process.

Optimization techniques in mechatronics design empower engineers to create highly efficient, reliable, and cost-effective mechatronic systems. From structural optimization to energy efficiency enhancement and multi-domain optimization, these techniques contribute significantly to the development of intelligent and adaptive mechatronic components and systems, advancing the field's capabilities in various industries. The integration of optimization in mechatronics design fosters innovation, enabling the creation of cutting-edge mechatronic systems that address complex engineering challenges and deliver sustainable and intelligent solutions.

4.4 Optimization in Mechatronics Control

Optimization techniques play a crucial role in enhancing the control strategies of mechatronic systems, enabling precise and efficient operation in diverse applications [7]. Mechatronics control optimization focuses on achieving optimal performance, adaptability, and robustness in controlling the motion, behavior, and response of mechatronic components and systems. This section explores the application of optimization techniques in mechatronics control and their impact on different aspects of control processes.

Model Predictive Control (MPC)

- Model Predictive Control (MPC) is a widely used optimization-based control strategy in mechatronics. MPC formulates a dynamic optimization problem, where the control inputs are computed by minimizing a cost function over a finite future time horizon while satisfying system dynamics and constraints. MPC allows for the incorporation of predictive models, system constraints, and performance objectives, making it suitable for controlling complex and time-varying mechatronic systems.

Trajectory Planning and Optimization

- Optimization techniques are employed in trajectory planning and optimization for mechatronic systems, such as robotic manipulators, autonomous vehicles, and drones. By optimizing the desired trajectories, control inputs, and motion profiles, mechatronic systems can achieve precise and smooth motion, reducing unnecessary energy consumption and minimizing wear and tear on mechanical components.

Motion Control Optimization

- Motion control optimization involves optimizing the control signals to achieve desired motion profiles, such as acceleration, velocity, and position control. Optimization techniques are used to minimize tracking errors, reduce settling times, and improve overall system performance in mechatronic systems. This is crucial in applications where precise motion control is essential, such as pick-and-place robots and CNC machining.

Adaptive Control Optimization

- Adaptive control optimization leverages optimization techniques to adapt control parameters and strategies based on system dynamics and uncertainties. Machine learning–based optimization approaches, such as reinforcement learning and adaptive control algorithms, enable mechatronic systems to learn and optimize their control strategies based on real-time feedback and changing operating conditions.

Fault-Tolerant Control Optimization

- Fault-tolerant control optimization focuses on ensuring the stability and performance of mechatronic systems even in the presence of faults or failures. Optimization techniques are applied to design control strategies that incorporate redundancy, reconfiguration, and fault detection mechanisms to maintain system operation under fault conditions.

Optimal Sensor and Actuator Placement

- In mechatronic systems with multiple sensors and actuators, optimization techniques are used to determine the optimal placement of these components. By strategically placing sensors and actuators, engineers can improve system observability, controllability, and performance, resulting in more efficient and responsive control strategies.

Real-Time Control Optimization

- Real-time control optimization involves optimizing control inputs and strategies in real time to respond to changing environmental conditions and uncertainties. Model-based optimization and predictive control techniques enable mechatronic systems to adapt and optimize their responses dynamically, ensuring robust and efficient performance.

Multi-Objective Control Optimization

- In some mechatronics applications, multiple control objectives need to be simultaneously considered, such as energy consumption, tracking accuracy, and safety. Multi-objective control optimization techniques help find trade-off solutions that optimize multiple performance criteria and facilitate decision-making in complex control scenarios.

Optimization techniques in mechatronics control play a pivotal role in achieving precise, efficient, and adaptive control of mechatronic components and systems. From MPC to adaptive control and fault-tolerant control, these techniques enable mechatronic systems to meet performance requirements, respond to dynamic operating conditions, and address uncertainties. The integration of optimization in mechatronics control strategies empowers engineers to design intelligent and responsive control systems, contributing to the advancement of autonomous systems, robotics, and automation in various industrial and consumer applications.

4.5 Optimization in Mechatronics Manufacturing

Optimization techniques play a significant role in improving the efficiency, productivity, and quality of mechatronics manufacturing processes [8]. Mechatronics manufacturing involves the integration of mechanical, electrical, and computer engineering in the production of complex systems and components. By applying optimization methodologies, engineers can streamline manufacturing processes, reduce costs, and enhance product quality. This section explores the application of optimization techniques in mechatronics manufacturing and their impact on different manufacturing processes.

Additive Manufacturing Optimization

- Additive manufacturing, commonly known as 3D printing, offers design freedom and the ability to produce complex geometries. Optimization techniques are used to optimize the design and orientation of components, minimize support material usage, and reduce printing time. Topology optimization is employed to create lightweight and structurally efficient components, maximizing the benefits of additive manufacturing.

CNC Machining Optimization

- Computer Numerical Control (CNC) machining is a widely used manufacturing process in mechatronics. Optimization techniques are applied to optimize tool paths, cutting parameters, and machining sequences. By minimizing machining time and tool wear while ensuring dimensional accuracy, optimization enhances the overall efficiency of CNC machining processes.

Robotic Assembly Line Optimization

- Optimization techniques are used to optimize the layout, sequencing, and task allocation in robotic assembly lines. By optimizing the placement of robotic arms and workstations, mechatronics engineers can minimize cycle times, reduce idle time, and increase throughput, resulting in more efficient and cost-effective assembly processes.

Supply Chain Optimization

- Supply chain optimization is crucial in mechatronics manufacturing to ensure smooth production processes and minimize inventory costs. Optimization techniques are employed to optimize production scheduling, material flow, and inventory management. By improving supply chain efficiency, manufacturers can respond quickly to changing demands and reduce lead times.

Quality Control Optimization

- Optimization techniques are utilized in quality control processes to optimize inspection strategies and detect defects early in the manufacturing process. Statistical process control and machine learning–based optimization enable manufacturers to identify potential quality issues, reduce scrap and rework, and enhance product quality.

Energy Efficiency Optimization

- Energy efficiency optimization is a critical aspect of mechatronics manufacturing, especially in energy-intensive processes. Optimization techniques are applied to minimize energy consumption while maintaining production targets. Energy-efficient mechatronics manufacturing reduces operational costs and contributes to sustainable manufacturing practices.

Multi-Objective Manufacturing Optimization

- Many manufacturing processes involve multiple conflicting objectives, such as minimizing production costs while maximizing product quality. Multi-objective optimization techniques are used to find trade-off solutions that balance these objectives. Manufacturers can make informed decisions based on the trade-offs identified by the optimization process.

Design for Manufacturing (DFM) Optimization

- Design for manufacturing (DFM) optimization aims to create designs that are easily manufacturable and cost-effective. Optimization techniques are employed to assess design feasibility, manufacturability, and cost implications. DFM optimization allows engineers to identify design changes that optimize production processes and reduce manufacturing costs.

Optimization techniques play a crucial role in mechatronics manufacturing, enabling engineers to enhance productivity, reduce costs, and improve product quality. From additive manufacturing and CNC machining to robotic assembly lines and supply chain optimization, these techniques contribute significantly to creating efficient and sustainable manufacturing processes. By embracing optimization in mechatronics manufacturing, industries can achieve increased competitiveness, innovation, and a more efficient and responsive manufacturing ecosystem.

4.6 Multi-Objective Optimization in Mechatronics

Multi-objective optimization (MOO) is a powerful approach that addresses complex mechatronics engineering problems involving multiple conflicting objectives. In mechatronics, systems often need to balance diverse and sometimes contradictory performance criteria, such as maximizing system efficiency while minimizing energy consumption or optimizing control precision while ensuring robustness. MOO techniques enable engineers to find a set of Pareto-optimal solutions, representing trade-offs between competing objectives [9, 10]. This section explores the application of MOO in mechatronics and its significance in achieving balanced and optimal solutions.

Pareto-Based Methods

- Pareto-based MOO techniques, such as Pareto frontiers or Pareto sets, aim to identify a set of solutions representing the best trade-offs between different objectives. These methods allow engineers to visualize and explore the trade-off space, making informed decisions based on the desired balance between objectives. Pareto frontiers help identify non-dominated solutions that are not inferior to any other solution in the set.

Epsilon-Constraint Methods

- Epsilon-constraint methods involve fixing one objective as the main objective while treating other objectives as constraints. The optimization process is repeated, each time optimizing a different objective while keeping the others within predefined limits. This approach generates a set of solutions, with each solution having a different main objective. Epsilon-constraint methods are effective when the designer has a preferred objective, and other objectives are considered secondary constraints.

Weighted Sum Method

- The weighted sum method converts MOO into a single-objective optimization problem by combining the objectives into a single objective function. The designer assigns weights to each objective based on their relative importance. The optimization process then seeks to maximize or minimize the weighted sum of the objectives. The choice of weights affects the final solution, allowing the engineer to emphasize different objectives.

Evolutionary Algorithms (EAs) in MOO

- Evolutionary algorithms, such as genetic algorithms and particle swarm optimization, are widely used in MOO problems in mechatronics. EAs are population-based search algorithms that iteratively evolve a population of candidate solutions to approximate the Pareto front. Their ability to handle multiple objectives, constraints, and nonlinearity makes them suitable for complex mechatronics engineering problems.

Applications in Mechatronics

- MOO finds application in various mechatronics domains, including mechatronic design, control, and system integration. In mechatronic design, MOO assists in optimizing components and systems for multiple criteria, such as weight, performance, and cost. In mechatronics control, MOO optimizes control strategies considering conflicting objectives, like tracking accuracy and energy efficiency. Moreover, MOO plays a crucial role in optimizing resource allocation, trajectory planning, and motion control in mechatronic systems.

Decision Support in Mechatronics Design

- MOO techniques provide valuable decision support in mechatronics design by offering a range of feasible solutions for engineers to choose from. The designer can make an informed decision based on the trade-offs between different objectives. MOO helps explore the design space comprehensively, leading to innovative and optimal solutions.

Sensitivity Analysis

- Sensitivity analysis is often performed in MOO to understand the impact of changes in objective weights on the resulting Pareto front. Sensitivity analysis helps engineers gain insights into the system's behavior and assists in making trade-offs between objectives based on the project's priorities.

Multi-Objective Reinforcement Learning

- Recent advances in multi-objective reinforcement learning have allowed for the optimization of control policies in mechatronic systems. Multi-objective reinforcement learning algorithms optimize the agent's behavior to achieve multiple objectives simultaneously, providing a more adaptable and robust control strategy in complex and dynamic environments.

Multi-objective optimization plays a critical role in addressing complex and conflicting objectives in mechatronics engineering. By providing a set of Pareto-optimal solutions, MOO allows engineers to make well-informed decisions in mechatronic design, control, and integration. These techniques empower engineers to achieve optimal trade-offs and design innovative, efficient, and adaptive mechatronic systems across various industrial applications.

4.7 Real-Time Optimization for Mechatronics

Real-time optimization is a dynamic approach that enables mechatronic systems to continuously adapt and optimize their performance in response to changing operating conditions and uncertainties. Real-time optimization plays a crucial role in enhancing the efficiency, adaptability, and robustness of mechatronic systems, particularly in applications that require rapid decision-making and response [11, 12]. This section explores the application of real-time optimization in mechatronics and its significance in achieving optimal performance in dynamic environments.

Model Predictive Control (MPC)

- Model Predictive Control (MPC) is a widely used real-time optimization technique in mechatronics. MPC formulates an optimization problem over a finite future time horizon and solves it repeatedly at each control interval. By considering future system behavior and constraints, MPC optimizes control inputs in real time, leading to precise and dynamic control of mechatronic systems. MPC is particularly effective in applications where system dynamics are subject to changes or uncertainties.

Real-Time Trajectory Planning

- Real-time trajectory planning involves continuously optimizing the desired trajectories and motion profiles for mechatronic systems, such as robotic manipulators and autonomous vehicles. By adapting trajectories in real time based on sensor feedback and environmental changes, mechatronic systems can navigate complex environments efficiently and safely.

Adaptive Control Strategies

- Real-time optimization is essential in adaptive control strategies, where control parameters are continuously adjusted based on system dynamics and uncertainties. Machine learning–based techniques, such as reinforcement learning, enable mechatronic systems to learn from real-time data and optimize their control strategies on the fly.

Optimal Resource Allocation

- In mechatronics applications with limited resources, such as energy, bandwidth, or processing power, real-time optimization is used to allocate resources optimally. This ensures that the available resources are efficiently utilized to meet performance objectives and adapt to changing demands.

Real-Time Fault Detection and Diagnosis

- Real-time optimization techniques are employed for fault detection and diagnosis in mechatronic systems. By continuously analyzing sensor data and comparing it with expected behavior, real-time fault detection algorithms can detect anomalies and trigger appropriate corrective actions, minimizing downtime and ensuring system reliability.

Dynamic Trajectory Tracking

- Real-time optimization is critical for dynamic trajectory tracking in mechatronic systems, such as drones or robotic arms. By continuously optimizing control inputs based on real-time feedback, these systems can precisely follow desired trajectories even in the presence of disturbances and uncertainties.

Real-Time Control in Autonomous Systems

- In autonomous mechatronic systems, real-time optimization is essential for decision-making and control in dynamic and unpredictable environments. Real-time optimization allows autonomous vehicles, drones, and robotic systems to adapt their behavior based on real-time sensor data, ensuring safe and efficient operation.

Networked Control Systems

- Real-time optimization is applied in networked control systems, where control decisions are distributed across multiple interconnected mechatronic components. By continuously optimizing control inputs in real time and exchanging information efficiently, networked control systems can achieve coordinated and optimal performance.

Real-time optimization is a fundamental aspect of mechatronics, enabling systems to achieve optimal performance and adaptability in dynamic environments. From real-time trajectory planning to adaptive control and fault detection, these techniques empower mechatronic systems to continuously optimize their behavior and respond to changing conditions. Real-time optimization is crucial for the advancement of autonomous systems, robotics, and automation, driving the development of intelligent and adaptive mechatronic solutions for various industries.

4.8 Challenges in Optimization for Mechatronics

The following are the challenges in optimization for mechatronics:

- *Computational Complexity*: Optimization problems in mechatronics can involve a large number of design variables, constraints, and objectives. Solving complex optimization problems in real time or within acceptable time frames can be computationally demanding, requiring advanced algorithms and powerful computing resources.
- *Model Uncertainty*: Real-world mechatronic systems often operate in uncertain environments, and accurate models may not be available. Uncertainties in system dynamics, parameters, and external disturbances can pose challenges in optimization, as the optimization process needs to be robust and adaptive to handle these uncertainties.
- *Real-Time Implementation*: In real-time applications, optimization algorithms need to be implemented efficiently to ensure timely decision-making and control. The time constraints of real-time control systems demand optimization algorithms that are fast and lightweight.
- *Multi-Objective Trade-Offs*: Mechatronic systems frequently involve multiple objectives that conflict with each other. Balancing these trade-offs and identifying the best compromise solution on the Pareto front can be challenging, requiring careful consideration of engineering priorities.
- *Dimensionality and Constraints*: High-dimensional optimization spaces and numerous constraints can make the optimization problem more challenging to solve, and

traditional optimization methods may struggle to handle such complexities.

4.9 Opportunities in Optimization for Mechatronics

The following are the opportunities in optimization for mechatronics:

- *Hybrid Optimization Techniques*: Combining different optimization methods, such as EAs with gradient-based methods or machine learning, can leverage the strengths of each approach and lead to more effective optimization solutions.
- *Machine Learning–Based Optimization*: The integration of machine learning with optimization techniques holds great promise. By using machine learning models to approximate complex functions or to guide the search process, optimization can be made more efficient and adaptable to uncertain and dynamic environments.
- *Real-Time Hardware and Sensors*: Advancements in hardware capabilities, sensors, and actuators provide opportunities to collect and process data in real time. Improved real-time data acquisition enhances the effectiveness of optimization algorithms in mechatronic systems.
- *Data-Driven Modeling*: Data-driven modeling techniques, such as system identification and surrogate modeling, can help develop accurate models and address model uncertainties in mechatronics optimization, leading to more reliable optimization results.
- *Machine Learning–Assisted Control Optimization*: Combining machine learning with control optimization allows for adaptive and self-learning control strategies. Mechatronic systems can continuously improve their control policies through real-time learning from data and experience.
- *Distributed and Networked Optimization*: For networked mechatronic systems, distributed optimization methods can be explored to achieve coordinated control and resource allocation across interconnected components, enhancing system performance and scalability.
- *Multi-Objective Reinforcement Learning*: Applying multi-objective reinforcement learning can optimize mechatronic

systems for multiple criteria simultaneously, leading to more versatile and adaptable control strategies.

4.10 Future Directions in Optimization for Mechatronics

The field of optimization for mechatronics is continuously evolving, driven by technological advancements and emerging research areas. As mechatronic systems become more complex and interconnected, optimization techniques must keep pace addressing new challenges and opportunities. Here are some key future directions in optimization for mechatronics:

Integration of Machine Learning and Optimization

- The integration of machine learning and optimization is a promising direction. Combining the strengths of both fields can lead to more efficient and adaptive optimization algorithms. Machine learning models can assist in solving complex optimization problems, improve real-time decision-making, and handle uncertainties, making optimization more robust and effective.

Reinforcement Learning for Control Optimization

- Reinforcement learning is gaining traction as a powerful approach for control optimization in mechatronics. Future research will likely explore more advanced and scalable reinforcement learning algorithms, enabling mechatronic systems to learn optimal control policies from experience in complex and dynamic environments.

Real-Time Implementation of Advanced Optimization Algorithms

- Advancements in computing technology will enable the real-time implementation of more sophisticated and computationally intensive optimization algorithms. Real-time optimization will become more feasible, allowing mechatronic systems to adapt and optimize their behavior on-the-fly in response to changing conditions.

Multi-Objective and Multi-Domain Optimization

- As mechatronic systems become more integrated and interconnected, multi-objective and multi-domain optimization will gain importance. Future optimization techniques will focus on achieving optimal solutions that balance conflicting objectives and consider the interdependencies between different domains in mechatronic systems.

Autonomous and Self-Optimizing Mechatronic Systems

- Autonomous mechatronic systems that can optimize their performance and adapt to varying conditions without human intervention will be a significant future direction. Such self-optimizing systems will continuously learn, evolve, and improve their behavior, leading to highly efficient and adaptable mechatronic solutions.

Hybrid Optimization Approaches

- The combination of different optimization techniques, such as EAs with gradient-based methods, holds great potential for solving complex mechatronics engineering problems. Hybrid optimization approaches can exploit complementary strengths to find better solutions and overcome the limitations of individual techniques.

Sustainable Design and Manufacturing Optimization

- As sustainability becomes increasingly important, future optimization techniques in mechatronics will focus on minimizing environmental impact. Sustainability-driven optimization will consider energy efficiency, materials usage, and recyclability, leading to greener and more environmentally friendly mechatronic systems.

Real-Time Multi-Objective Control

- Real-time multi-objective control optimization will become essential for mechatronic systems operating in uncertain and dynamic environments. These systems will continuously optimize control strategies to achieve multiple objectives simultaneously, enabling real-time adaptability and robustness.

The future of optimization for mechatronics is promising and dynamic. The integration of machine learning, the development of advanced real-time optimization algorithms, and the focus on multi-objective and autonomous systems will shape the next generation of mechatronic solutions. These advancements will contribute to the creation of intelligent, adaptive, and sustainable mechatronic systems that revolutionize various industries and improve the quality of human life.

4.11 Conclusion

Optimization techniques play a crucial role in enhancing the capabilities and efficiency of mechatronic systems across various domains, including design, control, and manufacturing. Mechatronics, as an interdisciplinary field, benefits greatly from the integration of advanced optimization methodologies, empowering engineers to address complex and conflicting objectives, adapt to dynamic environments, and achieve optimal performance. In mechatronics design, optimization allows for the creation of lightweight and structurally efficient components, optimal material selection, and energy-efficient systems. The ability to balance multiple objectives through MOO enables designers to make informed decisions and explore trade-offs in design solutions. In mechatronics control, optimization techniques ensure precise motion control, adaptive strategies, and robust performance. Real-time optimization enables mechatronic systems to continuously adapt to changing conditions, enhancing their responsiveness and adaptability in dynamic environments. In mechatronics manufacturing, optimization optimizes production processes, resource allocation, and quality control, leading to improved productivity, reduced costs, and enhanced product quality. Real-time optimization in manufacturing enables systems to adjust their operations in real time, making them more efficient and reliable. Despite the challenges related to computational complexity, model uncertainty, and real-time implementation, future developments in hybrid optimization, machine learning–based approaches, and real-time hardware advancements offer exciting opportunities to overcome these obstacles.

The future of optimization in mechatronics lies in the integration of machine learning with optimization techniques, enabling self-optimizing and adaptive mechatronic systems. Advances in real-time optimization will drive autonomous and intelligent mechatronic solutions that continuously learn, improve, and optimize their behavior. As optimization techniques evolve, mechatronics will play an increasingly vital role in shaping the future of various industries, from autonomous vehicles and robotics to smart manufacturing and sustainable technologies. The synergistic combination of optimization and mechatronics engineering will continue to drive innovation, efficiency, and sustainability, benefiting society and advancing our technological capabilities in the years to come.

Declarations

Conflict of Interest

The authors declare that they have no conflict of interest.

Ethics Approval and Consent to Participate

The article was submitted with the consent of all the authors to participate.

Consent for Publication

The article was submitted with the consent of all the authors and institutions for publication.

Competing Interests

The authors declare that they have no known competing financial interests or personal relationships that could have appeared to influence the work reported in this paper.

Open Access

Funding Statement

No funding was received.

References

1. Wang, B., Tao, F., Fang, X., Liu, C., Liu, Y., Freiheit, T., Smart manufacturing and intelligent manufacturing: A comparative review. *Engineering*, 7, 6, 738–757, 2021, https://doi.org/10.1016/j.eng.2020.07.017.
2. Szuster, M. and Hendzel, Z., Intelligent Control of Mechatronic Systems, in: *Intelligent Optimal Adaptive Control for Mechatronic Systems. Studies in Systems, Decision and Control*, vol. 120, Springer, Cham, 2018, https://doi.org/10.1007/978-3-319-68826-8.
3. Xiao, Y. and Liu, Q., Application of big data processing method in intelligent manufacturing. *Proceedings of 2019 IEEE International Conference on Mechatronics and Automation (ICMA)*, Tianjin, China, 2019 Aug 4–7, pp. 1895–1900, 2019, https://doi.org/10.1109/ICMA.2019.8816424.
4. Morales-Cru, C., Ceccarelli, M., Portilla-Flores, E.A., An Innovative Optimization Design Procedure for Mechatronic Systems with a Multi-Criteria Formulation. *Appl. Sci.*, 11, 19, 8900, 2021, https://doi.org/10.3390/app11198900.
5. Zhu, D., Mei, T., Luo, M., Adaptive sliding mode control for robots based on fuzzy support vector machines, in: *Proceedings of the IEEE International Conference on Mechatronics and Automation (ICMA 2009)*, Changchun, China, 9–12 August 2009, pp. 3469–3474, https://doi.org/10.1109/ICMA.2009.5246264.
6. Yixu, S., Hongbo, L., Zehong, Y., An adaptive modeling method for a robot belt grinding process. *IEEE/ASME Trans. Mechatron.*, 17, 309–317, 2012, https://doi.org/10.1109/TMECH.2010.2102047.

7. Zaitceva, I. and Andrievsky, B., Methods of Intelligent Control in Mechatronics and Robotic Engineering: A Survey. *Electronics*, 11, 2443, 2022, https://doi.org/10.3390/electronics11152443.
8. Bradley, D., Russell, D., Hehenberger, P., Azorin-Lopez, J., Watt, S., Milne, C., From Mechatronics to the Cloud, in: *Reinventing Mechatronics: Developing Future Directions for Mechatronics*, X.T. Yan, D. Bradley, D. Russell, P. Moore, (Eds.), pp. 17–33, Springer: Cham, Switzerland, 2020, https://doi.org/10.1007/978-3-030-29131-0_3.
9. Zeng, T., Ren, X., Zhang, Y., Li, G., Na, J., An Integrated Optimal Design for Guaranteed Cost Control of Motor Driving System With Uncertainty. *IEEE/ASME Trans. Mechatron.*, 24, 2606–2615, 2019, https://doi.org/10.1109/TMECH.2019.2937578.
10. Kayacan, E. and Chowdhary, G., Tracking Error Learning Control for Precise Mobile Robot Path Tracking in Outdoor Environment. *J. Intell. Robot. Syst.*, 95, 975–986, 2019, https://doi.org/10.1007/s10846-018-0916-3.
11. Cai, J., Kim, D., Jaramillo, R., Braun, J.E., Hu, J., A general multi-agent control approach for building energy system optimization. *Energy Build.*, 127, 337–351, 2016, http://manuscript.elsevier.com/S0378778816304091/pdf/S0378778816304091.pdf.
12. Filipescu, A. and Filipescu, A., Jr, Simulated hybrid model of an autonomous robotic system integrated into assembly/disassembly mechatronics line. *IFAC Proc.*, 47, 3, 9223–9228, 2014, https://doi.org/10.3182/20140824-6-ZA-1003.00556.

5

Reinforcement Learning for Adaptive Mechatronics Systems

D. Sathya[1]*, G. Saravanan[2] and R. Thangamani[1]

[1]Department of Artificial Intelligence, Kongu Engineering College, Tamil Nadu, India

[2]Department of Computer Science and Engineering, Erode Sengunthar Engineering College, Tamil Nadu, India

Abstract

Reinforcement Learning (RL) has emerged as a promising and powerful approach for developing adaptive mechatronics systems, capable of learning from interactions with their environment and dynamically adjusting their behavior to achieve desired objectives. This abstract explores the application of RL in the context of mechatronics, focusing on its ability to optimize control strategies, enhance system performance, and enable autonomous adaptation in response to changing conditions. The abstract begins with an introduction to mechatronics, emphasizing the need for adaptive systems that can continuously improve their performance and respond to uncertainties in real-world scenarios. It highlights the limitations of traditional control methods in handling complex and dynamic environments, underscoring the potential of RL to overcome these challenges. The core principles of RL are discussed, shedding light on its fundamental components, including the agent, environment, and reward system. RL algorithms [1], such as Q-learning, Deep Q-Networks (DQNs), and Proximal Policy Optimization (PPO), are explored, showcasing their ability to enable learning and decision-making in mechatronic systems. The integration of RL in adaptive mechatronics systems is elaborated, presenting real-world applications in various domains, such as robotics, autonomous vehicles, and industrial automation. RL-powered mechatronic systems have demonstrated remarkable capabilities, including path planning in dynamic environments, optimized motor control, and autonomous decision-making in uncertain conditions.

**Corresponding author*: sathya.ai@kongu.edu

Kolla Bhanu Prakash, Satish Kumar Peddapelli, Ivan C. K. Tam, Wai Lok Woo and Vishal Jain (eds.) Computational Intelligent Techniques in Mechatronics, (135–184) © 2024 Scrivener Publishing LLC

The abstract delves into the challenges and considerations in implementing RL in mechatronics, including the trade-off between exploration and exploitation, sample efficiency, and safety concerns. The exploration of safety measures to prevent RL-driven mechatronic systems from making critical mistakes is addressed. Case studies from different industries showcase successful RL implementations [2] in mechatronics, demonstrating how adaptive systems can improve operational efficiency, reduce energy consumption, and enhance overall system performance.

In conclusion, RL has proven to be a transformative tool for creating adaptive mechatronics systems, capable of autonomous learning and adaptation. By leveraging RL algorithms, mechatronic devices can navigate complex environments, optimize control strategies, and continuously improve their capabilities, making them increasingly indispensable in modern engineering applications. The abstract highlights the potential of RL for the future of adaptive mechatronics, fostering a new era of intelligent and versatile systems that have the ability to adapt, learn, and excel in diverse and dynamic scenarios.

***Keywords*:** Reinforcement learning (RL), adaptive mechatronics, autonomous adaptation, control strategies, system performance, Q-learning, deep Q-networks (DQNs), proximal policy optimization (PPO)

5.1 Introduction to Adaptive Mechatronics Systems

Adaptive mechatronics systems is an interdisciplinary [3] field that combines elements from mechanical engineering, electronics, control systems, and computer science to create intelligent and responsive machines and systems. The term "mechatronics" itself signifies the integration of mechanics and electronics, and the "adaptive" aspect refers to the ability of these systems to adjust and optimize their behavior based on changing conditions and requirements. In adaptive mechatronics systems, the goal is to develop machines and devices that can sense their environment, process information, and dynamically adjust their operations to achieve desired outcomes. This adaptability is particularly important in complex and uncertain environments where traditional control systems may fall short. The systems' ability to learn, evolve, and respond to new situations is a key characteristic that sets adaptive mechatronics apart from conventional systems.

Sensors are devices that collect data from the system's environment or its own internal state. These sensors can measure parameters such as temperature, pressure, position, velocity, and more. Actuators, on the other hand, are responsible for carrying out actions based on the system's decisions. They can include motors, servos, valves, and other mechanisms that generate movement, force, or other physical effects. Control algorithms are at the heart of adaptive mechatronics systems [4]. These algorithms process the data from sensors, make decisions based on predefined rules or learned behaviors, and generate control signals for the actuators. The algorithms can range from simple proportional-integral-derivative (PID) controllers to complex model-based controllers or machine learning–based controllers.

The feedback loop is a fundamental concept in adaptive systems. It involves continuously measuring the system's output, comparing it to the desired outcome, and adjusting the system's actions accordingly. This iterative process allows the system to adapt and improve its performance over time. Adaptive mechatronics systems incorporate mechanisms that allow them to change their behavior based on new information or changing conditions. This might involve modifying control parameters, adjusting setpoints, or even switching between different control strategies based on the system's state. Machine learning and artificial intelligence (AI) techniques play a significant role in adaptive mechatronics. These technologies enable systems to learn from data, recognize patterns, and make decisions that optimize their performance. Reinforcement Learning (RL), neural networks, and other AI methods can be employed to develop adaptive behaviors. Adaptive systems require decision-making logic that determines how the system responds to different situations. This logic can be rule-based, where predefined rules govern the system's behavior, or it can be more sophisticated, involving probabilistic reasoning or optimization algorithms. Interfaces that allow humans to interact with and monitor the system are crucial. These interfaces can include touch screens, graphical user interfaces, voice commands, or even haptic feedback systems. Human intervention might be necessary in certain scenarios or for fine-tuning the system's behavior. For systems that incorporate machine learning, a dataset for training and learning is essential. This dataset contains examples of various scenarios and the corresponding desired responses. The system learns from this data to generalize its behavior to new situations. Adaptive mechatronics [5] systems employ strategies to determine when and how to adapt. These strategies can be based on error thresholds, changing environmental conditions, or predefined triggers that indicate a need for adaptation. As adaptive systems operate in dynamic environments, safety mechanisms are vital. These mechanisms ensure that the system does not deviate from safe operating conditions even during adaptive changes in behavior.

In some cases, predictive modeling can help adaptive systems anticipate changes and adapt proactively. For example, an adaptive HVAC (heating, ventilation, and air conditioning) system might predict future occupancy levels to adjust temperature settings in advance. Reinforcement Learning (RL) is a powerful approach within the field of machine learning that can be applied to adaptive mechatronics systems to enable them to learn and optimize their behavior based on interactions with their environment. When combined with an existing mechatronics system [6], RL can enhance its adaptability and performance in dynamic and uncertain conditions. Define the problem you want the mechatronics system to solve using RL. This involves specifying the system's actions, the environment that it interacts with, and the goals that it aims to achieve. For example, consider an autonomous drone tasked with delivering packages efficiently while avoiding obstacles. Determine how to represent the system's current state and the information it needs to make decisions. This might involve sensor data (e.g., camera images and distance measurements) and internal variables (e.g., battery level and motor speeds). Define the actions that the mechatronics system can take in its environment. Actions could include motor commands, control inputs, or other changes in the system's behavior. Design a reward function that quantifies how well the system is performing with respect to its goals. The reward function provides feedback to the RL algorithm, guiding it to learn behaviors that lead to higher rewards. In the drone example, rewards could be given for successful deliveries and penalties for collisions. Choose an RL algorithm that suits your problem. Common algorithms include Q-learning, Deep Q-Networks (DQN), and Proximal Policy Optimization (PPO). The choice depends on factors like the complexity of the problem and the available data. Integrate the RL component into your existing mechatronics [7] system architecture. This might involve connecting the RL agent to the system's control inputs and receiving feedback from sensors. During the learning phase, the RL agent explores different actions to discover their effects on the environment and collects data to improve its policy. As learning progresses, the agent transitions from exploration to exploitation, using its learned policy to make decisions that maximize rewards. Train the RL agent using simulations or real-world interactions. The agent learns by iteratively updating its policy to maximize the cumulative rewards it receives over time. As the RL agent learns, it adapts its behavior to changing conditions and refines its actions to achieve better performance. This adaptability allows the mechatronics system to handle variations in the environment. Implement safety mechanisms to ensure that the learned behavior adheres to safety constraints. RL can sometimes lead to exploring risky actions, so careful design is necessary to prevent harmful outcomes. Continuously monitor the performance of the RL-enhanced mechatronics system and fine-tune parameters as needed. This might involve adjusting the learning rate, exploration strategies, or reward function. Incorporate a human–machine interface to allow human operators to intervene or provide guidance when necessary.

This ensures that the system remains under human control and safety is maintained.

Applying RL to adaptive mechatronics systems can lead to systems that not only respond to changing conditions but also actively learn and improve their behavior over time. It is important to note that RL requires careful consideration of factors such as reward shaping, exploration strategies, and generalization to ensure effective and safe learning.

5.2 Fundamentals of Reinforcement Learning

Reinforcement Learning (RL) constitutes a vital framework within the realm of machine learning, dedicated to instructing agents to make sequential decisions with the aim of maximizing cumulative rewards [8]. Departing from the conventional supervised learning paradigm, where labeled input–output pairs dictate the learning process, RL unfolds through an agent's interaction with an environment.

This interaction, driven by trial and error, empowers the agent to grasp optimal strategies by navigating the consequences of its actions.

At the core of RL lie several fundamental elements. The agent, representing the learner, interfaces with the environment, a dynamic entity that it seeks to manipulate. Actions, undertaken by the agent, elicit responses from the environment, leading to transitions between states. States encapsulate the encompassing conditions under which the agent operates, delineating its perception of the environment. Crucially, actions taken by the agent trigger rewards, numerical signals reflecting the desirability of the outcomes. These rewards serve as guides for the agent's learning, steering it toward decisions that yield greater cumulative rewards.

The agent's behavior hinges on its policy, a strategy mapping states to actions. The overarching goal of the agent is to learn a policy that maximizes the expected cumulative rewards over time. A pivotal challenge in RL is navigating the trade-off between exploration and exploitation.

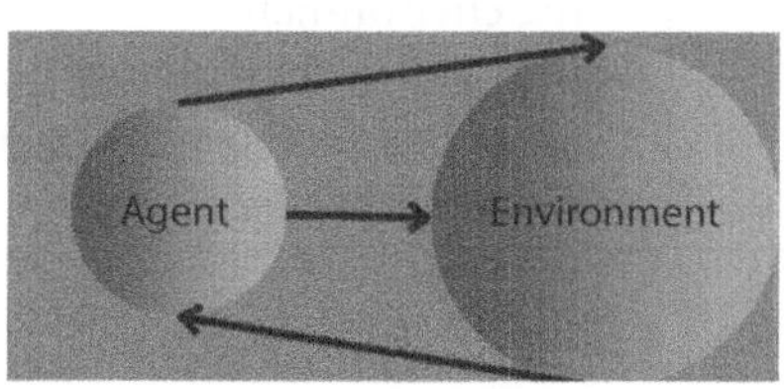

Figure 5.1 Fundamentals of RL are shown in this figure.

Exploration involves venturing into uncharted action spaces to uncover novel, potentially more rewarding paths, while exploitation entails utilizing the agent's learned knowledge to exploit immediate gains. Value functions assume a central role in RL. These functions estimate the expected cumulative rewards associated with specific states or state-action pairs. Value-based methods like Q-learning [9] utilize these functions to iteratively refine policies. Policy gradient methods, on the other hand, directly optimize policies to maximize expected rewards.

A panoply of RL algorithms exists, each tailored to specific scenarios. Q-learning, a foundational algorithm, employs iterative updates to estimate optimal state-action values. Policy gradient methods optimize policies through gradient descent, and actor-critic architectures blend policy-based and value-based approaches for robust learning. RL boasts a broad spectrum of applications, from robotics and autonomous systems to financial trading and recommendation systems [10]. By enabling machines to autonomously acquire knowledge and adapt to intricate, evolving environments, RL augments the capabilities of adaptive systems and fuels innovation across numerous domains.

Table 5.1 Fundamentals of RL in various fields are shown here.

Fundamentals	Description
1. Reinforcement Learning	A machine learning paradigm where agents learn by interacting with an environment to maximize cumulative rewards through sequential decision-making.
2. Agent and Environment	**Agent:** Learner that interacts with the environment. **Environment:** External system with which the agent interacts and learns.
3. Actions and States	**Actions:** Decisions taken by the agent to influence the environment. **States:** Represent the conditions under which the agent operates, conveying its perception of the environment.

(Continued)

Table 5.1 Fundamentals of RL in various fields are shown here. (*Continued*)

Fundamentals	Description
4. Rewards	Numerical feedback given by the environment to the agent in response to actions. Rewards indicate the desirability of outcomes and guide the agent's learning process.
5. Policy	Strategy that maps states to actions, guiding the agent's decisions. The agent aims to learn an optimal policy that maximizes the expected cumulative rewards over time.
6. Exploration vs. Exploitation	**Exploration:** Trying new actions to uncover better strategies and gather more information about the environment. **Exploitation:** Leveraging the agent's learned knowledge to exploit immediate gains and make informed decisions.
7. Value Functions	Functions that estimate the expected cumulative rewards associated with states or state-action pairs. Used to evaluate and refine policies. Value-based methods and algorithms like Q-learning leverage value functions.
8. Algorithms	**Q-learning:** An iterative algorithm that estimates optimal state-action values and refines policies. **Policy Gradient Methods:** Optimize policies directly using gradient descent. **Actor-Critic Architectures:** Blend policy-based and value-based methods.
9. Applications	RL finds application in various domains including robotics, autonomous systems, gaming, finance, healthcare, recommendation systems, and more. It enables autonomous learning and adaptive behaviors in complex and dynamic environments.
10. Impact	Reinforcement Learning empowers machines to learn from experience, adapt to evolving situations, and optimize decision-making, leading to improved efficiency, autonomy, and innovation across diverse industries.

5.3 Reinforcement Learning Algorithms for Mechatronics

Reinforcement Learning (RL) algorithms can be applied to mechatronics systems to enhance their adaptability, optimization, and autonomous decision-making capabilities. Here are a few RL algorithms that can be used for mechatronics applications:

1. **Q-Learning:** Q-learning is a foundational RL algorithm that is well suited for discrete action spaces. It is used to learn an optimal action-selection policy by estimating the value of state-action pairs. Q-learning [11] can be effective for mechatronics systems with a limited set of discrete actions, such as controlling robotic manipulators or autonomous vehicles' discrete maneuvers.
2. **Deep Q-Networks (DQNs):** DQN extends Q-learning to handle high-dimensional state spaces by using neural networks to approximate Q-values. This makes DQN suitable for mechatronics systems that involve raw sensor data like images or high-dimensional sensor readings. DQN has been applied to various tasks, including robotic control and game playing.
3. **Proximal Policy Optimization (PPO):** PPO is a policy-based algorithm designed to optimize policies directly. It iteratively refines policies while ensuring stable and controlled updates, making it suitable for mechatronics systems that require continuous control actions. PPO has been employed in applications such as robotic locomotion and control.
4. **Actor-Critic Methods:** Actor-critic algorithms combine value-based and policy-based methods. The "actor" learns a policy, whereas the "critic" estimates value functions. This approach can balance exploration and exploitation effectively, making it useful for mechatronics systems where both are crucial. Actor-critic algorithms have been applied to tasks like robotic manipulation and control.
5. **Trust Region Policy Optimization (TRPO):** TRPO is a policy optimization algorithm that ensures small policy updates to maintain stability. It is suitable for mechatronics systems where safety and stability are critical considerations. TRPO

has been applied to tasks involving robotic control and motion planning.

6. **Soft Actor-Critic (SAC):** SAC is an off-policy algorithm that emphasizes exploration and stochastic policies [12]. It is particularly well suited for continuous control tasks in mechatronics where exploration is essential. SAC has been

Table 5.2 Various algorithms of RL.

Algorithm	**Description**	**Suitability for mechatronics**
Q-Learning	Tabular algorithm for discrete action spaces.	Limited action choices, e.g., robotic manipulators
Deep Q-Networks (DQN)	Extends Q-learning to high-dimensional state spaces using neural networks	High-dimensional sensor data, e.g., image-based control
Proximal Policy Optimization (PPO)	Policy-based algorithm with stable updates	Continuous control tasks with smooth policy updates
Actor-Critic Methods	Combines policy-based and value-based approaches for stable learning	Balancing exploration and exploitation in various systems
Trust Region Policy Optimization (TRPO)	Policy optimization algorithm with safety guarantees	Critical systems where stability and safety are crucial
Soft Actor-Critic (SAC)	Off-policy algorithm that emphasizes exploration	Continuous control tasks with exploration requirements
Model-Based RL	Learns models of the environment to optimize actions	Complex mechatronics systems with intricate dynamics
Hierarchical RL	Learns policies at multiple levels of abstraction	Multi-agent systems or hierarchically structured systems

used in applications like robotic control and autonomous systems.

7. **Model-Based Reinforcement Learning:** While most RL algorithms are model-free, model-based RL uses learned models of the environment to plan and optimize actions. This approach can be useful for mechatronics systems that involve complex dynamics and require efficient decision-making. Model-based RL has applications in robotics and control.
8. **Hierarchical RL:** Hierarchical RL involves learning policies at multiple levels of abstraction. This can be beneficial for complex mechatronics systems with hierarchical structures, like multi-agent systems or systems with varying levels of control authority.

When selecting an RL algorithm for a mechatronics application, considerations include the complexity of the environment, the dimensionality of the state and action spaces, the need for exploration, safety requirements, and the availability of training data. It is often necessary to experiment with different algorithms to find the best fit for a specific mechatronics problem.

5.4 Adaptive Control Strategies in Mechatronics

Adaptive control strategies play a crucial role in mechatronics systems, allowing them to adjust and optimize their behavior in real time based on changing conditions [13]. These strategies ensure that the system maintains performance, stability, and robustness even when faced with uncertainties and variations. Here are some key adaptive control strategies commonly used in mechatronics:

1. **Model Reference Adaptive Control (MRAC):** MRAC adjusts the controller's parameters to match the response of the controlled system with a reference model. It is particularly effective when the system dynamics change over time or when modeling uncertainties are present.
2. **Self-Tuning Control:** Self-tuning control approaches, such as recursive parameter estimation techniques, continuously update model parameters to adapt to variations in the

system. These methods are well suited for mechatronics systems with time-varying dynamics.

3. **Adaptive Gain Control:** This strategy involves adjusting control gains based on system measurements or error signals. It enhances the controller's performance and robustness to variations in the plant's characteristics.
4. **Sliding Mode Control:** Sliding mode control generates control actions that force the system trajectory onto a predefined sliding surface. This technique is robust against uncertainties and disturbances, making it suitable for mechatronics systems operating in noisy environments.
5. **Model-Free Adaptive Control:** Unlike model-based approaches, model-free adaptive control does not require an accurate model of the system. Instead, it uses online learning techniques to adapt the controller based on real-time data.
6. **Reinforcement Learning:** As mentioned earlier, RL can be used to adapt control policies in mechatronics systems. RL algorithms learn optimal control strategies through trial and error, making them suitable for systems with complex dynamics and evolving environments.
7. **Neural Network–Based Control:** Neural networks can be employed to approximate complex system dynamics and control mappings. Adaptive neural network control adjusts network parameters to adapt to changes in the system.
8. **Fuzzy Logic Control:** Fuzzy logic control employs linguistic rules to adapt control actions based on imprecise or uncertain information. It is particularly effective for systems with nonlinear and uncertain dynamics.
9. **Adaptive Observers:** Adaptive observers estimate unmeasured state variables and system parameters, allowing the controller to adapt even when only partial information is available.
10. **Predictive Control:** Predictive control strategies optimize future control actions by considering predicted system responses and objectives. Adaptive predictive control adapts the predictions and control actions based on real-time data.

Adaptive control strategies are employed across a wide range of mechatronics applications, including robotics, manufacturing, automotive systems, aerospace, and more. The choice of strategy depends on factors such

Table 5.3 Adaptive control strategies employed in mechatronics systems.

Strategy	Description	Features	Applications
Model Reference Adaptive Control (MRAC)	Adjusts control parameters to match a reference model	Addresses varying dynamics and uncertainties	Aerospace systems and industrial robots
Self-Tuning Control	Continuously updates model parameters for adaptation	Real-time parameter estimation	Process control and chemical plants
Adaptive Gain Control	Adapts control gains based on system measurements	Improved performance and robustness	Automotive suspension systems and UAVs.
Sliding Mode Control	Generates control actions to ensure trajectory adherence.	Robust against uncertainties and disturbances.	Electric drives, mechanical systems.
Model-Free Adaptive Control	Adapts control based on real-time data without a model	No need for accurate system modeling	HVAC systems and traffic signal control
Reinforcement Learning	Learns optimal control strategies through trial and error	Handles complex dynamics and uncertainties	Robotic manipulation and autonomous systems.
Neural Network–based Control	Utilizes neural networks to approximate complex dynamics	Nonlinear system modeling and control	Process industries and motor control
Fuzzy Logic Control	Adapts control based on linguistic rules and uncertainty	Tolerant to imprecise information	Automotive systems and consumer electronics
Adaptive Observers	Estimates unmeasured variables and adapts control	Provides adaptation even with partial data	Biomedical systems and chemical reactors
Predictive Control	Optimizes control actions based on predicted responses	Considers future behaviors for adaptation	Building energy management and process optimization

as the system's complexity, the degree of uncertainty, available information, and desired performance criteria.

This comparison table offers insights into various adaptive control strategies, highlighting their distinctive features and their applicability to different mechatronics applications. The choice of strategy depends on factors such as the system's characteristics, the level of uncertainty, and the desired performance criteria.

5.5 Autonomous Decision-Making in Mechatronics

Autonomous decision-making in mechatronics [14] refers to the capability of mechatronic systems to make independent and intelligent choices without human intervention. These decisions are often based on data collected from sensors, knowledge of the system's environment, and pre-learned behaviors. This autonomy is crucial for enabling mechatronics systems to operate efficiently, adapt to changing conditions, and achieve their objectives. Here are key aspects and benefits of autonomous decision-making in mechatronics:

1. **Sensor Integration:** Mechatronic systems incorporate various sensors to perceive their environment. These sensors gather data about factors such as position, temperature, and pressure. Autonomous decision-making involves processing this sensor data to make informed choices.
2. **Data Fusion:** Autonomous mechatronics systems often employ data fusion techniques to combine information from multiple sensors. These integrated data provide a comprehensive understanding of the system's surroundings, improving decision accuracy.
3. **Environmental Awareness:** Mechatronics systems autonomously assess their surroundings to understand context. For example, autonomous vehicles use sensors to detect obstacles, road conditions, and traffic, allowing them to make decisions for safe navigation.
4. **Real-time Adaptation:** Autonomous decision-making enables mechatronics systems to adapt in real time to unforeseen circumstances. This agility is particularly beneficial in dynamic and unpredictable environments.
5. **Machine Learning and AI:** Many autonomous mechatronics systems utilize machine learning and AI algorithms.

These technologies allow systems to learn from data and experiences, improving decision-making accuracy over time.

6. **Path Planning and Navigation:** Autonomous mechatronics systems plan optimal paths and trajectories to reach a destination while avoiding obstacles. This is crucial for tasks like robotic exploration or autonomous vehicle navigation.
7. **Control Strategies:** Autonomous decision-making involves selecting appropriate control actions to achieve desired goals. This might involve adjusting actuators, controlling motor speeds, or modifying system parameters.
8. **Fault Detection and Recovery:** Autonomous mechatronics systems can detect faults or anomalies and make decisions to recover or ensure safe operation. This is vital for maintaining system reliability.
9. **Resource Optimization:** Autonomous decision-making optimizes resource utilization. For instance, in industrial automation, autonomous systems may allocate tasks to different machines for optimal production efficiency.
10. **Human Interaction:** Some autonomous mechatronics systems incorporate human interaction, allowing users to provide high-level instructions or intervene if needed. This human–machine interaction ensures a balance between autonomy and control.
11. **Benefits:** Autonomous decision-making enhances efficiency, reduces human intervention, enables operation in hazardous environments, increases system uptime, and allows systems to adapt to novel scenarios.
12. **Challenges:** Developing reliable autonomous decision-making systems requires addressing challenges such as handling uncertainties, ensuring safety, dealing with sensor noise, and managing complex decision spaces.

In summary, autonomous decision-making in mechatronics empowers [7] systems to operate independently and intelligently. It is a critical component for achieving advanced functionalities, efficiency improvements, and adaptability in a wide range of applications, including robotics, manufacturing, transportation, and more.

5.6 Optimization and Energy Efficiency in Mechatronics

Optimization and energy efficiency are paramount considerations in mechatronics, aimed at designing systems that maximize performance while minimizing resource consumption.

Table 5.4 Overview of optimization and energy efficiency in mechatronics.

Aspect	Optimization in mechatronics	Energy efficiency in mechatronics
Definition	Optimization involves finding the best solutions to design and operational problems. It aims to maximize performance, minimize costs, and meet specific constraints.	Energy efficiency focuses on minimizing energy consumption while maintaining desired performance levels
Importance	- Enhances system performance and functionality	- Reduces operational costs associated with energy use
	- Improves resource utilization	- Extends the lifespan of components and systems
	- Reduces waste and inefficiencies	- Contributes to environmental sustainability
Methods	- Mathematical optimization techniques (e.g., linear programming and nonlinear optimization)	- Efficient component selection to minimize power consumption
	- Heuristic algorithms (e.g., genetic algorithms and particle swarm optimization)	- Implementing power management strategies like sleep modes and dynamic voltage scaling
	- Sensitivity analysis to identify critical design parameters	- Regenerative braking and energy harvesting from environmental sources.

(*Continued*)

Table 5.4 Overview of optimization and energy efficiency in mechatronics. (*Continued*)

Aspect	Optimization in mechatronics	Energy efficiency in mechatronics
	- Simulation-based optimization using computer-aided engineering tools.	- Implementing predictive control algorithms to optimize energy usage
	- Multi-objective optimization considering conflicting design goals.	- Integrating energy-efficient actuators and sensors
Examples	- Optimal design of mechatronic systems like robotic arms for maximum payload and minimum power consumption	- Energy-efficient HVAC systems for buildings
	- Parameter optimization of a control algorithm for stability and response time	- Smart manufacturing systems with optimized production processes
	- Optimal allocation of resources in a production line for maximum throughput and minimal energy usage	- Energy-efficient transportation systems like electric vehicles and hybrid engines
	- Optimal trajectory planning for mobile robots to conserve energy.	- Renewable energy integration in mechatronic systems, like solar-powered devices

Here is a breakdown of key aspects and strategies related to optimization and energy efficiency in mechatronics:

1. **Design Optimization:**

   - Structural optimization to achieve optimal mechanical designs.
   - Material selection for components that balance strength and weight.
   - Topology optimization to determine optimal material distribution for lightweight designs.
   - Shape optimization for components that minimize stress and improve efficiency.

2. **Control Strategies for Efficiency:**

    - Advanced control algorithms to optimize dynamic behavior.
    - Model-based predictive control to optimize performance while meeting constraints.
    - Feedback control loops for precise and energy-efficient motion control.
    - Adaptive control to adjust parameters based on varying operating conditions.

3. **Energy-Efficient Actuators and Motors:**

    - Selection of energy-efficient actuators and motors.
    - Variable frequency drives for controlling motor speed and energy consumption.
    - Permanent magnet motors for high efficiency and power density.

4. **Energy Recovery Systems:**

    - Regenerative braking systems in mechatronic vehicles to recover energy during deceleration.
    - Energy recovery from vibration and motion in industrial systems.

5. **Smart Sensors and Sensor Fusion:**

    - Utilization of energy-efficient sensors.
    - Sensor fusion techniques to reduce redundant measurements and optimize accuracy.

6. **Efficient Algorithms and Software:**

    - Algorithm optimization for faster computation and reduced processing requirements.
    - Coding practices that minimize energy consumption in embedded systems.

7. **Predictive Maintenance:**
    - Condition monitoring and predictive maintenance to prevent energy waste due to malfunctions.
    - Detection of abnormal behaviors in mechatronic systems to avoid unnecessary energy consumption.
8. **Lifecycle Assessment:**
    - Lifecycle analysis to consider energy consumption during manufacturing, operation, and disposal.
    - Design decisions that optimize for overall lifecycle energy efficiency.
9. **Hybrid and Electric Drivetrains:**
    - Integration of hybrid or electric drivetrains in mechatronic vehicles.
    - Optimization of power distribution between different energy sources.
10. **Efficiency in Manufacturing Processes:**
    - Automation and robotics to streamline manufacturing processes and reduce energy consumption.
    - Design for manufacturability to minimize material waste during production.
11. **Energy-Efficient Control System Design:**
    - Low-power electronics and efficient microcontrollers for control systems.
    - Use of sleep modes and power management strategies in embedded systems.
12. **Benefits and Sustainable Practices:**
    - Reduced operational costs due to energy savings.
    - Environmental benefits by reducing energy consumption and carbon footprint.
    - Compliance with energy efficiency regulations and standards.

13. **Challenges and Future Trends:**
    - Balancing energy efficiency with performance requirements.
    - Integration of renewable energy sources in mechatronic systems.
    - Embracing emerging technologies like energy harvesting and IoT for efficiency gains.

Optimization and energy efficiency [15] in mechatronics embody a holistic approach that encompasses design, control, materials, and system behavior. These principles enable the creation of sustainable and high-performance mechatronic systems across various domains, from industrial automation to automotive engineering.

5.7 Safety and Robustness in Reinforcement Learning

Safety and robustness are critical considerations when applying RL algorithms, especially in domains where the learned behaviors interact with real-world systems. Ensuring that RL agents operate safely and reliably is essential to prevent unintended consequences and potential risks. Here is how safety and robustness are addressed in RL:

1. **Reward Engineering:** The reward function plays a pivotal role in RL. Designing a well-defined reward function that aligns with the desired behavior is crucial. Improperly designed rewards can lead to undesirable outcomes, so careful engineering is essential to guide the agent toward safe behaviors.
2. **Constraint Handling:** Introducing constraints to RL can prevent the agent from taking actions that could lead to unsafe states or situations. Constraint-based RL ensures that the agent adheres to safety limits throughout the learning process.
3. **Exploration Strategies:** During exploration, RL agents might take risky actions to gather data. Employing cautious exploration strategies helps strike a balance between exploring new behaviors and avoiding harmful actions.
4. **Model-Based RL:** Incorporating models of the environment allows for safer learning in simulated environments before

deploying to the real world. Model-based RL lets agents experiment with potentially hazardous situations in a controlled setting.

5. **Adversarial Testing:** Testing RL agents against adversarial scenarios can help identify vulnerabilities and weaknesses in learned policies. These tests simulate situations where the agent might face unexpected challenges.
6. **Off-Policy Learning:** Off-policy learning allows agents to learn from previously collected data, reducing the risk of causing damage during the learning process. This can prevent the agent from making mistakes in a live environment.
7. **Human-in-the-Loop:** Introducing human oversight and intervention can enhance safety. Human operators can intervene to correct the agent's actions if it deviates from safe behavior.
8. **Value Alignment:** Ensuring that the RL agent's objectives align with human values and goals is crucial. Mismatches between the agent's objectives and human intentions could lead to unsafe outcomes.
9. **Transfer Learning and Sim-to-Real:** Transferring knowledge from simulated environments to real-world situations can mitigate risks. This approach allows agents to learn in a controlled setting before encountering actual physical systems.
10. **Benchmarking and Standards:** Establishing benchmarks and safety standards for RL algorithms can help ensure that agents meet minimum safety criteria before deployment.
11. **Continuous Monitoring and Updates:** RL agents should be continuously monitored in real-world scenarios. Updates to the learned policies can be deployed based on real-time observations to adapt to changing conditions.
12. **Legal and Ethical Considerations:** Incorporating legal and ethical guidelines into RL system design helps ensure responsible behavior. Compliance with regulations is crucial to avoid legal liabilities.

Incorporating safety and robustness measures into RL systems is essential for their successful deployment in critical applications, including robotics, autonomous vehicles, medical devices, and industrial automation. Combining these considerations with effective training, testing, and validation processes can result in RL agents that operate reliably and safely in complex environments.

5.8 Real-World Applications and Case Studies

Reinforcement Learning (RL) has found numerous applications in mechatronics, enhancing the adaptability and performance of systems in various domains. Here are some real-world applications and case studies that showcase the synergy between RL and mechatronics:

1. **Robotic Manipulation:**
   - **Case Study:** A robot arm learns to grasp and manipulate objects in cluttered environments using RL. The agent learns to adapt its grasping strategy based on object shapes and positions.

   Robotic manipulation involves teaching robots to interact with and manipulate objects in their environment. Reinforcement Learning (RL) plays a significant role in enabling robots to learn optimal grasping strategies for various objects. One of the critical challenges in robotic manipulation is adapting the robot's grasp to different object shapes, sizes, and orientations.

Case Study: Dex-Net 2.0 - Grasping with Deep Reinforcement Learning

Overview: Dex-Net 2.0 is a case study that showcases how RL is used to teach robots effective grasping strategies for a wide range of objects. This study aims to improve the grasping success rate by enabling robots to learn from their interactions with objects.

Methodology:

1. **Data Collection:** Dex-Net 2.0 collects a vast dataset of 3D object models and associated grasp qualities, including force closure and friction. This dataset serves as the foundation for training the RL agent.
2. **Grasp Quality Prediction:** A deep neural network is trained to predict the probability of successful grasping based on input images of objects and their poses. This neural network becomes a critical component of the RL agent.
3. **Reinforcement Learning:** The RL agent learns grasp policies through trial and error. It interacts with objects in

simulation and learns to optimize grasp quality prediction. The agent takes actions (grasping poses) in the simulation, and, based on the predicted grasp quality, it receives rewards.
4. **Policy Optimization:** The RL agent uses policy optimization algorithms to adjust its grasp strategies based on the rewards received during training. The goal is to learn grasp policies that maximize the predicted grasp quality.

Results: The Dex-Net 2.0 case study demonstrates impressive results in terms of improving robotic grasping capabilities:

- The RL-trained agent shows a significant improvement in grasping success rates compared to traditional grasp planning methods.
- The agent is able to generalize its grasp policies to various object shapes and sizes that were not explicitly seen during training.
- The learned grasping strategies exhibit adaptability to novel objects, highlighting the agent's ability to generalize from its training data.

Benefits:

- Improved Grasping Performance: RL enables robots to learn effective grasping strategies, enhancing their success rates in picking up and manipulating objects.
- Generalization: The learned policies generalize across a diverse range of objects, reducing the need for explicit programming for each object.
- Adaptation: The agent's ability to adapt to novel objects showcases the potential of RL in addressing complex and dynamic manipulation scenarios.

Observation: The Dex-Net 2.0 case study illustrates how RL empowers robots to autonomously learn and adapt their grasping strategies. By combining deep learning for grasp quality prediction with RL for policy optimization, robots become capable of interacting with objects in a more versatile and efficient manner. This advancement in robotic manipulation has implications for industries such as manufacturing, logistics, and automation.

- **Benefits:** Improved object manipulation, adaptability to different objects, and robustness in changing environments.

2. **Autonomous Vehicles:**
   - **Case Study:** RL is used to train self-driving cars to navigate complex urban scenarios. The agent learns to make safe and efficient driving decisions, including lane changes and avoiding pedestrians.

Autonomous vehicles represent a transformative technology that aims to enable vehicles to navigate and make driving decisions without human intervention. Reinforcement Learning (RL) has been explored as a tool to develop decision-making algorithms for self-driving cars, allowing them to adapt to various traffic scenarios and road conditions.

Case Study: End-to-End Autonomous Driving with Deep RL

Overview: This case study focuses on using RL to develop an end-to-end autonomous driving system capable of making driving decisions in complex urban environments. The RL agent learns to control the vehicle's steering, throttle, and brake to navigate safely and efficiently.

Methodology:

1. **Data Collection:** Real-world driving data, including images from cameras, sensor measurements, and vehicle control inputs, are collected. This dataset serves as the foundation for training the RL agent.

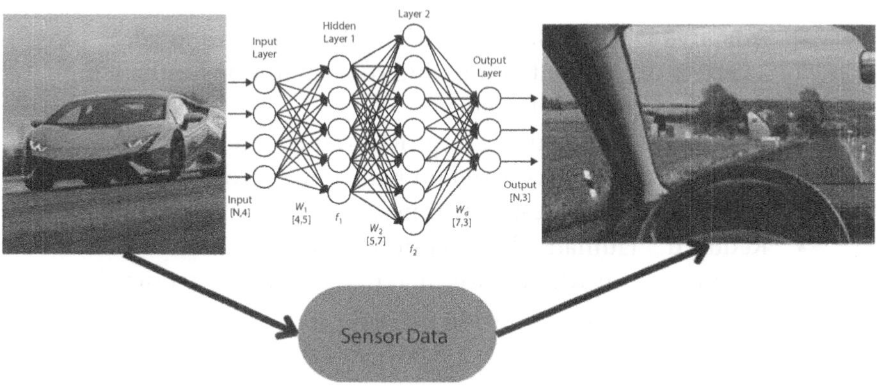

Figure 5.2 End-to-end autonomous driving with deep RL.

2. **Image Processing:** Deep Convolutional Neural Networks process the camera images to extract features and relevant information about the environment, such as lane markings, vehicles, and pedestrians.
3. **Reinforcement Learning:** The RL agent learns to control the vehicle by taking actions (steering, throttle, and brake) based on the processed images. The agent interacts with the environment (simulated or real-world) and receives rewards based on its driving behavior.
4. **Policy Optimization:** The RL agent optimizes its driving policy using algorithms like PPO. It learns to make driving decisions that maximize rewards while adhering to safety constraints.

Results: The end-to-end autonomous driving case study demonstrates the potential of RL in developing self-driving systems:

- The RL-trained agent is capable of navigating through complex urban scenarios, including lane changes, intersections, and traffic lights.
- The agent's driving behavior becomes more refined over time as it learns to respond to different road conditions and traffic patterns.
- RL-driven autonomous vehicles exhibit adaptability to dynamic environments and are capable of making real-time decisions.

Benefits:

- Adaptive Driving Behavior: RL enables vehicles to adapt to a variety of traffic situations and road conditions, enhancing driving safety and efficiency.
- Real-Time Decision-Making: RL-trained agents can make immediate decisions based on real-time sensor inputs, contributing to faster response times.
- Reduced Human Intervention: Autonomous vehicles equipped with RL decision-making algorithms require less human intervention during driving.

Challenges and Considerations:

- Safety: Ensuring that RL-trained agents drive safely and avoid accidents is of paramount importance.
- Generalization: RL algorithms need to generalize across diverse road scenarios, weather conditions, and driving environments.
- Verification and Validation: Thorough testing, simulation, and validation are essential to guarantee the safety and reliability of autonomous driving systems.

Observation: The end-to-end autonomous driving case study illustrates how RL can contribute to the development of decision-making algorithms for self-driving cars. By training agents to learn driving policies from data, autonomous vehicles can navigate through complex urban environments and contribute to the advancement of safer and more efficient transportation systems.

o **Benefits:** Enhanced navigation, adaptive responses to traffic dynamics, and improved safety.

3. Industrial Automation:
o **Case Study:** In manufacturing, RL optimizes the control of robotic arms in tasks like assembly and welding. The agent learns efficient trajectories and task sequences, improving productivity.

Industrial automation involves using technology to automate manufacturing processes, improve efficiency, and reduce human labor. Reinforcement Learning (RL) has applications in optimizing robotic assembly tasks within manufacturing environments, enhancing productivity and adaptability.

Case Study: Robotic Assembly Line Optimization with RL

Overview: This case study focuses on using RL to optimize a robotic assembly line for manufacturing tasks such as picking, placing, and assembling components. The goal is to improve the speed and efficiency of the assembly process while adapting to changes in the production line.

Methodology:

1. **Task Definition:** Define the assembly tasks, such as picking up parts, aligning them, and placing them correctly.

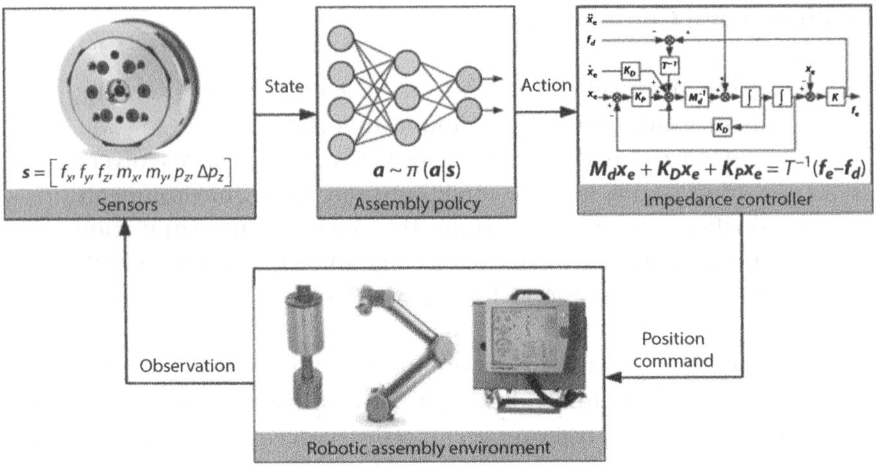

Figure 5.3 Robotic assembly line optimization with RL.

These tasks are specified as the actions the RL agent needs to learn.

2. **Simulation Environment:** Create a simulation environment that models the assembly line and robot's interactions with parts. The RL agent learns within this virtual environment before transferring its learned policies to the real-world setting.
3. **Reinforcement Learning:** Train the RL agent to perform assembly tasks efficiently. The agent learns by attempting different actions and receiving feedback (rewards) based on task completion and accuracy.
4. **Policy Optimization:** Utilize RL algorithms like TRPO or PPO to optimize the agent's policy. The agent learns to perform actions that maximize rewards over time.

Results: The robotic assembly line optimization case study demonstrates the benefits of using RL in industrial automation:

- The RL-trained agent learns efficient strategies for picking, placing, and assembling components, reducing cycle times and enhancing productivity.
- The agent adapts to variations in assembly conditions and learns to handle variations in part sizes, orientations, and positioning.

Benefits:

- Increased Efficiency: RL-driven assembly lines achieve faster task completion, leading to higher production rates.
- Adaptability: RL agents adapt to changes in production requirements and variations in the assembly process.
- Reduced Programming Effort: Traditional programming for robot assembly tasks can be time-consuming and challenging. RL reduces the need for explicit programming by enabling robots to learn from experience.

Challenges and Considerations:

- Safety: Ensuring that RL-trained robots perform assembly tasks without causing damage to components or equipment is crucial.
- Real-World Transfer: Transferring RL policies from simulation to the physical environment requires careful validation to ensure performance consistency.
- Learning Complexity: Complex assembly tasks might require longer training times and more advanced RL techniques.

Observation: The robotic assembly line optimization case study highlights the potential of RL in improving industrial automation processes. By training robots to autonomously learn optimal assembly strategies, manufacturing systems can achieve higher efficiency, adaptability to changes, and reduced manual intervention. This synergy between RL and industrial automation contributes to streamlining manufacturing processes and enhancing overall productivity.

- **Benefits:** Increased manufacturing efficiency, reduced cycle times, and minimized errors.

4. Energy Management:

- **Case Study:** RL is applied to control HVAC systems in buildings. The agent learns to adjust temperature and ventilation settings to optimize comfort while minimizing energy consumption.

Energy management involves optimizing the use of energy resources to achieve desired outcomes while minimizing wastage and costs. Reinforcement Learning (RL) can be applied to energy management systems, enabling smarter and more efficient control of energy-consuming devices such as HVAC systems.

Case Study: RL-Based HVAC Energy Optimization

Overview: This case study focuses on using RL to optimize the energy consumption of HVAC systems in commercial buildings. The goal is to maintain comfortable indoor conditions while minimizing energy usage.

Methodology:

1. **State and Action Definition:** Define the state of the HVAC system, including indoor and outdoor temperatures, humidity, occupancy, and setpoint temperatures. Actions include adjusting temperature setpoints and fan speeds.
2. **Reward Function:** Develop a reward function that balances comfort and energy efficiency. The reward function guides the RL agent's decision-making process.
3. **Reinforcement Learning:** Train the RL agent to learn control policies that dynamically adjust HVAC settings to achieve optimal comfort and energy efficiency. The agent interacts with a simulation or real building environment.
4. **Policy Optimization:** Use RL algorithms such as DQNs or PPO to optimize the control policies. The agent learns to take actions that maximize the long-term cumulative reward.

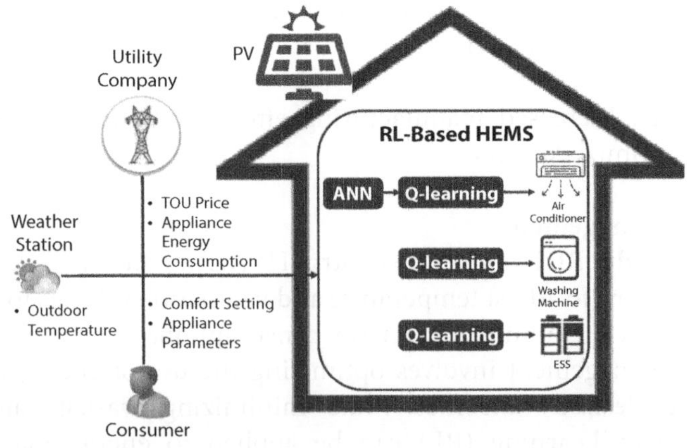

Figure 5.4 RL-based HVAC energy optimization.

Results: The HVAC energy optimization case study demonstrates the potential of RL in energy management:

- The RL-trained agent learns to adapt HVAC settings based on occupancy, outdoor conditions, and time of day to optimize comfort and energy consumption.
- Energy savings are achieved without compromising occupant comfort levels, showcasing the effectiveness of RL-driven control strategies.

Benefits:

- Energy Savings: RL-driven HVAC control reduces energy consumption and associated costs.
- Adaptive Control: The agent adapts to changing occupancy patterns, weather conditions, and time of day.
- Enhanced Comfort: Optimized HVAC settings maintain comfortable indoor conditions for occupants.

Challenges and Considerations:

- Balance Between Comfort and Energy: Designing a reward function that balances occupant comfort and energy savings is crucial.
- Generalization: RL algorithms need to generalize across diverse building types and occupancy patterns.
- Real-World Validation: Transferring RL policies to real building environments requires validation to ensure energy savings are achieved reliably.

Observation: The HVAC energy optimization case study demonstrates how RL can be leveraged to enhance energy management in buildings. By training agents to make optimal control decisions, energy-intensive systems like HVAC can operate efficiently while maintaining occupant comfort. This synergy between RL and energy management contributes to sustainability and cost savings in commercial and residential buildings.

- o **Benefits:** Energy savings, improved occupant comfort, and reduced environmental impact.

5. Healthcare Robotics:

- **Case Study:** RL guides robotic exoskeletons for physical therapy. The agent adjusts assistance levels based on patient movements, facilitating personalized rehabilitation.

Healthcare robotics involves the integration of robotic systems into medical applications to improve patient care, diagnostics, and treatment. Reinforcement Learning (RL) is being explored in healthcare robotics to develop adaptive and personalized solutions, especially in the field of rehabilitation.

Case Study: RL-Driven Robotic Rehabilitation

Overview: This case study focuses on using RL to develop robotic rehabilitation systems that assist patients in their recovery process. The goal is to provide tailored and effective rehabilitation therapies that adapt to each patient's progress.

Methodology:

1. **Patient Assessment:** Assess the patient's condition, limitations, and rehabilitation goals. This information guides the design of the RL-based rehabilitation program.
2. **Robotic Rehabilitation Device:** Develop a robotic device that assists patients in performing rehabilitation exercises. The device is equipped with sensors to monitor patient movements and progress.
3. **Reinforcement Learning:** Train the RL agent to provide personalized assistance to patients during their rehabilitation sessions. The agent adapts its assistance level based on the patient's performance and progress.
4. **Feedback and Reward:** Define a reward mechanism that encourages patients to perform exercises correctly and make progress. The RL agent provides feedback and rewards to reinforce desired behaviors.
5. **Policy Optimization:** Utilize RL algorithms like TRPO or Deep Deterministic Policy Gradients to optimize the assistance policies. The agent learns to provide assistance that maximizes patient improvement.

Results: The robotic rehabilitation case study demonstrates the potential of RL in healthcare robotics:

- The RL-driven rehabilitation system adapts to each patient's abilities, ensuring that exercises are appropriately challenging and tailored to their progress.
- Patients experience more engaging and effective rehabilitation sessions, leading to improved recovery outcomes.

Benefits:

- Personalized Rehabilitation: RL-driven systems provide personalized therapy plans based on each patient's needs and progress.
- Adaptive Assistance: The system adapts its assistance to match the patient's current abilities, promoting continuous improvement.
- Engaging Therapy: RL-driven rehabilitation enhances patient engagement and motivation, leading to better adherence to therapy routines.

Challenges and Considerations:

- Patient Safety: Ensuring that RL-driven rehabilitation devices operate safely and do not cause harm to patients is of utmost importance.
- Clinical Validation: Rigorous clinical studies are necessary to validate the efficacy and safety of RL-driven rehabilitation therapies.
- Ethical Considerations: Balancing the autonomy of the RL agent with the patient's needs and preferences is a crucial ethical consideration.

Observation: The robotic rehabilitation case study exemplifies how RL can revolutionize healthcare robotics by delivering personalized and effective rehabilitation therapies. By tailoring assistance to individual patients' needs, RL-driven systems contribute to faster and more successful recoveries. This integration of RL and healthcare robotics has the potential to enhance patient care and outcomes in rehabilitation settings.

- **Benefits:** Tailored therapy, adaptive assistance, and efficient patient recovery.

6. Quadcopter Flight Control:

- **Case Study:** RL is employed to stabilize and control quadcopters in dynamic and windy environments. The agent learns to adjust motor speeds for stable flight.

Quadcopters, also known as quadrotor drones, are versatile unmanned aerial vehicles with applications in surveillance, exploration, and entertainment. Reinforcement Learning (RL) can be applied to quadcopters to develop autonomous flight control systems that enable them to navigate complex environments and perform tasks efficiently.

Case Study: Autonomous Quadcopter Navigation with RL

Overview: This case study focuses on using RL to develop an autonomous quadcopter navigation system that enables the drone to navigate and make decisions in complex and dynamic environments.

Methodology:

1. **State and Action Space:** Define the state of the quadcopter, including its position, orientation, velocity, and sensor measurements. Actions include adjusting motor speeds and controlling pitch, roll, and yaw.
2. **Reward Function:** Develop a reward function that incentivizes desired behaviors, such as stable flight, obstacle avoidance, and efficient path following.
3. **Reinforcement Learning:** Train the RL agent to learn control policies that enable the quadcopter to navigate through

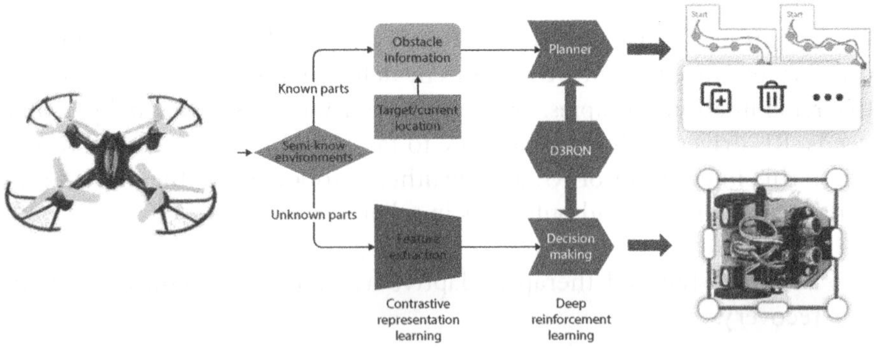

Figure 5.5 Autonomous quadcopter navigation with RL.

challenging scenarios. The agent interacts with a simulated or real-world environment.

4. **Policy Optimization:** Utilize RL algorithms like PPO or TRPO to optimize the control policies. The agent learns to take actions that maximize the long-term cumulative reward.

Results: The autonomous quadcopter navigation case study demonstrates the potential of RL in aerial robotics:

- The RL-trained agent learns to navigate through dynamic environments, avoid obstacles, and maintain stable flight.
- The quadcopter can perform tasks such as exploration, surveillance, or even package delivery in urban or outdoor settings.

Benefits:

- Autonomous Operation: RL-driven quadcopters [16] can operate autonomously, reducing the need for manual piloting.
- Complex Environments: The agent can navigate through environments that might be challenging for manual control, such as cluttered spaces or hazardous areas.
- Adaptive Flight: RL enables the quadcopter to adapt its flight behavior to changing conditions and obstacles.

Challenges and Considerations:

- Safety: Ensuring that the RL-trained quadcopter navigates safely and avoids collisions is critical.
- Real-World Transfer: Transferring RL policies to real-world quadcopter systems requires validation to ensure consistent performance.
- Dynamic Environments: RL algorithms need to handle dynamic scenarios, such as moving obstacles and changing weather conditions.

Observation: The autonomous quadcopter navigation case study illustrates the potential of RL in aerial robotics for autonomous navigation and decision-making [17]. By training agents to learn effective control policies, quadcopters can autonomously navigate through challenging environments and perform tasks that benefit various domains, including surveillance, exploration, and disaster response. This synergy between RL and quadcopter flight control contributes to the advancement of aerial robotics applications.

- **Benefits:** Stable flight under varying conditions, improved maneuverability, and reduced risk of crashes.

7. **Supply Chain Optimization:**
   - **Case Study:** RL is used to optimize inventory management in warehouses. The agent learns inventory replenishment strategies to balance stock levels and demand.
   - **Benefits:** Reduced stockouts, efficient use of storage space, and cost savings.

8. **Multi-Robot Systems:**
   - **Case Study**: RL coordinates a swarm of robots for tasks like exploration or surveillance. The agents learn to collaborate and adapt to changing team dynamics.

   - **Benefits:** Efficient exploration, improved coverage, and collaboration in complex environments.

9. **Mechatronic Systems Testing:**
   - **Case Study:** RL is utilized to perform stress testing on mechatronic systems. The agent learns to simulate extreme conditions and identify potential weaknesses.

Case Study: Autonomous Drone Landing System Testing using RL

Scenario: Imagine a mechatronic system involving an autonomous drone landing system [13]. The goal is to create an efficient and reliable system that enables a drone to autonomously land on a moving platform (such as a moving vehicle or a ship) in various environmental conditions.

Challenge: Designing and testing such a system is complex due to the interactions between the drone's dynamics, the platform's movement, wind conditions, and other variables. Traditional testing methods might not cover all possible scenarios.

Solution: Integrate RL into the testing process to optimize the drone's landing control policies and ensure robust performance.

Steps:

1. **Simulation Setup:** Develop a physics-based simulation environment that accurately models the drone, the landing platform, and various environmental factors (wind, turbulence, etc.). This simulation will be used as the training ground for the RL agent.

2. **RL Agent Training:**

   - Define the RL agent's action space: Actions could involve adjusting the drone's altitude, pitch, roll, and yaw to control its landing.
   - Define the RL agent's observation space: Observations might include the drone's position, velocity, orientation, and the platform's movement.
   - Define the reward function: Design a reward function that encourages safe and precise landing. For example, a higher reward for successful landings and penalties for crashes or overshooting the landing zone.

3. **Training Process:**

   - Use RL algorithms (e.g., PPO and DQN) to train the agent in the simulated environment.
   - The agent learns optimal control policies by repeatedly attempting landings and adapting its strategies based on the rewards received.

4. **Testing and Validation:**

   - Deploy the trained RL agent into the simulation to perform autonomous landings under various conditions: different platform speeds, wind strengths, and initial drone positions.
   - The RL agent's performance is evaluated based on successful landings, precision, and overall safety.

5. **Transfer to Real System:**

   - After successful testing in the simulation, transfer the learned control policies to the real drone landing system with necessary adjustments.

- o Continue testing and fine-tuning in a controlled real-world environment.

6. **Continuous Monitoring and Adaptation:**

   - o Deploy the drone landing system in operational scenarios.
   - o Continuously monitor its performance and adapt the RL control policies based on real-time data.
   - o This adaptive process helps the system handle unforeseen challenges and changing conditions.

Benefits:

- **Efficiency:** RL enables the drone to learn effective control strategies faster than manual tuning.
- **Robustness:** The learned policies are more likely to handle a wide range of scenarios, contributing to a robust landing system.
- **Adaptation:** RL enables the system to adapt to changing conditions and learn from real-world experiences.
- **Reduced Risk:** Rigorous testing in simulation reduces the risk of accidents during real-world testing.

Challenges:

- **Simulation Fidelity:** Ensuring that the simulation accurately represents real-world dynamics and uncertainties.
- **Safety:** Ensuring that RL-trained policies prioritize safety and avoid risky behaviors.
- **Real-World Transfer:** Ensuring that the policies learned in simulation work effectively on the real system.

This case study illustrates how combining mechatronic system testing with RL can streamline the development and testing of complex systems, leading to more efficient, adaptive, and reliable outcomes.

- o **Benefits:** Comprehensive testing, improved system reliability, and reduced failure rates.

10. **Aerospace Systems:**
    - **Case Study:** RL is employed to optimize trajectory planning for spacecraft re-entry. The agent learns to minimize heat load while ensuring a safe landing.

Autonomous Aircraft Control using RL

Scenario: In the context of aerospace systems, consider an autonomous aircraft tasked with flying through changing weather conditions and avoiding obstacles while maintaining stable flight [18]. Traditional control systems might struggle to handle the complexity of such scenarios.

Challenge: Designing a control system that can adapt to dynamic environments and make optimal decisions in real time.

Solution: Integrate RL into the control system to enable the aircraft to learn adaptive flight control policies.

Steps:

1. **Simulation Setup:** Create a high-fidelity flight simulator that accurately models the aircraft's dynamics, environmental conditions, and potential obstacles. This simulator will serve as the training environment for the RL agent.
2. **RL Agent Training:**
    - Define the RL agent's action space: Actions could involve adjusting the aircraft's pitch, roll, yaw, throttle, and flaps.
    - Define the RL agent's observation space: Observations might include the aircraft's altitude, airspeed, orientation, wind conditions, and obstacle proximity.
    - Define the reward function: Design a reward function that encourages safe flight, obstacle avoidance, and stable control. For instance, a high reward for smooth flight and avoiding obstacles, and penalties for erratic behavior.
3. **Training Process:**
    - Utilize RL algorithms (such as PPO or TRPO) to train the agent in the simulated environment.

  - The agent learns optimal control policies by repeatedly flying through scenarios and adapting its behavior based on the rewards it receives.

4. **Testing and Validation:**

   - Evaluate the RL-trained agent's performance in various scenarios, including changing weather conditions, obstacle-laden environments, and emergency situations.
   - Measure the success rate of obstacle avoidance, stability in turbulent conditions, and adherence to safe flight behavior.

5. **Transfer to Real Aircraft:**

   - Once validated in simulation, transfer the learned control policies to a real aircraft's flight control system.
   - Implement necessary safety measures to ensure that the RL control policies prioritize safe flight.

6. **Real-World Testing:**

   - Conduct real-world flight tests in controlled environments with safety mechanisms in place.
   - Monitor the aircraft's behavior and make adjustments if needed.

7. **Continuous Learning and Adaptation:**

   - Incorporate a feedback loop that allows the RL agent to continue learning from real-world flight experiences.
   - Adapt the control policies based on data collected during actual flights.

Benefits:

- **Adaptive Control:** RL allows the aircraft to adapt to changing conditions and unforeseen challenges.
- **Optimal Decision-Making:** The RL agent learns to make decisions that maximize safety and performance objectives.

Table 5.5 Real-world applications, case studies, and corresponding descriptions.

Application	**Case study**	**Description**
Industrial Robotics	Optimizing Robotic Welding Arms	A car manufacturing plant used optimization techniques to minimize the energy consumption of robotic welding arms while ensuring precise welding. This resulted in substantial energy savings and improved production efficiency.
Renewable Energy Systems	Solar Panel Tracking System	A solar power plant implemented a solar tracking mechanism that adjusts the angles of solar panels optimally based on the sun's position. This led to an increase in energy output by up to 25%, contributing to enhanced energy efficiency.
HVAC Systems	Smart HVAC Controls in Buildings	An office building integrated smart HVAC controls that adapt temperature and ventilation according to occupancy and time of day. This approach optimized energy use while maintaining occupant comfort.
Autonomous Vehicles	Energy-Efficient Autonomous Drones	A drone delivery company developed autonomous drones with optimized flight paths and energy-efficient navigation algorithms. This allowed longer flight times and increased delivery efficiency with minimized energy consumption.
Manufacturing Processes	Energy-Optimized CNC Machining	A machining facility employed optimization techniques to determine the best toolpaths and cutting parameters for CNC machines. This resulted in reduced machining time, lower energy usage, and improved product quality.
Smart Grid Systems	Demand Response and Smart Grid	An energy utility implemented smart grid systems and demand response strategies to balance energy supply and demand dynamically. This optimization helped prevent grid overload and reduced the need for additional power generation during peak times.
Transportation Systems	Energy-Efficient Electric Vehicles	An electric vehicle manufacturer designed energy-efficient electric drivetrains and regenerative braking systems. These optimizations extended the driving range and enhanced overall energy efficiency of the vehicles.

- **Reduced Human Intervention:** The aircraft can autonomously handle complex situations, reducing the need for constant human intervention.
- **Improved Efficiency:** RL enables the aircraft to fly more efficiently by adjusting control parameters based on real-time conditions.

Challenges:

- **Safety:** Ensuring that the RL control policies prioritize safety at all times.
- **Real-World Transfer:** Verifying that the policies learned in simulation perform well on the real aircraft.
- **Continuous Learning:** Designing mechanisms for the RL agent to learn from real-world data while avoiding unsafe behavior.

This case study demonstrates how RL can revolutionize aerospace systems by enabling aircraft to autonomously navigate complex and dynamic environments, contributing to safer and more efficient flight operations.

o **Benefits:** Enhanced re-entry accuracy, reduced thermal stress, and mission success.

These case studies demonstrate the versatility of RL in mechatronics, enabling systems to learn and adapt to diverse and complex environments [19]. The synergy between RL algorithms and mechatronics principles drives innovation, efficiency, and improved performance across a wide range of applications.

5.9 Challenges and Future Directions

Combining mechatronics with RL presents exciting opportunities, but it also comes with its own set of challenges and potential future directions [20]. Here are some challenges and future directions for the integration of mechatronics and RL:

Challenges:

1. **Sample Efficiency:** RL algorithms often require a large number of interactions with the environment to learn

effective policies. In mechatronic systems, real-world interactions can be time-consuming, expensive, and potentially risky. Developing techniques to improve sample efficiency is crucial.

2. **Safety and Risk:** In mechatronics, especially when dealing with physical systems, safety is paramount. RL agents might explore risky actions that could lead to damage or harm. Ensuring safe exploration and risk mitigation is a critical challenge.
3. **Real-Time Requirements:** Many mechatronic systems, such as robotics or autonomous vehicles, operate in real-time environments. RL algorithms might have computational demands that make them unsuitable for real-time decision-making. Balancing the computational load with real-time requirements is essential.
4. **Transferability:** Learned policies in simulation might not transfer well to real-world systems due to differences in dynamics, noise, and uncertainties. Bridging the gap between simulated and real-world performance is a significant challenge.
5. **Robustness to Changing Conditions:** Mechatronic systems often encounter dynamic and changing conditions. Ensuring that RL policies remain effective under varying conditions, including disturbances and uncertainties, is challenging.
6. **Curriculum Learning:** Teaching RL agents to learn progressively from simpler to more complex tasks could be challenging in mechatronics, where the task hierarchy might not be obvious.

Future Directions:

1. **Safe RL and Human–Robot Interaction:** Developing safe RL algorithms that can handle uncertainty and unexpected situations is crucial for real-world mechatronic systems. Integrating human expertise and preferences into RL decision-making processes could enhance the safety and reliability of interactions.
2. **Domain Adaptation and Transfer Learning:** Exploring techniques that allow RL policies to adapt to changes between simulation and reality can lead to more robust mechatronic systems.

3. **Hybrid Control Approaches:** Integrating RL with traditional control methods can combine the strengths of both approaches. Hybrid control strategies can provide stability guarantees while harnessing the adaptability of RL.
4. **Meta-Learning:** Meta-learning involves training RL agents to learn new tasks quickly with limited data. In mechatronics, where each system might be unique, meta-learning could facilitate faster adaptation to new environments.
5. **Multi-Agent Systems:** Many mechatronic systems involve multiple interacting components. Developing RL techniques that enable coordinated and cooperative behavior among multiple agents is an important direction.
6. **Inverse RL and Imitation Learning:** Instead of learning from scratch, using inverse RL or imitation learning to learn from human or expert demonstrations could expedite the learning process and ensure safe behaviors.
7. **Online and Lifelong Learning:** Enabling RL agents to continually learn and adapt over time as they operate in the field is crucial for maintaining optimal performance in mechatronic systems.
8. **Human-Centric RL:** Designing RL algorithms that can collaborate with and learn from human operators can enhance the usability and acceptance of mechatronic systems.

The integration of mechatronics and RL holds immense promise for developing intelligent, adaptable, and efficient systems. Overcoming the challenges and pushing forward with innovative research in these future directions will contribute to the advancement of both fields and the realization of more capable mechatronic systems.

5.10 Ethical and Societal Implications

Integrating RL into mechatronics introduces a range of ethical and societal implications that need careful consideration [21]. Here are some key ethical and societal aspects to think about when combining RL and mechatronics:

Ethical Implications:

1. **Safety and Liability:** If RL-based mechatronic systems fail to perform as intended, then determining liability can

be complex. Who is responsible if an autonomous system causes harm? Clear guidelines for accountability and liability are essential.

2. **Bias and Fairness:** If the training data used for RL contains biases, then the learned policies could also exhibit bias. Ensuring fairness in decision-making and avoiding reinforcement of existing societal biases is critical.
3. **Human Well-Being:** Implementing RL in mechatronics should prioritize the well-being of humans and society. Systems must be designed to prevent harm and respect human values.
4. **Privacy:** RL systems might need access to large amounts of data, including personal information. Ensuring the privacy and security of individuals' data is important.
5. **Transparency and Interpretability:** RL algorithms can be complex and difficult to interpret. Ensuring that the decision-making process of mechatronic systems is transparent and understandable is vital.
6. **Adversarial Attacks:** Just like other AI systems, RL-based mechatronic systems might be vulnerable to adversarial attacks that manipulate their behavior. Safeguarding against such attacks is crucial.

Societal Implications:

1. **Job Disruption and Workforce:** The integration of RL and mechatronics might lead to automation of tasks traditionally performed by humans. This could have an impact on employment and require workforce reskilling.
2. **Unequal Access:** Access to RL technology and advanced mechatronic systems might be unevenly distributed, exacerbating societal inequalities.
3. **Dependency and Autonomy:** As mechatronic systems become more capable, there is a risk of over-reliance on these technologies, potentially leading to loss of individual skills and autonomy.
4. **Economic Impact:** The deployment of RL-based mechatronics can have economic implications, affecting industries, markets, and supply chains.
5. **Regulation and Governance:** Establishing appropriate regulations and governance frameworks for RL-based

mechatronics is essential to ensure their responsible development and deployment.

Mitigation Strategies:

1. **Ethics by Design:** Embed ethical considerations into the design process of RL-based mechatronic systems from the outset.
2. **Testing and Validation:** Rigorous testing and validation processes can help identify potential biases, safety risks, and performance issues before deployment.
3. **Explainability:** Strive to make the decision-making process of RL systems transparent and interpretable to stakeholders.
4. **Human Oversight:** Maintain a level of human oversight and intervention in critical decision-making processes, particularly in high-stakes applications.
5. **Collaborative Development:** Involve interdisciplinary teams including ethicists, sociologists, policymakers, and end-users to ensure a holistic approach.
6. **Public Discourse:** Engage in public discussions and consultations to gather diverse perspectives and incorporate societal values into technology development.
7. **Education and Awareness:** Educate the public, policymakers, and stakeholders about the capabilities, limitations, and potential impacts of RL-based mechatronic systems.

The integration of RL and mechatronics offers remarkable technological advancements, but ethical and societal considerations must be central to their development and deployment. Addressing these implications responsibly ensures that the benefits of these technologies are realized while minimizing potential risks and negative consequences for society as a whole.

5.11 Conclusion

In conclusion, the convergence of RL and mechatronics holds immense potential to revolutionize various industries by creating intelligent, adaptive, and efficient systems [10]. This fusion brings together the power of AI-driven decision-making and the precision of mechatronic engineering, enabling systems to autonomously navigate complex environments, optimize tasks, and learn from experiences. Throughout this discussion, we have explored the various facets of this synergy, from the foundational concepts to practical applications and the challenges they entail.

As mechatronic systems become more intricate and interconnected, RL offers a dynamic approach to control and optimization [22]. By enabling machines to learn from interactions with their environment, RL-driven mechatronics can adapt to changing conditions, optimize performance, and even discover novel solutions that might be challenging to achieve through traditional engineering approaches alone.

However, with this fusion comes a responsibility to address ethical, societal, and safety considerations. Ensuring transparency, fairness, and accountability in the decision-making processes of RL-driven mechatronic systems is paramount. As these technologies become more integrated into our lives, it is crucial to engage in thoughtful discourse, involve diverse perspectives, and establish regulatory frameworks that balance innovation with societal well-being.

The journey of integrating RL with mechatronics is a continuous one, characterized by collaboration between engineers, researchers, ethicists, policymakers, and society at large. As this field evolves, the potential for advancements in automation, robotics, aerospace, healthcare, and beyond is substantial. With careful attention to the challenges and opportunities, RL-driven mechatronics can shape a future where intelligent systems coexist harmoniously with humanity, augmenting our capabilities and enhancing our quality of life.

References

1. Acar, M. and Parkin, R.M., Engineering education for mechatronics. *IEEE Trans. Ind. Electron.*, 43, 106–112, 1996, [Google Scholar] [CrossRef][Green Version].
2. Camposaragna, M., Chatterton, S., Righettini, P., Riva, R., Strada, R., Zappa, B.R., Design and development of a Cartesian plotter as a way to teach mechatronics to mechanical engineering students, in: Proceedings of the IEEE Industrial Electronics (IECON 2006), Paris, France, pp. 4474–4479, 7–10 November 2006, [Google Scholar].
3. Habib, M.K., Mechatronics: A unifying interdisciplinary and intelligent engineering paradigm. *IEEE Ind. Electron. Mag.*, 1, 12–24, 2007, [Google Scholar] [CrossRef].
4. Habib, M.K., Interdisciplinary mechatronics: Problem solving, creative thinking and concurrent design synergy. *Int. J. Mechatron. Manuf. Syst.*, 1, 264–269, 2008, [Google Scholar].

5. Stockmans-Daou, C., Cruz-Martin, A., Fernandez-Madrigal, J.A., A heterogeneity-enabled development system for educational mechatronics, in: *Proceedings of the IEEE International Conference on Mechatronics (ICM'09)*, Malaga, Spain, pp. 1–6, 14–17 April 2009, [Google Scholar].
6. Stark, B., Li, Z., Smith, B., Chen, Y.Q., Take-home mechatronics control labs: A low-cost personal solution and educational assessment, in: *Proceedings of the ASME 2013 International Design Engineering Technical Conferences & Computers and Information in Engineering Conference (IDETC/CIE 2013)*, Portland, OR, USA, pp. 294–301, 4–7 August 2013, [Google Scholar].
7. Félix-Herrán, L.C., Rendon-Nava, A.E., Nieto Jalil, J.M., Challenge based learning: An I-semester for experiential learning in Mechatronics Engineering. *Int. J. Interact. Des. Manuf.*, 13, 1367–1383, 2019, [Google Scholar] [CrossRef].
8. Darling-Hammond, L., Barron, B., Pearson, P.D., Schoenfeld, A.H., Stage, E.K., Zimmerman, T.D., Cervetti, G.N., Tilson, J.L., *Powerful Learning: 67 What We Know about Teaching for Understanding*, JosseyBass, San Francisco, CA, USA, 2008, [Google Scholar].
9. Herrera, F. and Herrera-Viedma, E., Linguistic Decision Analysis: Steps for Solving Decision Problems under Linguistic Information. *Fuzzy Sets Syst.*, 115, 67–82, 2000, [Google Scholar] [CrossRef].
10. Brooks, R.A., A robust layered control system for a mobile robot. *IEEE J. Robot. Autom.*, 2, 14–23, 1986, [Google Scholar] [CrossRef][Green Version].
11. Quint, J. and Condliffe, B., Project-Based Learning: A Promising Approach to Improving Student Outcomes; ERIC Number: ED580907, Institute of Education Sciences (IES): Washington, DC, USA, 2018, Available online: https://eric.ed.gov/?id=ED580907 (accessed on 22 January 2021).
12. Dole, S., Bloom, L., Kowalske, K., Transforming pedagogy: Changing perspectives from teacher-centered to learner-centered. *Interdiscip. J. Probl. Based Learn*, 10, 2016, [Google Scholar] [CrossRef][Green Version].
13. Benli, E., Motai, Y., Rogers, J., Human behavior-based target tracking with an omni-directional thermal camera. *IEEE Trans. Cogn. Dev. Syst.*, 11, 36–50, 2019, [Google Scholar] [CrossRef].
14. Abdulla, A., Maria, J., Francisco, M., Arturo, E., Jose, M., An Appearance-Based Tracking Algorithm for Aerial Search and Rescue Purposes. *Sensors*, 19, 652, 2019, [Google Scholar] [CrossRef][Green Version].
15. Reynolds, M., Wild frontiers–reflections on experiential learning. *Manage. Learn.*, 40, 387–392, 2009, [Google Scholar] [CrossRef].
16. Aydogmus, Z., A real-time robust fuzzy-based level control using programmable logic controller. *Elektron. Elektrotech.*, 21, 1, 13–17, 2015, 10.5755/j01.eee.21.1.7812.
17. Experion overview, Honeywell Int. Inc, May 5, 2006, Release 300.1.
18. Morari, M. and Zafiriou, B., Robust Process Control, Prentice Hall, New Jersey, 1989. Mousa, H.M., Koutb, M.A., El-Araby, S.M., Elsayed, H.M.A.,

Design of optimal fuzzy controller for water level of U-Tube steam generator in nuclear power station. *J. Am. Sci.*, 7, 4, 629–637, 2011.
19. Yordanova, S., Intelligent approaches to real time level control. Int. J. Intell. Sys. Appl., 7, 10, 19–27, 2015, 10.3233/IFS-141242.
20. Thieme, C., Sodium carbonates, in: *Ullmann's Encyclopedia of Industrial Chemistry*, pp. 299–316, Wiley-VCH Verlag GmbH & Co. KGaA, Weinheim, Germany, 2012, https://onlinelibrary.wiley.com/doi/10.1002/14356007.a24_299, 33.
21. Kahraman, C. and Kabak, O., *Fuzzy Statistical Decisio-Making: Theory and Applications*. Studies in Fuzziness and Soft Computing, Springer, Berlin/Heidelberg, Germany, 2018, ISBN 9783319817934, [Google Scholar].
22. Terano, T., Asal, K., Sugeno, M., *Fuzzy Systems Theory and Its Applications*, 1st, Academic Press, San Diego, CA, USA, 1992, [Google Scholar].

Further Reading

Ahmad, S., Ali, S., Tabasha, R., The design and implementation of a fuzzy gain-scheduled PID controller for the Festo MPS PA compact workstation liquid level control. *Eng. Sci. Technol.*, 23, 307–315, 2020, 10.1016/j.jestch.2019.05.014.

Akdogan, E., Topuz, V., Akbas, A., An education tool study on mechatronics: Emulation of stepper motor driving systems by using a micro controller based system interface, in: *Proceedings of the IEEE International Conference on Mechatronics (ICM'04)*, Istanbul, Turkey, 3–5 June 2004, pp. 509–512,[Google Scholar].

Chabni, F., Taleb, R., Benbouali, A., Bouthiba, M.A., The application of fuzzy control in water tank level using Arduino. *Int. J. Adv. Comput. Sci. Appl.*, 7, 4, 261–265, 2016, 10.14569/IJACSA.2016.070432.

Clark, J. and White, G.W., Experiential learning: A definitive edge in the job market. *Am. J. Bus. Educ.*, 3, 115–118, 2010, Available online: http://journals.cluteonline.com/index.php/AJBE/article/view/390/379 (accessed on 22 January 2021). [CrossRef].

Darling-Hammond, L. and Adamson, F., *Beyond Basic Skills: The Role of Performance Assessment in Achieving 21st Century Standards of Learning*, Stanford University, Stanford Center for Opportunity Policy in Education, Stanford, CA, USA, 2010, [Google Scholar].

Fuzzy logic toolbox: User's guide for use with MATLAB, TheMathWorks, Inc, Natick, MA, 1998.

Garay-Rondero, C.L., Calvo, E.Z.R., Salinas-Navarro, D.E., Experiential learning at Lean-Thinking-Learning Space. *Int. J. Interact. Des. Manuf. (IJIDEM)*, 13, 1129–1144, 2019, [Google Scholar] [CrossRef][Green Version].

Grover, R., Krishnan, S., Shoup, T., Khanbaghi, M., A competition-based approach for undergraduate mechatronics education using the arduino platform, in:

Proceedings of the 2014 4th Interdisciplinary Engineering Design Education Conference (IEDEC), Santa Clara, CA, USA, pp. 78–83, 3 March 2014, [Google Scholar].

Habib, M.K., Human adaptive and friendly mechatronics (HAFM), in: *Proceedings of the IEEE International Conference on Mechatronics and Automation*, Takamatsu, Japan, 5–8 August 2008, pp. 61–65, [Google Scholar].

Helbing, D., Farkas, S., Vicsek, T., Simulating Dynamic Features of Escape Panic. *Nature*, 407, 487–490, 2020, [Google Scholar] [CrossRef] [PubMed] [Green Version].

Huang, T.-C., Seeing creativity in an augmented experiential learning environment. *Univ. Access Inf. Soc*, 18, 301–313, 2019, [Google Scholar][CrossRef].

Jantzen, J., Foundations of Fuzzy Control, John Wiley and Sons, New York, 2007, Online ISBN: 9780470061176, doi: 10.1002/9780470061176.

Kanagasabai, N. and Jaya, N., Fuzzy gain scheduling of PID controller for a MIMO process. *Int. J. Comput. Appl.*, 91, 10, 13–20, 2014, 10.5120/15916-4803.

Kasabanov, N. and Kozma, R., *Neuro-Fuzzy Techniques for Intelligent Information Systems*, vol. 30, p. 449, Springer, Berlin/Heidelberg, Germany, 1999, ISBN1 -10 3790811874. ISBN2 -13 978-3790811872, [Google Scholar].

Kitahara, N., Nagata, F., Otsuka, A., Sakakibara, K., Watanabe, K., Habib, M.K., A proposal of experimental education system of mechatronics, in: *Proceedings of the 17th International Symposium on Artificial Life and Robotics (AROB 17th '12)*, Beppu-Oita, pp. 166–169, Japan, 19–21 January 2012, [Google Scholar].

Kolb, D.A., Experiential Learning: Experience as the Source of Learning and Development, Prentice-Hall, Englewood Cliffs, NJ, USA, 1984, [Google Scholar].

Kolbari, H., Sadeghnejad, S., Parizi, A.T., Rashidi, S., Baltes, J.H., Extended fuzzy logic controller for uncertain teleoperation system, in: *Proceedings of the 4th International Conference on Robotics and Mechatronics, Tehran, Iran*, 26–28 October 2016, pp. 78–83, [Google Scholar].

Krajcik, J.S. and Shin, N., Project-based learning, in: *The Cambridge Handbook of the Learning Sciences*, 2nd, R.K. Sawyer, (Ed.), pp. 275–297, Cambridge University Press, New York, NY, USA, 2014, [Google Scholar].

Lee, G.H. and Jung, S., Design and control of an inverted pendulum system for intelligent mechatronics system control education, in: *Proceedings of the 2008 IEEE/ASME International Conference on Advanced Intelligent Mechatronics*, Xian, China, pp. 1254–1259, 2–8 July 2008, [Google Scholar].

Leal-Rodríguez, A.L. and Albort-Morant, G., Promoting innovative experiential learning practices to improve academic performance: Empirical evidence from a Spanish Business School. *J. Innov. Knowl.*, 4, 97–103, 2019, [Google Scholar] [CrossRef].

Liliana, D. and Florina, P.S., Education, knowledge and innovation from a mechatronics perspective. *Proc. Soc. Behav. Sci.*, 203, 205–209, 2015, [Google Scholar] [CrossRef][Green Version].

Minea, M., Dumitrescu, C., Costea, I., Chiva, I., Semenescu, A., Developing a Solution for Mobility and Distribution Analysis Based on Bluetooth and Artificial Intelligence. *Sensors*, 20, 7327, 2020, [Google Scholar] [CrossRef] [PubMed].

Mioduser, D., Framework for the study of cognitive and curricular issues of technological problem solving. *Int. J. Technol. Des. Educ.*, 8, 167–184, 1998, [Google Scholar] [CrossRef].

Neshkov, T., Yordanova, S., Topalova, I., Process Control and Production Automation, Technical University of Sofia, Sofia, 2007.

Pratumsuwan, P., A conceptual framework and its application for project developing in mechatronics education, in: *Proceedings of the 2014 2nd International Conference on Technology, Informatics, Management, Engineering, and Environment (TIME-E)*, Bandung, Indonesia, pp. 222–227, 19–21 August 2014, [Google Scholar].

Precup, R.E., David, R.C., Petriu, E.M., Radac, M.B., Fuzzy logic-based adaptive gravitational search algorithm for optimal tuning of fuzzy-controlled servo systems. *IET Control Theory Appl.*, 7, 1, 99–107, 2013, 10.1049/iet-cta.2012.0343.

Roboaca, S., Dumitrescu, C., Mantan, I., Aircraft Trajectory Tracking using Radar Equipment with Fuzzy Logic Algorithm. *Mathematics*, 8, 207, 2020, [Google Scholar] [CrossRef][Green Version].

Sell, R., Seiler, S., Ptasik, D., Microcontroller based intelligent platform for research and education in mechatronics, Mechatronics (MECATRONICS), in: *Proceedings of the 9th France-Japan & 7th Europe-Asia Congress on and Research and Education in Mechatronics (REM)*, Paris, France, pp. 188–193, 21–23 November 2012, [Google Scholar].

Shome, A. and Ashok, D., Fuzzy logic approach for boiler temperature and water level control. *Int. J. Sci. Eng. Res.*, 3, 6, 1–6, 2012.

Steels, L. and Brooks, R., *The Artificial Life Route to Artificial Intelligence: Building Embodied, Situated Agents*, Routledge, Oxford, UK, 2018, [Google Scholar].

Tan, W., Water level control for a nuclear steam generator. *Nucl. Eng. Des.*, 241, 5, 1873–1880, 2011, 10.1016/j.nucengdes.2010.12.010.

Tanaka, K. and Wang, H.O., *Fuzzy Control Systems Design and Analysis: A Linear Matrix Inequality Approach*, John Wiley and Sons, New York, 2001.

Tokuyasu, T., Installation of mechatronics education using the MindStorms for dept. of mechanical engineering. *O.N.C.T. Mechatron. Syst. Appl.*, 339–352, 2010, [Google Scholar] [CrossRef][Green Version].

Versaci, M., Calcagno, S., Cacciola, M., Morabito, F., Palamara, I., Pellicanò, D., Innovative Fuzzy Techniques for Characterizing Defects in Ultrasonic Nondestructive Evaluation, in: *Ultrasonic Nondestructive Evaluation Systems*, P. Burrascano, S. Callegari, A. Montisci, M. Ricci, M. Versaci, (Eds.), Springer: Cham, Switzerland, 2015, ISBN 978-3-319-10565-9, [Google Scholar] [CrossRef].

Widiastuti, I. and Budiyanto, C.W., Applying an Experiential Learning Cycle with the Aid of Finite Element Analysis in Engineering Education. *J. Turk. Sci. Educ.*, 15, 97–103, 2018, [Google Scholar].

Wolf, D., Portfolio assessment: Sampling pupil's work. Educ. Leadersh., 45, 35–39, 1989, [Google Scholar].

Yilmaz, O. and Tuncalp, K., A mixed learning approach in mechatronics education. *IEEE Trans. Educ.*, 54, 294–301, 2011, [Google Scholar] [CrossRef].

Yordanova, S., Design of fuzzy supervisor-based adaptive process control systems, in: *New Approaches in Intelligent Control: Techniques, Methodologies and Applications*, K. Nakamatsu and R. Kountchev (Eds.), Springer Int. Publ., Switzerland, 2016, DOI: 10.1007/978-3-319-32168-4_1.

Yordanova, S., *Methods for Design of Fuzzy Logic Controllers for Robust Process Control*, 2011, KING, Sofia, Bulgarian.

Yordanova, S., Slavov, M., Gueorguiev, B., Parallel distributed compensation for improvement of level control in carbonisation column for soda production. *Contr. Eng. Pract.*, 71, 53–60, 2018, 10.1016/j.conengprac.2017.10.003, https://www.aspentech.com/en/products/msc/aspen-infoplus21.

Zadeh, L., Fuzzy logic = computing with words. *IEEE Trans. Fuzzy Syst.*, 4, 103–111, 1996, [Google Scholar] [CrossRef][Green Version].

6

Application of PLC in the Mechatronics Industry

Harpreet Kaur Channi[1*], Pulkit Kumar[1] and Arvind Dhingra[2]

[1]*Department of Electrical Engineering, Chandigarh University, Mohali, Punjab, India*

[2]*Department of Electrical Engineering, Guru Nanak Dev Engineering College, Ludhiana, Punjab, India*

Abstract

Industrial automation and control systems have been revolutionized by the mechatronics sector's widespread use of programmable logic controllers (PLCs). Many industries are adopting mechatronics, merging mechanical, electrical, and computer engineering to improve efficiency, accuracy, and production. This study delves into the widespread use of PLCs in mechatronics applications, covering their underlying concepts, programming methods, and seamless integration with mechatronics gadgets. The first section of the article serves as an introduction to PLCs, discussing their history, architecture, and programming languages. It highlights the benefits of using PLCs in mechatronics systems and how well they interface with other devices and parts. The fundamentals of PLC programming, including ladder logic, sequential function charts, and function block diagrams, are also covered in depth. The importance of sensors and actuators for monitoring physical variables and regulating mechanical motion is discussed in this study, which focuses on PLC-based mechatronics systems. It highlights the effective deployment of PLCs in various industrial contexts by showcasing real-world case studies, including illustrations of PLC-controlled automated assembly lines, robotic arms, and computer numerical control (CNC) machine tools. The importance of motion control in mechatronics and the role that PLC-based motion control approaches play in obtaining precise motion are discussed. In addition, the study explores human–machine interaction/PLC integration, guiding how to create intuitive user interfaces for mechatronics. The research, conducted as interest in Industry 4.0 grows, looks at PLCs and how they may be used to facilitate smart production by connecting to the Internet of

**Corresponding author*: harpreetchanni@yahoo.in

Kolla Bhanu Prakash, Satish Kumar Peddapelli, Ivan C. K. Tam, Wai Lok Woo and Vishal Jain (eds.) Computational Intelligent Techniques in Mechatronics, (185–210) © 2024 Scrivener Publishing LLC

Things. It also covers security and safety concerns, defining guidelines for protecting PLC-based mechatronics systems. Future developments in PLC technology, integration with machine learning and artificial intelligence, and potential applications and prospects are discussed to round out the presentation. Overall, the study shows how important PLCs are in advancing the mechatronics sector and enabling more advanced forms of industrial automation.

***Keywords*:** Programmable logic controllers (PLCs), mechatronics, motion control, Industry 4.0

6.1 Introduction

Programmable logic controllers (PLCs) have come a long way from their early versions as basic relay controllers and are now capable of completing complicated automation tasks with pinpoint accuracy and high throughput as shown in Figure 6.1. They are crucial in contemporary manufacturing because of their capacity to process data from a wide variety of sources, exchange information with other systems, and carry out logical operations. Advanced equipment, robotics, and automated processes have all benefited from mechatronics because of the field's focus on the integration of mechanical and electrical components with sophisticated control systems. As a result of this integration, operational efficiency, downtime, and product quality have all been increased across a wide range of mechatronic applications, such as factory automation, robotics, automotive systems, and smart factories. The constant cooperation between PLCs and mechatronics is moving the industry toward increasingly complex and linked systems that are crucial in creating the future of automation and industrial technology.

A programmable logic controller (PLC) is a multipurpose computer device used to automate and control a wide range of operations in manufacturing and other industrial contexts. Control logic modules, or PLCs, are built to take sensor data as input, process it using logic that has already been coded, and then send commands to actuators and other devices. Engineers may readily make changes to their functioning because of the programming versatility they provide. PLCs are an essential part of today's mechatronics systems because of their significant impact on improving productivity, accuracy, and safety in industrial automation. Regarding automating factories and other industrial processes, mechatronics is a cutting-edge multidisciplinary discipline that combines mechanical engineering, electrical engineering, computer science, and control systems. This approach to interdisciplinary engineering aims to improve industrial efficiency, accuracy, and production by developing complex and intelligent

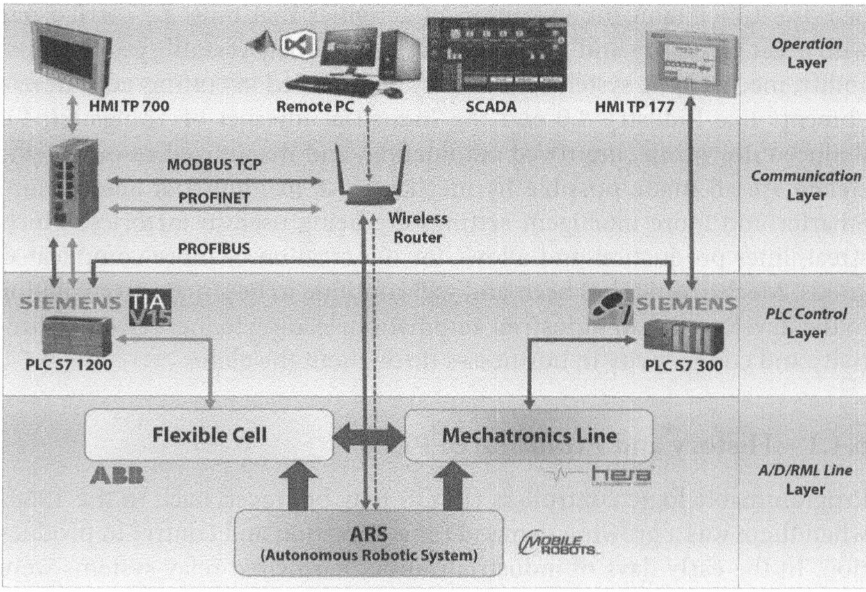

Figure 6.1 PLC-based autonomous robotic system [1] (CCBY 4.0).

systems. Smart machines and systems with cutting-edge sensors and actuators are the backbone of mechatronics in industrial automation. Sensors like this constantly monitor environmental factors, including temperature, pressure, speed, and location. In contrast, actuators take commands from smart controllers and turn them into precisely executed mechanical motions. When sensors and actuators work together, mechatronic systems may improve their operation in response to dynamic conditions. Intelligent controllers, such as microcontrollers and PLCs, are essential to mechatronics in industrial automation. These controllers take in information from sensors and, using algorithms written in advance, make calculated judgments about how to command the actuators. Improved product quality and less waste result from more accurate and reactive industrial operations management. Mechatronics is important in industrial automation, especially when incorporating human–machine interaction (HMI). HMIs make it easy for operators and engineers to work with mechatronic systems by providing a streamlined interface. This let them keep tabs on real-time operations, get critical information, and exert effective machine control. HMIs help businesses run more smoothly and make better decisions because of their user-friendly interfaces. Increased levels of precision and accuracy in manufacturing processes are largely made possible by mechatronics in industrial automation. Increased productivity is the result because it facilitates the execution of activities that must be carried out with great regularity and precision. Because of their versatility and adaptability, mechatronic systems may easily be included in cutting-edge developments like Industry 4.0 and the Industrial Internet of Things (IIoT). Reduced downtime, improved automation, and maximized resource efficiency are all made possible by mechatronics in industrial automation. Smarter and more intelligent settings are being used in factories, which streamlines production and allows for the creation of more complicated goods. Mechatronics has been and will continue to be a major contributor to the development of industrial automation, leading to increased productivity and connectivity in businesses throughout the globe.

6.1.1 History and Evolution of PLCs

Programmable logic controllers (PLCs) may be traced back to the 1960s when there was a growing demand for automation and control in production. In the early days of industrialization, hardwired relay systems were used to operate factory machines, but these systems were difficult to adapt to new uses. The seminal change that initiated PLC formation occurred in the late 1960s. The first modern PLC was created in 1968 by a team of GM engineers headed by Richard E. Morley and dubbed the "Modicon 084." This cutting-edge tool is a programmable alternative to traditional hardwired relay systems. It used digital logic and stored programming instructions, enabling engineers to readily alter control functions without having to rewire the hardware.

Because of its adaptability, user-friendliness, and dependability, the Modicon 084 inspired the name "programmable logic controller" and rapidly became a standard in many other fields. The PLC was a game-changer in industrial automation because it offered a flexible and inexpensive method of managing complex systems. PLCs grew in strength and sophistication during the 1970s and 1980s. Increases in processing speed and memory capacity brought about by advances in microprocessor technology have made PLCs capable of performing sophisticated control tasks and facilitating interoperability with other devices. Standardized programming languages and communication protocols became possible with the advent of "open" PLCs in the 1990s. With these standards in place, PLCs and other industrial devices could communicate, which paved the way for more efficient system integration. More networked and interconnected automation systems were built in the 2000s when PLCs were combined with newer technologies like Ethernet and the Internet. Modular PLC systems, whereby separate modules may be added or swapped out to increase or decrease the controller's capabilities, also gained popularity during this period. PLCs have developed and adapted in recent years to fulfill the needs of Industry 4.0 and IIoT. PLCs are increasingly widely used in smart manufacturing, where they facilitate the sharing of information and the making of well-informed decisions that improve production efficiency. From its origins as a substitute for hardwired relay systems to their current status as the backbone of contemporary mechatronics and industrial control systems, the evolution of PLCs is a testimony to their importance in industrial automation. Their ongoing development has revolutionized manufacturing

6.1.2 Literature Review

Martell, F., & Sanchez, I. Y. (2024), discussed that the control systems are essential components in mechatronic products as well as in the automation of production systems. The design of control systems is generally required for the development of multidisciplinary engineered systems. Mechatronic design methodologies rely heavily on system modeling and real time computer simulation, which are fundamental engineering tools for the design and verification of control systems [2]. Iskandar, H. *et al.* (2024) analyzed various actions to achieve R.I. competency 4.0, namely through mechatronics training as a learning transformation. To obtain research data through (1) analysis of basic needs, (2) creation of project-based training program schemes, and (3) evaluation of mechatronics training schemes and training media. The results of the basic needs analysis were carried out to see the dominant material, namely (1) mechanics 43%, (2) electricity 30%, (3) informatics 27% as well as training media needs and time allocation for the training process including; (1) preparation 10%, (2) analysis and design 45%, (3) process management 30%, (4) evaluation 15%, while for assessing the feasibility of training schemes and training media all indicators received a 'valid' interpretation. Based on these results, the training scheme and training media are suitable for use for mechatronics training in the Mechatronics Laboratory as a form of learning transformation in the R.I. 4.0. era [3]. Guo, L., Fang, D., & Tian, F. (2023) described that PLC output power is getting higher and higher, creating more favorable conditions for PLC control coil. With the maturity of sensing technology, the level of manipulator assembly operation is enhanced. The application of electromechanical intelligent (EI) control technology based on PLC technology is studied and analyzed in this paper. It takes YL-235A optical EI equipment as the research object, introduces its basic working process, introduces the hardware composition, electrical control composition, software required for work and use requirements of YL-235A optical EI training equipment in detail, and finally discusses the overall working process and requirements of the equipment in detail, and discusses the application of EI intelligent control technology of PLC technology [4].

Kaswan *et al.* (2022) aimed to propose a framework to integrate Green Lean Six Sigma (GLSS) and Industry 4.0 to improve organizational sustainability. The integration of GLSS and Industry 4.0 is proposed based on theoretical facets of the individual approaches. A generic, conceptual framework of an integrated GLSS–Industry 4.0 approach is then proposed using the application of different tools and techniques of GLSS and Industry 4.0 at different stages of the realization of a project. Findings of both approaches have common facets related to enablers and barriers, and the integrated application of tools and techniques of each approach supplements the common focus of both related to sustainability enhancement. The development of cultured meat, the full application of by-products, and the emergence of new technologies provide opportunities for the application of 3D printing in the meat industry [5]. Dong *et al.* (2022) highlight the current challenges and opportunities for the application of 3D printing in meat to provide new ideas for the development of 3D printing. The use of natural surfactants can reduce the generation of pollutants from the use of synthetic detergents or chemical solvents without sacrificing economic gains for the oil industry [6]. Filho *et al.* (2023) aimed to provide an overview of many existing studies claiming that the widespread application of digital twins will shift the industry to a principally new level of development [7].

Khalyasmaa *et al.* (2023) provide an extensive overview of the industrial application experience of digital twin technologies for solving the problems of modern power systems with a particular focus on the task of high-voltage power equipment life-cycle management [8]. Mo *et al.* (2023) presented a design of digital-twin human–machine interface sensor (DT-HMIS) with intelligent finger gesture recognition.

The design of a DT-HMIS is proposed. This is a digital-twin sensor that can meet the demands of human–machine automation collaboration in Industry 5.0. Augmented reality (AR) technology can skilfully blend virtual information with the real world, so it can greatly help the development of the manufacturing industry [9]. Zhao *et al.* (2023) studied the introduction to the application of AR technology in the digital manufacturing process. The AR technology is mainly focused on 3D laser projection technology [10].

6.1.3 Scope and Objectives

The scope of implementing mechatronics systems in various industries is multifaceted and aimed at achieving specific goals. Mechatronics, as an interdisciplinary field that integrates mechanical, electronic, and computer engineering, offers a wide range of applications and objectives. It encompasses the seamless integration of mechanical, electronic, and computer engineering components to create systems that are more efficient, precise, and adaptable. Mechatronics systems are designed to automate complex processes, control machinery, and enable autonomous operation, improving productivity.

They focus on achieving high levels of precision and accuracy in tasks such as manufacturing, quality control, and medical diagnostics. The application of mechatronics can lead to improved energy efficiency and resource utilization, reducing waste and costs. Additionally, mechatronics systems often include safety features and redundancy to enhance system reliability and protect personnel and equipment. They drive innovation in product development, enabling the creation of advanced technologies and products, such as self-driving cars, robotic surgeries, and smart manufacturing systems. Mechatronics systems are designed to be adaptable and easily reconfigurable to meet changing requirements and tasks.

The objectives of implementing PLCs with mechatronics systems are multifaceted and centered around enhancing the efficiency, control, and functionality of mechatronic applications. PLCs are integral components of mechatronics systems, and their objectives include:

- One of the core objectives is to achieve high levels of precision and accuracy in mechatronic processes. PLCs provide fine-grained control over critical parameters, enabling systems to perform tasks with minimal errors or deviations.
- PLCs contribute to improved energy efficiency and resource utilization by optimizing the operation of mechatronic systems. They help reduce energy consumption, material wastage, and overall operational costs.
- Safety is a paramount objective when integrating PLCs with mechatronics systems. PLCs facilitate the implementation of safety interlocks, emergency stop systems, and fault tolerance mechanisms to ensure the well-being of personnel and the protection of equipment

6.2 Role of PLC in Mechatronics System Integration

In order to orchestrate the smooth combination of mechanical and electrical parts with smart control systems, the area of mechatronics relies heavily on the incorporation of PLCs. Integrated systems rely heavily on

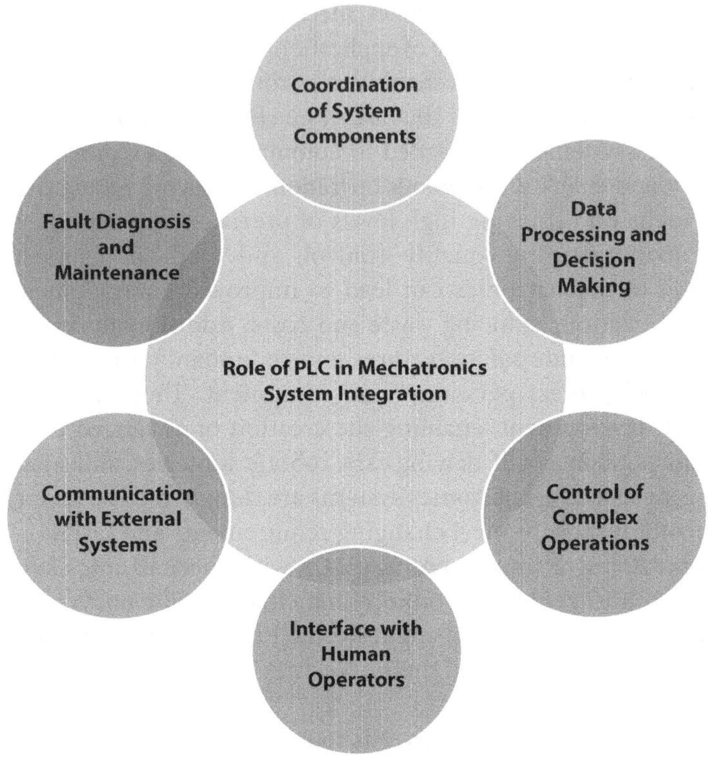

Figure 6.2 Role of PLC in mechatronics system integration.

PLCs, which operate as the system's "brain" and coordinate everything from motors and sensors to actuators and computers. These controllers are responsible for handling real-time data, making quick judgments based on predetermined algorithms, and carrying out many activities simultaneously, all in an effort to keep complicated processes running smoothly. Moreover, PLCs act as a bridge between the mechatronic system and its human operators by delivering user-friendly interfaces for data visualization, control, and monitoring. Moreover, because they can talk to other systems, including supervisory control and data acquisition (SCADA) and manufacturing execution systems (MES), they can be easily integrated into larger industrial processes and data management infrastructures.

PLCs are crucial to the smooth operation and integration of contemporary mechatronic systems because they facilitate problem detection and expedite maintenance, which, in turn, reduces downtime, increases system flexibility, and maximizes overall productivity. Programmable logic controllers (PLCs) play a crucial role in ensuring the smooth operation of sophisticated automated systems in the field of mechatronics as shown in Figure 6.2. Programmable logic controllers (PLCs) serve as the nerve center, coordinating the activities of mechanical parts, sensors, actuators, and computers. PLCs play many crucial parts in the integration of mechatronics systems.

6.2.1 Integration of PLC with Mechanical Systems

Fundamental to mechatronics is the coordination and control of diverse mechanical components in automated settings via the use of PLCs. Actuators like as motors, hydraulic systems, and pneumatic cylinders are precisely controlled and regulated by PLCs via their interface with these devices. Furthermore, PLCs oversee complex tasks like speed regulation, trajectory planning, and synchronized movement of multiple axes through their sophisticated motion control capabilities, allowing for the harmonious coordination of mechanical systems in uses ranging from robotics to CNC machines. PLCs are able to make instantaneous modifications to account for environmental changes or variations in operations because they include feedback mechanisms like as sensors and encoders, which monitor the state of mechanical components. When it comes to safeguarding workers and expensive machinery, PLCs play a crucial role in handling emergency stops, safety gates, and other essential duties. Within the larger mechatronics framework, PLCs contribute to increased operational efficiency, enhanced precision, and overall safety by executing complex control algorithms for tasks such as temperature regulation, pressure control, and material handling.

6.2.2 Integration of PLC with Electrical Systems

Integrating PLCs with electrical systems is crucial in the field of mechatronics because it allows for the streamlined coordination and accurate control of various electrical components in automated settings. PLCs play a significant role in maintaining reliable power distribution, which is crucial for the proper operation of machinery and the prevention of damage from power fluctuations. In addition, PLCs may be programmed to communicate with various sensors and switches, enabling continuous monitoring of voltage, current, and resistance. With this capacity, electrical components may be controlled and fine-tuned in real time to ensure optimum performance at all times. In addition, PLCs provide for the exact control and regulation of motor speed in applications that need it, thanks to their ability to regulate and control electrical drives like variable frequency drives and servo drives. By enforcing safety procedures such as the usage of circuit breakers, fuses, and emergency shut-off mechanisms, PLCs help to make a facility safer for its employees and the equipment they work with. To further improve operational efficiency and safety in a wide range of mechatronic applications, PLCs establish seamless communication with control panels and operator interfaces, providing a user-friendly platform for operators to monitor, control, and troubleshoot electrical processes.

6.2.3 Integration of PLC with Computing Systems

To facilitate the smooth convergence of physical components and digital interfaces, contemporary mechatronics relies heavily on the integration of PLCs with computer systems. Informed decision-making and precise control over the integrated mechatronic processes are made possible by this integration's seamless data exchange and processing. PLCs provide monitoring and control throughout the mechatronic landscape via their strong interconnection with industrial networks such as Ethernet, Fieldbus systems, and wireless settings [11]. PLCs work in tandem with computers to accomplish complicated control algorithms and logic functions, so ensuring the proper execution of difficult tasks and boosting efficiency in the intertwined mechanical and electrical domains. PLCs and computer systems with user-friendly HMIs offer a robust foundation for efficient system monitoring, control, and data visualization, increasing the mechatronic setup's accessibility and usability. In addition, the recording and reporting capabilities made possible by this interface allow for complete data management, which, in turn, allows performance assessment and predictive maintenance planning, both of which improve operating efficiency and expedite maintenance schedules [12]. This seamless integration of PLCs and computer systems exemplifies the characteristics of a data-driven, efficient, and interconnected mechatronic ecosystem.

The use of PLCs for system integration in the field of mechatronics has several advantages that boost the productivity and reliability of manufacturing processes. PLCs' primary benefit is the streamlined coordination that they provide between the many mechanicals, electrical, and computational parts of the system. With everything under one roof, it is easier to share information and coordinate actions, leading to better overall performance. Further, PLCs allow for real-time monitoring and modifications, which improves operational flexibility and response to environmental changes or production needs in real time [12]. The system can be quickly adapted to new production requirements and operating environments because of its programmability and flexibility, which also allows for easy reconfiguration. PLCs also improve system dependability by allowing for more thorough safety measures and diagnostics to be put in place, which reduces the likelihood of malfunctions and keeps machinery and people safe. Across a wide range of industries, PLCs are used in system integration because of the critical roles that they play in ensuring the efficiency, productivity, and safety of mechatronic systems [13].

6.3 PLC Applications in Mechatronics Industry

Figure 6.3 shows the potential applications of PLCs in the mechatronics industry are diverse and continually expanding. Mechatronics, which harmoniously combines mechanical, electronic, and computer engineering, leans heavily on PLCs for orchestrating and fine-tuning intricate systems across a multitude of domains. From manufacturing automation, where PLCs are the linchpin for controlling machinery, conveyors, and robots with impeccable precision, to the realm of robotics and industrial automation, where PLCs dictate the movements and tasks of robotic arms, even in hazardous environments. In packaging and material handling systems, PLCs take charge of conveyor belts and sorting mechanisms, ensuring that

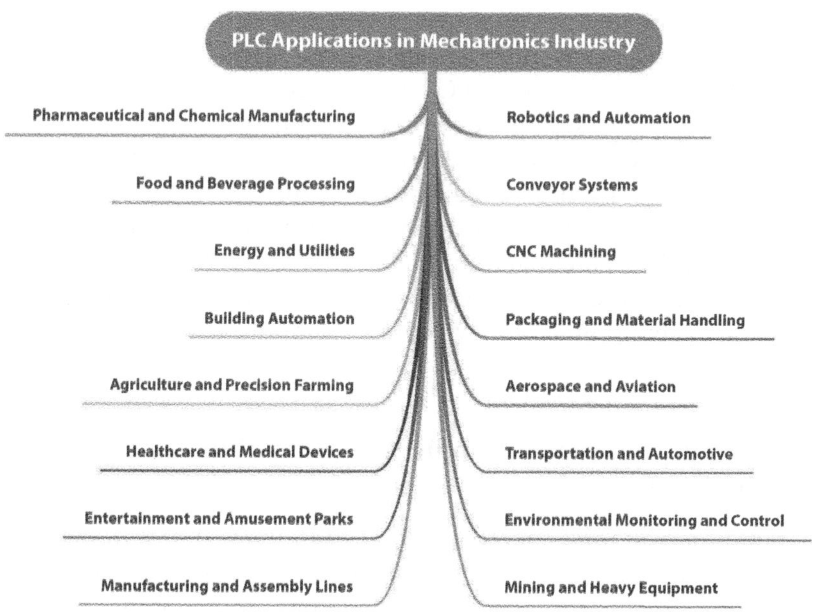

Figure 6.3 PLC applications in mechatronics industry.

products are efficiently processed and prepared for distribution. In CNC machining and 3D printing, PLCs deliver precise tool and print head control, yielding components and products of unparalleled accuracy. In the aerospace industry, PLCs govern aircraft functions, from landing gear to wing flaps, and underpin the assembly of aircraft components. In pharmaceutical and chemical manufacturing [14], PLCs manage complex processes, ensuring precision, repeatability, and regulatory compliance. The food and beverage industry relies on PLCs for cooking, mixing, filling, and packaging, upholding product consistency and hygiene standards. Within the energy sector, PLCs are pivotal in controlling power generation, distribution, and transmission, maintaining electrical grids, water treatment facilities, and renewable energy installations. Building automation systems depend on PLCs to oversee HVAC, lighting, security, and fire safety, enhancing energy efficiency and occupant comfort. PLCs are instrumental in modern agriculture, automating tasks like precision planting, irrigation, and monitoring, ultimately increasing crop yields and resource efficiency. In healthcare, PLCs are integral to medical devices, guaranteeing accuracy and reliability in diagnostic equipment, patient monitoring systems, and laboratory instruments [15].

Transportation and automotive sectors benefit from PLCs that manage traffic lights, railway systems, public transportation, and engine control units for vehicles. PLCs also contribute to environmental monitoring and control by supervising air quality, water treatment, and waste management systems, ensuring environmental safety. Even in the realm of entertainment and amusement parks, PLCs play a crucial role in ride safety and synchronization, overseeing ride sequences, safety systems, and lighting effects. Lastly, PLCs are instrumental in the mining industry, controlling heavy machinery, drilling equipment, and conveyor systems, optimizing resource extraction and transportation. These applications merely scratch the surface of PLC potential in mechatronics, as PLCs continue to evolve and expand their role in automation, control, and precision across various industries, enabling increasingly sophisticated and efficient mechatronic systems [16].

6.3.1 Programming and Implementation of PLC in Mechatronics

Programming and implementing PLCs within the realm of mechatronics requires a systematic approach that encompasses various crucial stages. First and foremost, grasping the essentials of PLC programming languages, including ladder logic, function block diagrams, and structured text, lays the foundation for crafting logical programs that govern the intricate processes and components within the mechatronic system. Employing specialized PLC software development tools and simulation software is essential for creating, testing, and refining PLC programs prior to their deployment, ensuring seamless integration and optimal functionality within the mechatronic environment. Moreover, giving due consideration to design aspects such as system requirements, hardware specifications, and interface compatibility is imperative for the successful implementation of PLCs. Selecting appropriate PLC hardware and peripherals that align with the specific demands of the mechatronic system is vital for achieving a cohesive and efficient integration process [17].

Finally, conducting thorough testing and validation of the PLC program within the mechatronic setup is paramount to guaranteeing its smooth operation and adherence to predefined control logic and safety protocols. By following these systematic steps, engineers and technicians can navigate the complexities of programming and implementing PLCs in mechatronics, ultimately contributing to the optimization of industrial processes and the seamless coordination of diverse mechanical, electrical, and computing components within the system. Programming and implementing PLCs in the field of mechatronics is a critical process that involves several key steps and considerations [18].

6.4 PLC in Mechatronics System Design

6.4.1 Integration of PLCs in Mechatronics Systems

Integration of PLCs into mechatronics systems involves utilizing PLCs as a central control unit to manage and coordinate various components and processes within the mechatronic system. These PLCs serve as the central intelligence, responsible for orchestrating the operation of mechanical, electrical, and electronic elements. They interface with sensors to gather real-time feedback from the environment, enabling precise monitoring and adjustment of system parameters [14]. PLCs execute programmed logic, ranging from simple conditional statements to complex algorithms, to make decisions based on sensor inputs. Their flexibility allows for reprogramming to adapt to changes in system requirements, making them suitable for dynamic environments. Additionally, many PLCs come equipped with built-in diagnostic features for detecting faults or malfunctions, enhancing system reliability and safety [19].

6.4.2 Mechatronics System Components

Mechatronics systems consist of a combination of mechanical, electrical, electronic, and software components that collectively perform a specific task or function. Mechanical components, including actuators, motors, gears, and structural elements, handle physical movement and manipulation within the system. Electrical components encompass power supplies, relays, switches, and circuitry, providing electrical energy and controlling the flow of electricity. Electronic components, such as microcontrollers, sensors, transducers, and amplifiers, are responsible for signal processing, feedback, and interfacing with the PLC. Sensors capture data about the environment or system state, including position, temperature, and pressure. Actuators, like motors, pneumatic cylinders, and solenoids, convert electrical signals into physical action. The control software and logic, embedded in the PLC and other control units, determine how the system responds to various inputs and conditions, ensuring the effective coordination and operation of the mechatronics system [20].

6.4.3 PLC Hardware Selection

Selecting the right PLC hardware is crucial for optimizing the performance and reliability of a mechatronics system. Considerations in hardware selection include the required processing power and speed, based on the complexity of control tasks. Assess the number and types of I/O points needed to interface with sensors and actuators, ensuring compatibility with their specifications. Verify that the PLC hardware supports necessary communication protocols, such as Ethernet, Fieldbus, or Modbus, for interfacing with other system components. Depending on the operational environment, choose hardware that can withstand environmental factors like temperature extremes, humidity, and vibration. Evaluate scalability and expandability, accounting for potential future modifications or expansions. Balancing performance with budget constraints is essential, and higher-end PLCs with advanced features may come at a higher cost. Vendor reputation, technical support, and compatibility with existing equipment and software tools are also vital considerations. For safety-critical applications, prioritize PLC hardware that complies with relevant safety standards and offers safety features. Lastly, opt for hardware with user-friendly programming tools and ample documentation and training resources to facilitate integration and personnel training [21].

6.5 Safety in Mechatronics Systems

Safety in mechatronics systems refers to the implementation of measures and practices to ensure that the system operates without harm to people, property, or the environment as shown in Figure 6.4. This involves identifying potential hazards, minimizing risks, and establishing protocols for handling emergencies and faults [22].

6.5.1 Safety Standards and Regulations

Safety standards and regulations are integral to guaranteeing the secure operation of mechatronics systems. These standards, established by government authorities and industry associations, provide essential guidelines and prerequisites that manufacturers and operators must adhere to in order to uphold safety. They serve as a structured framework for the design, operation, and maintenance of mechatronics systems, with the primary objective of minimizing risks and averting potential accidents.

Moreover, safety standards are often industry-specific, tailored to the unique requirements and potential hazards of various sectors. For instance, ISO 26262 is pertinent to the automotive industry, delineating functional safety standards, whereas ANSI/RIA R15.06 governs the safety of industrial robots in manufacturing. By emphasizing risk mitigation, these standards specify safety features, mandate risk assessments, and outline performance criteria [23].

Compliance with safety standards is not just a best practice but often a legal mandate, ensuring that mechatronics systems are developed and operated with paramount safety considerations. Non-compliance can lead to legal repercussions and regulatory penalties, making adherence crucial for

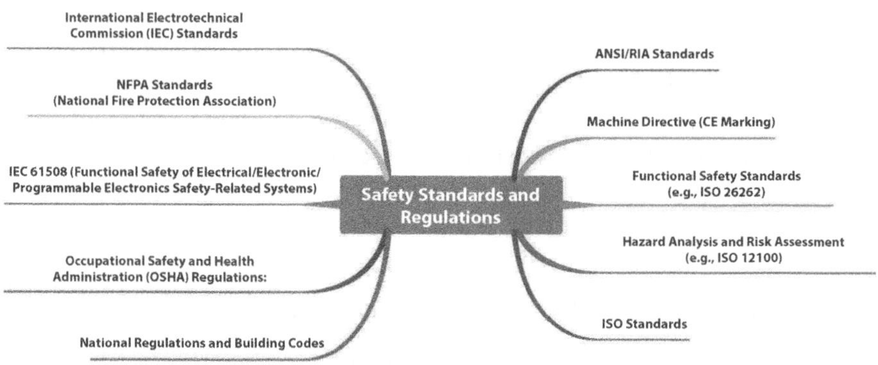

Figure 6.4 Safety standards and regulations for mechatronics systems.

both safety and legal obligations. Additionally, in some cases, systems may necessitate certification to validate their conformity with safety standards, thereby providing an extra layer of assurance that the system effectively meets safety requirements and follows best practices in risk management and hazard mitigation. These measures collectively contribute to creating safer and more reliable mechatronics systems, fostering trust and confidence in their operation within diverse industries [24].

Relevant safety standards and regulations for mechatronics systems encompass a range of sources and guidelines. The International Electrotechnical Commission (IEC) sets global standards that address electrical equipment safety for machines, exemplified by IEC 60204-1, which covers aspects like electrical safety and control circuits. In the United States, the Occupational Safety and Health Administration enforces workplace safety regulations that extend to mechatronics systems in industrial settings, including machine guarding and lockout/tag-out procedures [20]. The International Organization for Standardization (ISO) provides various safety standards, with ISO 13849 focusing on safety guidelines for control systems in mechatronics applications. In the automotive sector, ISO 26262 establishes functional safety standards, ensuring the safe design and testing of mechatronics systems in vehicles. Additionally, the American National Standards Institute (ANSI) and the Robotic Industries Association (RIA) developed ANSI/RIA R15.06, outlining safety standards for industrial robots, which is vital for mechatronics systems involving robots [25].

In the European Union, the Machinery Directive mandates specific safety requirements for machinery, including mechatronics systems, indicated by the Conformité Européenne (CE) marking for compliance with EU safety standards. National governments often contribute to this landscape with their unique regulations and building codes applicable to mechatronics systems across various applications, such as construction, transportation, and healthcare. The choice of which standards and regulations to adhere to is context-specific, depending on the industry, application, and geographical location. Manufacturers and system integrators must diligently navigate and comply with these standards and regulations to guarantee the safety and reliability of mechatronics systems, recognizing that non-compliance carries legal consequences and regulatory penalties and, most crucially, jeopardizes human safety [26].

6.5.2 Safety Interlocks and Emergency Stop Systems

Safety interlocks and emergency stop systems serve as indispensable safety features within mechatronics systems, playing a pivotal role in accident prevention and the safeguarding of individuals, equipment, and the surrounding environment. Safety interlocks consist of mechanisms and sensors designed to verify that specific safety conditions are met prior to permitting system operation. For instance, a safety door interlock ensures that a machine cannot initiate its operation if the safety door remains open, thereby averting accidents arising from unauthorized access or unsafe conditions [27].

In parallel, emergency stop (E-Stop) systems provide a rapid and definitive means to halt system functions during emergency scenarios, and they can be triggered by pressing a button, pulling a cord, or activating a safety switch. The deployment of E-Stop systems is instrumental in promptly terminating system operation to forestall injuries or damage in situations posing imminent threats. Moreover, safety interlocks and E-Stop systems often encompass safety monitoring functions, continually scrutinizing the status of safety devices and relaying feedback to the control system. In cases where a safety device is compromised, these systems can promptly enact appropriate measures to preserve safety. The seamless integration of safety interlocks and E-Stop systems with the control system, allowing them to interface with the PLC or control unit, guarantees that these safety features can effectively halt system processes or modify their behavior as necessary. Consequently, these robust safety measures contribute to the overall protection and dependability of mechatronics systems, instilling confidence in their secure operation across diverse applications [28].

6.5.3 Fault Detection and Tolerance

Fault detection and tolerance mechanisms play a crucial role in the realm of mechatronics systems, as they are implemented to recognize system failures, deviations from standard operation, or potentially hazardous conditions. The paramount objective of these mechanisms is to minimize risks and maintain the safe operation of the system. Mechatronics systems encompass a variety of approaches for fault detection, which includes continuous monitoring of sensor readings, actuator responses, and the overall system performance. These mechanisms may rely on sensor data, redundancy, or analytical algorithms to pinpoint deviations from expected behavior, providing early warning of potential issues. In the event that a fault is identified, the system is programmed to respond judiciously. Responses may entail the shutdown of specific processes or components, triggering alarms, displaying error messages, or even seamlessly transitioning to backup systems to ensure safety. This becomes even more critical in safety-critical applications, where the detection and handling of faults are of paramount importance [29].

Redundancy and self-checking features are commonly integrated to guarantee that the system can maintain safe operation, even in the presence of specific component or sensor failures. The continuous monitoring of faults is an integral part of the system's ongoing evaluation processes, ensuring that the system remains secure and reliable over time. This proactive approach to fault detection significantly reduces the likelihood of accidents and injuries. By comprehensively implementing safety standards, safety interlocks, emergency stop systems, and fault detection mechanisms, mechatronics systems significantly minimize risks, protect human safety, and ensure that the system operates within a safe and controlled environment. These proactive safety measures are indispensable for averting accidents, complying with legal prerequisites, and upholding the overall integrity of the system, fostering trust and confidence in its operation [30].

6.6 Case Studies for Mechatronics Systems Using PLCs

Let us delve into real-world case studies that exemplify the versatility and impact of mechatronics systems using PLCs across various industries.

6.6.1 Automotive Manufacturing

In the automotive industry, PLC-driven mechatronics systems are instrumental in ensuring efficient, precise, and safe vehicle assembly. Case in point, the Tesla Gigafactory employs a myriad of robots and mechatronics systems controlled by PLCs for tasks ranging from welding to painting. These systems ensure consistency in vehicle quality and production efficiency. When a car's chassis moves along the assembly line, PLCs communicate with robotic arms to weld specific parts together with extraordinary precision. Sensors continuously provide feedback to PLCs, enabling real-time adjustments if deviations are detected. This level of automation and control is fundamental to achieving the high production standards required by the automotive industry [31].

6.6.2 Bottling and Packaging Industry

The bottling and packaging industry is another domain where mechatronics systems using PLCs excel. A case study can highlight a beverage bottling plant that relies on PLC-controlled systems for the seamless production of bottled drinks. PLCs manage the entire bottling process, from filling and capping to labelling and packaging. Sensors positioned at various points along the assembly line monitor bottle positions, product quality, and system performance, sending data to the PLC. Based on these data, the PLC ensures that the bottles are filled to the correct level, caps are sealed tightly, and labels are applied accurately. In the event of any discrepancies, the PLC can halt the process and trigger alarms, preventing subpar products from reaching consumers [32].

6.6.3 Aircraft Landing Gear Control

Safety is paramount in the aviation industry, and mechatronics systems using PLCs play a pivotal role in ensuring the safety of aircraft during take-off and landing. A case study can examine the landing gear control system of a commercial airliner. When a plane prepares for landing, the PLC orchestrates the precise extension of landing gear, ensuring a safe touchdown. Sensors located within the landing gear provide feedback to the PLC, confirming that each gear is correctly deployed. In the event of an emergency, such as a sudden need to abort a landing, the PLC rapidly retracts the landing gear, allowing the plane to ascend and make another attempt. This responsiveness and precision are critical for safe air travel [33].

6.6.4 E-Commerce Warehouse Automation

In the fast-paced world of e-commerce, warehouse automation is a competitive advantage. A case study can delve into a state-of-the-art e-commerce distribution center that relies on mechatronics systems controlled by PLCs to expedite order processing. These systems coordinate a complex network of conveyor belts, robotic pickers, and automated storage and retrieval systems to efficiently manage inventory and fulfill orders. The PLCs, in this context, are responsible for orchestrating the movement of goods and ensuring that products are picked, packed, and shipped accurately and swiftly. Real-time feedback from sensors allows the PLC to optimize order fulfillment, track inventory levels, and promptly respond to fluctuations in demand [34].

6.6.5 CNC Machining Centers

The precision and versatility of mechatronics systems using PLCs shine in CNC machining centers. A case study could focus on a high-precision manufacturing facility where these systems are integral to producing complex components with microscopic tolerances. CNC machining centers are guided by PLCs that control the movement of cutting tools, such as drills and milling machines, in intricate patterns on workpieces. These systems meticulously follow programmed instructions, ensuring that each component is manufactured with unparalleled accuracy. Additionally, sensors continually monitor the position and condition of tools and workpieces, providing real-time feedback to the PLC. This feedback allows the system to adapt to changing conditions, such as tool wear, guaranteeing consistent component quality throughout the production process [35].

6.6.6 Precision Agriculture

Agriculture has witnessed a transformation through mechatronics systems controlled by PLCs. A case study could showcase a modern agricultural operation employing autonomous tractors and robotic equipment guided by PLCs. These systems undertake tasks such as ploughing, planting, and harvesting crops with remarkable precision and efficiency. Sensors positioned on tractors and equipment monitor environmental conditions, soil quality, and crop growth [36].

6.7 Challenges and Future Trends

6.7.1 Challenges in Implementing Mechatronics Systems

Implementing mechatronics systems presents a combination of opportunities and challenges that demand careful consideration. Figure 6.5 shows these challenges encompass intricate integration involving multiple disciplines, the need for a team with interdisciplinary expertise to bridge gaps in knowledge, cost management amid the advantages of heightened automation, the complexity of maintaining and ensuring the reliability of increasingly sophisticated systems, the mounting concern of cybersecurity threats as these systems become more network-connected, the absence of universal standards posing interoperability and consistency issues, and the constant need to keep up with the rapid evolution of technology in the mechatronics field to maintain system relevance and competitiveness. Addressing these challenges is crucial to ensure the successful deployment of mechatronics systems and unlock their full potential in various industries and applications [37].

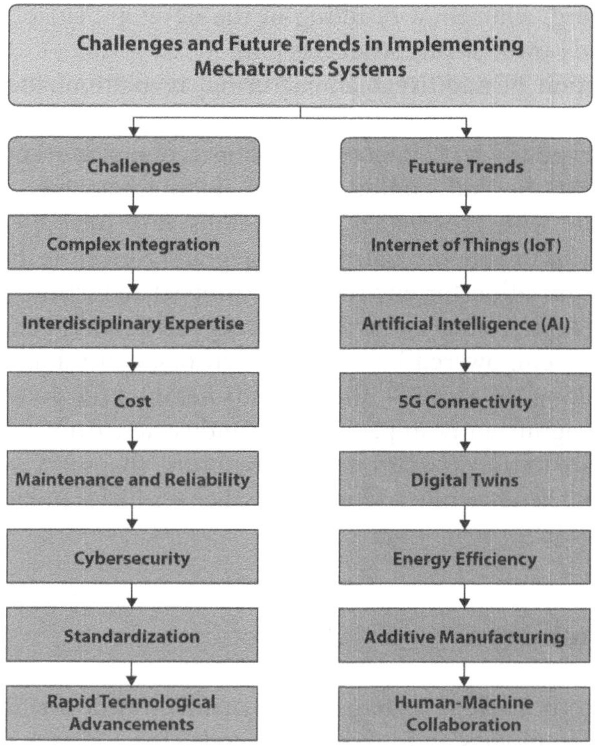

Figure 6.5 Challenges and future trends in implementing mechatronics systems.

6.7.2 Emerging Technologies and Trends in Mechatronics

Mechatronics is on a dynamic trajectory, driven by a wave of emerging technologies and transformative trends. These advancements are reshaping the landscape of mechatronics, propelling it toward a future of unprecedented innovation. Among the key trends and technologies shaping this evolution are the Internet of Things (IoT), which facilitates real-time data collection and predictive maintenance; the integration of artificial intelligence (AI) and machine learning, empowering mechatronics systems with data-driven decision-making and optimization; the rollout of 5G networks, promising low-latency communication for enhanced control and remote operation; the advent of digital twins, enabling comprehensive modeling and simulation to reduce development costs and time; and an escalating focus on energy efficiency, resulting in the development of eco-friendly actuators and power-saving strategies [38].

The adoption of additive manufacturing, revolutionizing component production and reducing lead times; the rise of human–machine collaboration, exemplified by collaborative robots (cobots) working in tandem with humans; the burgeoning field of biomechatronic, revolutionizing healthcare with prosthetics, exoskeletons, and wearable devices; the embrace of sustainable mechatronics, integrating green technologies and resource optimization for environmental impact reduction; and the proliferation of autonomous systems, including self-driving vehicles, drones, and more, are empowered by advanced sensors, control algorithms, and decision-making capabilities. These trends herald a future where mechatronics will significantly impact diverse industries, from manufacturing and healthcare to transportation and agriculture, shaping a world of innovation and efficiency while addressing challenges like standardization and cybersecurity [39].

6.8 Conclusion

The mechatronics field is undergoing a dynamic transformation, driven by emerging technologies and innovative trends. The IoT is making mechatronic systems more interconnected, enabling real-time data analysis, predictive maintenance, and remote monitoring. Artificial intelligence (AI) and machine learning are augmenting decision-making and automation within these systems. The rollout of 5G connectivity promises low-latency communication for real-time control and remote operation. Digital twins are revolutionizing development by enabling virtual modeling and simulation. Energy efficiency is taking center stage, with a focus on eco-friendly components and power-saving strategies. Additive manufacturing is revolutionizing component production. Human–machine collaboration is exemplified by collaborative robots working alongside humans. Biomechatronics is revolutionizing healthcare through prosthetics, exoskeletons, and wearable devices. Sustainable mechatronics is integrating green technologies and resource optimization. Autonomous systems are on the rise, from self-driving vehicles to drones, driven by advanced sensors and decision-making capabilities.

In conclusion, the application of PLC in the mechatronics industry is a testament to human innovation and engineering excellence. These systems, guided by PLCs, are changing the way of manufacturing, automation, and operation across various industries. They offer precision, efficiency, and safety, and they address real-world challenges, from intricate integration to cybersecurity threats. Looking into the future, it is evident that mechatronics will continue to evolve, offering new opportunities and solutions. The key lies in embracing emerging technologies and trends, from IoT and AI to 5G and sustainable practices. These innovations will shape the next generation of mechatronics systems, making them even more efficient, responsive, and capable of addressing the evolving demands of a rapidly changing world.

References

1. Ionescu, D. *et al.*, Communication and Control of an Assembly, Disassembly and Repair Flexible Manufacturing Technology on a Mechatronics Line Assisted by an Autonomous Robotic System. *Inventions*, 7, 2, 43, 2022.
2. Martell, F. and Sanchez, I. Y., *Practical control engineering for mechatronics and automation*. CRC Press, 2024.
3. Iskandar, H., Triono, S., Efendi, H., Project-based mechatronics training scheme and media: Case study in a mechatronics laboratory. In *Proceedings of the 5th International Conference on Innovation in Education, Science, and Culture, ICIESC 2023*, 24 October 2023, Medan, Indonesia, 2024, January.
4. Guo, L., Fang, D., Tian, F., Mechatronic intelligent control technology based on PLC technology. In *International Conference on Computer Graphics, Artificial Intelligence, and Data Processing (ICCAID 2022)*, vol. 12604, pp. 551–557, SPIE, 2023, May.
5. Kaswan, M.S. *et al.*, Integrating Green Lean Six Sigma and industry 4.0: a conceptual framework. *J. Manuf. Technol. Manage.*, 34, 1, 87–121, 2022.
6. Dong, H. *et al.*, 3D printing based on meat materials: Challenges and opportunities. *Curr. Res. Food Sci.*, 6, 100423, 2022.
7. Selva Filho, A.A.P. *et al.*, Biosurfactants as Multifunctional Remediation Agents of Environmental Pollutants Generated by the Petroleum Industry. *Energies*, 16, 3, 1209, 2023.
8. Khalyasmaa, A.I. *et al.*, Review of the Digital Twin Technology Applications for Electrical Equipment Lifecycle Management. *Mathematics*, 11, 6, 1315, 2023.
9. Mo, D.-H. *et al.*, Design of Digital-Twin Human-Machine Interface Sensor with Intelligent Finger Gesture Recognition. *Sensors*, 23, 7, 3509, 2023.
10. Zhao, Z., Jia, Z., Li, Y., Introduction to the application of augmented reality technology in the digital manufacturing process, in: *Ninth Symposium on Novel Photoelectronic Detection Technology and Applications*, SPIE, 2023.
11. Azarmipour, M. *et al.*, Plc 4.0: A control system for industry 4.0, in: *IECON 2019-45th Annual Conference of the IEEE Industrial Electronics Society*, IEEE, 2019.

12. Shi, Z., Wei, H., Zhu, J., Edge computing-empowered task offloading in PLC-wireless integrated network based on matching with quota. *Comput. Commun.*, 182, 110–116, 2022.
13. Baek, S., System integration for predictive process adjustment and cloud computing-based real-time condition monitoring of vibration sensor signals in automated storage and retrieval systems. *Int. J. Adv. Manuf. Technol.*, 113, 955–966, 2021.
14. Guo, L., Fang, D., Tian, F., Mechatronic intelligent control technology based on PLC technology, in: *International Conference on Computer Graphics, Artificial Intelligence, and Data Processing (ICCAID 2022)*, SPIE, 2023.
15. Mašek, V., Fait, D., Čermák, R., Transfer of learning in mechatronics education for industry 4.0, in: *INTED2023 Proceedings*, IATED, 2023.
16. Jensen, S.F. *et al.*, Digital product passports for a circular economy: Data needs for product life cycle decision-making. *Sustain. Prod. Consumption*, 37, 242–255, 2023.
17. Suyetno, A., Kustono, D., Sudjimat, D.A., Pneumatic control system simulation based a PLC micro in mechatronics courses, in: *AIP Conference Proceedings*, AIP Publishing, 2016.
18. Alizadeh, T. and Mosadeghzad, M., A hands-on course on mechatronics, based on modular production systems, in: *2019 IEEE International Conference on Mechatronics and Automation (ICMA)*, IEEE, 2019.
19. Dosoftei, C.-C., Simulation Power vs. Immersive Capabilities: Enhanced Understanding and Interaction with Digital Twin of a Mechatronic System. *Appl. Sci.*, 13, 11, 6463, 2023.
20. Isermann, R., Designs and Specification of Mechatronic Systems, in: *Springer Handbook of Automation*, pp. 287–313, 2023.
21. Milik, A., On hardware synthesis and implementation of PLC programs in FPGAs. *Microprocess. Microsyst.*, 44, 2–16, 2016.
22. Lin, S. *et al.*, A new function-topology-based method for assessing passive safety of mechatronics systems. *IEEE Access*, 8, 9312–9324, 2020.
23. Mhenni, F., Safety analysis integration in a systems engineering approach for mechatronic systems design, Ecole Centrale, Paris, 2014.
24. Kaltjob, P.O., *Control of mechatronic systems: model-driven design and implementation guidelines*, John Wiley & Sons, Hoboken, New Jersey, 2020.
25. Bruzzone, L., New trends in the control of robots and mechatronic systems. *Appl. Sci. MDPI*, Basel, Switzerland, 13, 5, 3112, 2023.
26. Staessens, T., Lefebvre, T., Crevecoeur, G., Optimizing Cascaded Control of Mechatronic Systems through Constrained Residual Reinforcement Learning. *Machines*, 11, 3, 402, 2023.
27. Kago, M. *et al.*, System Design of Accelerator Safety Interlock for the XFEL/SPRING-8. *Proceedings of IPAC*, pp. 10, 2010.
28. Hong, J. *et al.*, The evolution of the interlocks for KSTAR. *Fusion Eng. Des.*, 194, 113871, 2023.

29. Bhuiyan, E.A. *et al.*, A survey on fault diagnosis and fault tolerant methodologies for permanent magnet synchronous machines. *Int. J. Autom. Comput.*, 17, 763–787, 2020.
30. Cordoneanu, D. and Nițu, C., A review of fault Diagnosis in mechatronics systems, in: *Proceedings of the International Conference of Mechatronics and Cyber-MixMechatronics–2018*, pp. 2, Springer, 2019.
31. Shirley, J. *et al.*, A mechatronics and material handling systems laboratory: experiments and case studies. *Int. J. Electr. Eng. Educ.*, 48, 1, 92–103, 2011.
32. Caballero-García, A., González-Altamirano, M. N., Velázquez-Velázquez, J. E., Galván-Guerra, R., Diseño e Implementación de Controladores: Retos en la integración. *Pädi Boletín Científico de Ciencias Básicas e Ingenierías del ICBI*, 11, Especial4, 147–156, 2023..
33. Jean-Charles, M. and Jian, F., Review on signal-by-wire and power-by-wire actuation for more electric aircraft. *Chin. J. Aeronaut.*, 30, 3, 857–870, 2017.
34. Ojea, J.A. and Solowjow, E., Case Study: Siemens–Flexible Robot Grasping with Deep Neural Networks, in: *Springer Handbook of Automation*, pp. 1467–1473, Springer, 2023.
35. Filipescu, A. *et al.*, Multifunctional Technology of Flexible Manufacturing on a Mechatronics Line with IRM and CAS, Ready for Industry 4.0. *Processes*, 9, 5, 864, 2021.
36. Amertet, S. *et al.*, Assessment of Smart Mechatronics Applications in Agriculture: A Review. *Appl. Sci.*, 13, 12, 7315, 2023.
37. Durfee, W. K., Mechatronics for the masses: A hands-on project for a large, introductory design class. *Int. J. Eng. Educ.*, 19, 4, 593–596, 2003.
38. Tantawi, K. *et al.*, Mechatronics Education-Current and Future Trends, in: *2023 11th International Symposium on Digital Forensics and Security (ISDFS)*, IEEE, 2023.
39. Song, D. *et al.*, Flexible sensors for mechatronic engineering education. *Sens. Int.*, 4, 100236, 2023.

7

Fuzzy Logic and Its Applications in Mechatronic Control Systems

D. Sathya[1]*, G. Saravanan[2] and R. Thangamani[1]

[1]Department of Artificial Intelligence, Kongu Engineering College, Tamil Nadu, India
[2]Department of Computer Science and Engineering, Erode Sengunthar Engineering College, Tamil Nadu, India

Abstract

Fuzzy logic is a computing paradigm that deals with reasoning and decision-making in systems that involve uncertain or imprecise information. Unlike classical binary logic, which deals with true or false values (0 or 1), fuzzy logic allows for degrees of truth between 0 and 1, representing various levels of membership or similarity. Fuzzy logic finds numerous applications in various fields, and one of its significant applications is in mechatronic control systems. Mechatronics is an interdisciplinary field that combines mechanical engineering, electronics, computer science, and control engineering. Fuzzy logic is well suited for mechatronic control systems because it can handle the inherent complexities and uncertainties in these systems. Fuzzy logic can be used to design fuzzy control systems, which are control systems that use linguistic variables rather than precise numerical values. Fuzzy control allows engineers to create control rules based on human expertise and experience, making it easier to control systems that are challenging to model mathematically.

Fuzzy logic–based controllers often exhibit robustness in the face of uncertainties and external disturbances. This is particularly useful in mechatronic systems where variations in the environment, material properties, or component characteristics can occur. Mechatronic systems can be highly nonlinear, making traditional control approaches less effective. Fuzzy logic can handle nonlinearities better, making it an attractive choice for controlling such systems. Fuzzy logic controllers [1] can adapt to changing conditions and system dynamics without needing a complete model of the system. This adaptivity is advantageous in

**Corresponding author*: sathya.ai@kongu.edu

Kolla Bhanu Prakash, Satish Kumar Peddapelli, Ivan C. K. Tam, Wai Lok Woo and Vishal Jain (eds.) Computational Intelligent Techniques in Mechatronics, (211–242) © 2024 Scrivener Publishing LLC

mechatronic systems, which may experience varying operating conditions. Fuzzy logic provides a way to incorporate human-like reasoning and decision-making in control systems. It allows control rules to be expressed in natural language and mimics human decision processes. Mechatronic systems often rely on data from multiple sensors. Fuzzy logic can be used for sensor fusion, combining information from different sensors and handling uncertainties associated with sensor data [2]. Fuzzy logic can be used to develop expert systems for mechatronic control, where the system's behavior is guided by expert knowledge and rules. Fuzzy logic has applications in path planning and navigation tasks in mechatronics, such as autonomous robots or vehicles navigating in complex environments. It enables the development of intelligent control strategies that can effectively handle real-world scenarios and lead to improved performance and robustness. However, it is worth noting that other control techniques, such as neural networks and evolutionary algorithms, are also used in combination with fuzzy logic or as alternatives for specific mechatronic control applications.

Keywords: Fuzzy logic, mechatronic systems, control systems, intelligent control, adaptive systems, rule-based systems

7.1 Introduction

In the intricate landscape of modern engineering, the fusion of mechanical, electrical, and computer sciences has given birth to mechatronics, a multidisciplinary field [3] that aims to design and create intelligent systems [4] with precise control capabilities. However, the inherent complexities, uncertainties, and nonlinear behaviors present in many mechatronic systems challenge conventional control methodologies. This is where the innovative concept of fuzzy logic steps onto the stage, offering a more adaptable and human-like approach to control systems. Fuzzy logic, a cornerstone of soft computing, tackles the dilemma of dealing with imprecise or uncertain information. Unlike traditional binary logic, where statements are either true or false, fuzzy logic acknowledges and quantifies the degrees of membership or truthfulness within a continuum of possibilities. This ability to handle vagueness and partial truths makes fuzzy logic particularly well suited for real-world systems that involve nuances and shades of accuracy. At the heart of fuzzy logic lies the concept of linguistic variables and fuzzy sets. These allow engineers to represent inputs, outputs, and control rules in linguistic terms, akin to human reasoning. Fuzzy logic controllers process these linguistic variables to produce control signals that navigate the complexities of mechatronic systems [5]. This human-like decision-making ability makes fuzzy logic an invaluable tool for managing uncertain environments and nonlinear behaviors that often characterize mechatronic domains.

In the realm of robotics [3], where robots interact with dynamic and unpredictable surroundings, fuzzy logic emerges as a guiding light. It empowers robots to make decisions based on sensor inputs that may be noisy or imprecise. This allows robots to navigate through cluttered environments, handle objects of varying shapes, and adapt their actions to diverse scenarios. Fuzzy logic finds its way into automotive control systems, enhancing safety and efficiency. Adaptive cruise control systems use fuzzy logic to adjust vehicle speed based on distance and relative velocity to the preceding vehicle. Fuzzy engine control optimizes fuel efficiency and emissions across changing driving conditions.

Process control systems in industries grapple with variables that are not always crisp. Fuzzy logic provides a robust solution by accommodating imprecise sensor readings and ensuring stability even in uncertain conditions. This is vital for maintaining the integrity of industrial processes. Household appliances like washing machines and air conditioners leverage fuzzy logic for intelligent decision-making. Fuzzy washing machine controllers adjust wash cycles based on load size and dirtiness, whereas fuzzy air conditioner systems optimize temperature and humidity levels for comfort and energy efficiency [6]. Medical devices, such as infusion pumps, rely on fuzzy logic to deliver precise dosages of medication tailored to individual patient conditions. Fuzzy controllers adapt to varying patient responses and sensor readings, ensuring patient safety and effective treatment. In the skies, where aerospace systems navigate the complexities of flight dynamics, fuzzy logic aids in flight control systems. It helps manage aerodynamic uncertainties [7] and nonlinear behaviors, contributing to smoother flight paths and stable maneuvers. Fuzzy logic plays a role in maintaining optimal conditions in homes. Heating, ventilation, and air conditioning systems utilize fuzzy controllers to balance factors like temperature, humidity, and occupancy, ensuring comfort and energy efficiency. Fuzzy logic transforms traffic management by optimizing traffic light timings in real-time. It considers variables such as traffic density, pedestrian flow, and road conditions to reduce congestion and enhance overall traffic flow. In the intricate dance of mechatronics, where precision meets complexity, fuzzy logic emerges as a valuable partner. Its ability to handle uncertainties, adapt to nonlinear behaviors, and make decisions akin to human cognition has propelled it into the forefront of mechatronic control systems. The applications across diverse industries, from robotics to healthcare, automotive to aerospace, underscore its versatility and transformative potential. As mechatronic systems continue to evolve, fuzzy logic stands as a beacon of adaptable control, ensuring that machines can navigate the uncertain and nonlinear paths of our world with finesse and intelligence.

In the present era, mechatronic systems are pervasive, from industrial automation [8] and robotics to automotive advancements and healthcare devices. These systems comprise intricate networks of sensors, actuators, and computational elements that work in tandem to achieve specific tasks. However, the real world is inherently uncertain, with imprecise data, changing environments, and nonlinear behaviors. Traditional control methods, which rely on precise mathematical models, struggle to cope with such complexities and uncertainties.

In response, engineers are turning to innovative paradigms that can accommodate these challenges. Fuzzy logic, a pillar of soft computing, has gained prominence for its ability to deal with imprecision and uncertainty. Unlike traditional binary logic, fuzzy logic operates in a realm of partial truths, allowing for a more nuanced and human-like approach to decision-making. By leveraging fuzzy logic, engineers can create control systems that exhibit adaptability, robustness, and effectiveness in the face of uncertain and changing conditions.

Fuzzy logic provides a framework that mirrors human reasoning, allowing systems to process imprecise or vague information. It introduces the concept of linguistic variables and fuzzy sets, which enable the representation of data in linguistic terms rather than fixed values. These variables and sets can capture the nuances of real-world scenarios, making fuzzy logic a valuable asset in mechatronics.

In the context of mechatronic control systems, fuzzy logic offers a versatile approach. It can model and manage nonlinear behaviors, adjust to changing environments, and handle uncertainties arising from sensor noise and system dynamics. This adaptability is particularly crucial in scenarios where traditional control methods fall short.

7.1.1 Applications of Fuzzy Logic in Mechatronic Control Systems

The applications of fuzzy logic in mechatronics span across industries and sectors are shown in Table 7.1:

1. **Robotics:** Fuzzy logic enables robots to navigate through dynamic environments, interact with objects of varying shapes, and adapt to unforeseen situations.

Table 7.1 Applications of fuzzy logic in mechatronic control systems.

Application domain	Specific applications
Robotics	Autonomous navigation, manipulation
Automotive Systems	Engine control, adaptive cruise control
Industrial Automation	Process control, quality control
Healthcare Devices	Medical imaging, surgical robots
Aerospace Systems	Flight control, unmanned aerial vehicles
Consumer Electronics	Home automation, appliances
Manufacturing	CNC machines, assembly lines

2. **Automotive Systems:** Fuzzy logic enhances adaptive cruise control and engine management, optimizing performance under diverse driving conditions.
3. **Industrial Automation:** Fuzzy logic plays a key role in process control, ensuring stable and efficient operations in the presence of uncertainties.
4. **Consumer Electronics:** Household appliances equipped with fuzzy logic adapt to user preferences and changing conditions, optimizing energy usage and performance.
5. **Healthcare Devices:** Fuzzy logic–based medical devices adjust treatment parameters based on real-time patient feedback and sensor data.
6. **Aerospace Systems:** Fuzzy logic aids in flight control, enabling smoother maneuvers and precise responses in complex flight dynamics.
7. **Home Automation:** Fuzzy logic optimizes heating, cooling, and lighting systems to create comfortable and energy-efficient living environments.

7.2 Fuzzy Control Systems

7.2.1 Bridging Precision and Flexibility

In the realm of control engineering, where the goal is to manipulate systems to achieve desired outcomes, the traditional approach often relies on precise mathematical models. However, in scenarios where systems exhibit uncertainties, complexities, and nonlinear behaviors, a more adaptable and nuanced approach is required. This is where fuzzy control systems come into play, offering a bridge between the rigidity of traditional control methods and the fluidity of real-world dynamics.

7.2.2 Understanding Fuzzy Control Systems

Fuzzy control systems are a subset of control engineering that leverage fuzzy logic—a form of mathematical reasoning that mimics human decision-making processes. Unlike classical control systems that require exact numerical inputs and outputs, fuzzy control systems operate in a realm of linguistic variables and degrees of membership. They enable control decisions based on imprecise, vague, or uncertain information, allowing for more human-like and flexible responses.

At the core of fuzzy control systems are linguistic variables and fuzzy sets. These linguistic variables define input and output values using linguistic terms (e.g., "high," "low," "warm," and "cold") rather than precise numbers. Fuzzy sets assign degrees of membership to these linguistic terms, allowing for gradual transitions between categories. By applying fuzzy logic rules to these variables and sets, a fuzzy control system generates control actions that accommodate uncertainty and imprecision are shown in Figure 7.1.

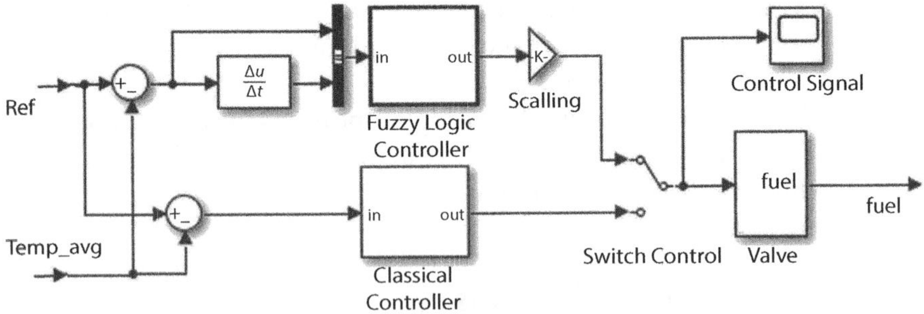

Figure 7.1 Fuzzy control system.

7.2.3 Applications of Fuzzy Control Systems

The adaptability of fuzzy control systems finds applications across various domains:

1. **Temperature Control:** Fuzzy control systems excel in maintaining stable temperatures in environments with fluctuations. They adjust heating or cooling systems based on linguistic variables like "too hot" or "slightly cold."
2. **Industrial Processes:** Fuzzy control optimizes complex industrial processes where variables are difficult to quantify exactly. It ensures smooth operations in scenarios like chemical reactions, where precise measurements may not be feasible.
3. **Traffic Signal Timing:** Fuzzy control enhances traffic signal timing by considering factors such as traffic density and pedestrian flow, leading to more efficient traffic management.
4. **Consumer Electronics:** Appliances like washing machines and air conditioners use fuzzy control to adapt to varying load sizes or environmental conditions, resulting in energy savings and improved performance.
5. **Robotic Systems:** Fuzzy control allows robots to navigate through dynamic environments, make decisions based on sensor inputs, and handle objects of different shapes and sizes.
6. **Financial Trading:** In the realm of algorithmic trading, fuzzy control can adapt strategies based on market volatility and risk tolerance.
7. **Healthcare Devices:** Medical devices, such as insulin pumps, use fuzzy control to adjust dosages based on patient conditions and sensor feedback.

7.2.4 Benefits and Challenges

Table 7.2 show strengths of fuzzy control systems lie in their ability to handle uncertainties and nonlinearities that often challenge traditional control methods. They provide a systematic approach to integrating human expertise into control algorithms, making them more robust and adaptable.

However, implementing fuzzy control systems requires careful consideration of linguistic variable definitions, fuzzy set memberships, and

Table 7.2 Benefits and challenges of fuzzy control systems.

Advantage	Explanation
Adaptability	Handling uncertain and changing conditions
Nonlinear Behavior Handling	Addressing complex and nonlinear system responses
Expert Knowledge Integration	Incorporating human expertise into control strategies
Uncertainty Management	Dealing with imprecise and uncertain sensor data
Transparency and Interpretability	Making control decisions understandable and explainable

rule base design. Tuning these elements effectively demands both domain knowledge and an understanding of fuzzy logic principles.

Fuzzy control systems stand at the crossroads of control engineering and cognitive science, offering a method to bridge the gap between the precision of mathematical models and the uncertainties of the real world. As technology advances and systems become more intricate, the adaptive nature of fuzzy control systems positions them as valuable tools for achieving effective control in dynamic and uncertain environments.

7.2.5 Advantages of Fuzzy Logic in Mechatronic Control

Fuzzy logic offers a range of advantages that make it a valuable tool in mechatronic control systems, where precise and adaptable control is essential. These advantages include the following:

1. **Handling Uncertainty:** Mechatronic systems often operate in uncertain environments with imprecise sensor data. Fuzzy logic's ability to work with linguistic variables and degrees of membership allows it to process and respond to uncertain information, ensuring robust control even in the presence of noise.
2. **Adaptability:** Fuzzy logic excels in systems where conditions change rapidly or unpredictably. Its flexible decision-making process allows systems to adapt smoothly to

varying situations, making it suitable for dynamic mechatronic applications.

3. **Nonlinear Behavior:** Mechano-electrical systems frequently exhibit nonlinear behaviors that challenge traditional control methods. Fuzzy logic's inherent capacity to handle nonlinear relationships enables accurate control across a wide range of system responses.
4. **Human-Like Decision Making:** Fuzzy logic mirrors human cognitive processes by using linguistic terms and gradual transitions. This human-like approach simplifies control rule formulation, allowing engineers to translate expert knowledge into intuitive control strategies.
5. **Domain Expertise Integration:** Engineers can incorporate domain expertise directly into the control system through fuzzy logic. By defining linguistic variables and membership functions, experts contribute valuable insights that enhance the system's performance.
6. **Real-World Context:** Mechatronic systems often encounter real-world scenarios where crisp numerical values may not adequately represent conditions. Fuzzy logic's linguistic approach enables systems to make decisions that align with how humans perceive and respond to situations.
7. **Reduced Mathematical Modeling:** Traditional control methods require precise mathematical models of systems, which can be complex and time-consuming to develop. Fuzzy logic reduces the reliance on precise modeling, allowing control systems to be designed based on qualitative input.
8. **Smooth Control Transitions:** Fuzzy logic's gradual transitions between linguistic terms facilitate smooth and gradual changes in control actions. This feature is advantageous in systems where abrupt control changes could lead to instability.
9. **Interpretability:** Fuzzy logic's linguistic rules and variables provide transparency in control decisions. Engineers can understand and explain how the system is making decisions, enhancing trust and facilitating troubleshooting.
10. **Reduced Sensitivity to Noise:** Fuzzy logic's inherent ability to tolerate imprecise inputs and filter out noise contributes to stable control, even in scenarios where sensor data may fluctuate.

11. **Customization:** Fuzzy control systems can be fine-tuned easily to achieve specific control goals. Engineers can adjust membership functions and rules to optimize performance based on the system's requirements.

In mechatronic control systems, where the interplay between mechanical and electronic components demands precise yet adaptable control, the advantages offered by fuzzy logic make it a valuable approach. By harnessing its capacity to manage uncertainty, handle nonlinearities, and incorporate human insights, engineers can design control systems that excel in complex and ever-changing environments.

7.3 Fuzzy Logic Applications in Mechatronic Control Systems

Fuzzy logic finds diverse and impactful applications in mechatronic control systems, enhancing their performance, adaptability, and precision. Some notable applications include the following:

1. **Robotics and Automation:** Fuzzy logic is extensively used in robotic control systems to navigate through complex environments, avoid obstacles, and adapt to varying terrains. It enables robots to make real-time decisions based on sensor inputs, allowing them to handle uncertainties and changing scenarios. Fuzzy logic also plays a crucial role in the coordination of multiple robots in collaborative tasks.
2. **Automotive Systems:** In modern vehicles, fuzzy logic is employed in engine control units to optimize fuel efficiency and reduce emissions based on driving conditions. Fuzzy-based adaptive cruise control adjusts vehicle speed according to traffic conditions, enhancing safety on highways.
3. **Industrial Processes:** Fuzzy logic is applied in industrial automation to regulate processes with nonlinear behaviors and uncertain inputs. It ensures stable operation and efficient control in areas such as chemical processing, manufacturing, and energy management.
4. **Consumer Electronics:** Fuzzy logic is used in consumer appliances like washing machines to adapt wash cycles based on load size and dirtiness. It optimizes air conditioning systems by adjusting cooling or heating settings according to comfort preferences and room conditions.

5. **Medical Devices:** Fuzzy logic enhances the control of medical devices like anesthesia machines and infusion pumps. It adjusts dosages and flow rates based on patient conditions and real-time feedback, ensuring patient safety and optimal treatment.
6. **Aerospace Systems:** Fuzzy logic contributes to flight control systems by handling uncertain aerodynamic conditions and providing stable responses in the presence of external disturbances. It assists in flight path optimization and reduces pilot workload.
7. **Home Automation:** Fuzzy logic is utilized in smart home systems to control lighting, temperature, and security systems. It adapts settings based on user preferences and environmental conditions, creating comfortable and energy-efficient living spaces.
8. **Traffic Management:** Fuzzy logic optimizes traffic signal timings by considering traffic density, pedestrian flow, and road conditions. This leads to improved traffic flow, reduced congestion, and enhanced road safety.
9. **Renewable Energy Systems:** Fuzzy logic aids in the control of renewable energy sources like solar panels and wind turbines. It adjusts energy production based on weather conditions and energy demand, maximizing efficiency and grid integration.
10. **Precision Machinery:** Fuzzy logic ensures precise control in precision machinery applications, such as CNC (computer numerical control) machining and robotic arms. It compensates for mechanical tolerances and variations, achieving accurate and consistent results.

In essence, fuzzy logic's applications in mechatronic control systems span various industries and domains. Its ability to handle uncertainty, nonlinear behaviors, and human-like decision-making makes it a versatile and valuable tool for designing control systems that excel in complex and dynamic environments [7].

7.4 Fuzzy Expert Systems in Mechatronics

7.4.1 Enhancing Decision-Making and Control

Fuzzy expert systems stand as a potent fusion of fuzzy logic and artificial intelligence, revolutionizing decision-making and control processes in the realm of mechatronics. Mechatronic systems, characterized by their multidisciplinary nature and intricate interactions, benefit greatly from the incorporation of fuzzy expert systems, which bring adaptability, precision, and human-like reasoning to the forefront.

7.4.2 Understanding Fuzzy Expert Systems

Fuzzy expert systems combine the knowledge-driven approach of expert systems with the adaptability of fuzzy logic. They are designed to capture and emulate the expertise of human specialists in a particular domain. By incorporating fuzzy logic's ability to handle imprecise and uncertain data, these systems offer nuanced decision-making akin to human cognition. In a fuzzy expert system, domain-specific knowledge is represented using fuzzy linguistic variables, membership functions, and a rule base. These elements encapsulate the expertise of human professionals and experts, making the system capable of making context-sensitive and informed decisions, even in complex scenarios.

7.4.3 Applications in Mechatronics

Fuzzy expert systems find extensive applications in mechatronic systems, enhancing control, diagnosis, and decision-making across diverse domains:

1. **Control Systems:** Fuzzy expert systems optimize control strategies in mechatronic systems, adapting to changing conditions and uncertainties. They adjust parameters based on expert-derived rules and linguistic variables, ensuring effective control even in complex dynamics.
2. **Fault Detection and Diagnosis [8]:** These systems play a pivotal role in diagnosing faults and anomalies in mechatronic components. By integrating expert knowledge and fuzzy logic, they identify deviations from normal behavior, enabling timely maintenance and minimizing downtime.
3. **Quality Control:** In manufacturing, fuzzy expert systems contribute to quality control by assessing product characteristics and deviations. They determine whether a product meets predefined standards based on expert-defined rules, membership functions, and input data.
4. **Robot Path Planning:** Fuzzy expert systems guide robot path planning in dynamic environments. They enable robots to make decisions on the optimal path, considering both safety and efficiency, and adjusting their trajectory based on changing conditions.

5. **Process Optimization:** These systems enhance process optimization in mechatronic manufacturing, adjusting parameters for optimal performance. By integrating domain expertise with real-time data, they fine-tune processes to achieve desired outcomes.
6. **Human-Machine Interaction:** Fuzzy expert systems facilitate natural and intuitive interactions between humans and mechatronic systems. They interpret human inputs, such as gestures or voice commands, and adjust system behavior accordingly.
7. **Resource Allocation:** In energy-efficient mechatronic systems, fuzzy expert systems allocate resources like power or energy based on expert-defined criteria. This ensures optimal resource utilization while considering varying load conditions.

7.4.4 Benefits and Challenges

Fuzzy expert systems offer several advantages, including the ability to incorporate human expertise, handle uncertainty, and make context-aware decisions. They enhance system adaptability, improve performance, and contribute to more efficient and effective control processes. However, designing and implementing fuzzy expert systems require domain-specific expertise and meticulous rule formulation. The accuracy of the system heavily depends on the quality of the knowledge base and rule sets, demanding careful attention to detail during development. Fuzzy expert systems are a formidable asset in the mechatronics toolkit, amalgamating the strengths of fuzzy logic and expert knowledge. Their applications extend across control, diagnostics, and decision-making, enhancing the performance of mechatronic systems in diverse industries. By imbuing systems with expert-derived, context-sensitive reasoning, fuzzy expert systems pave the way for more intelligent, adaptive, and human-like interactions between machines and their environments.

7.5 Fuzzy Logic and Machine Learning in Mechatronics

7.5.1 A Synergistic Approach to Intelligent Control

In Figure 7.2 the intricate landscape of mechatronics, where mechanical, electrical, and computational elements converge, the synergy of fuzzy logic

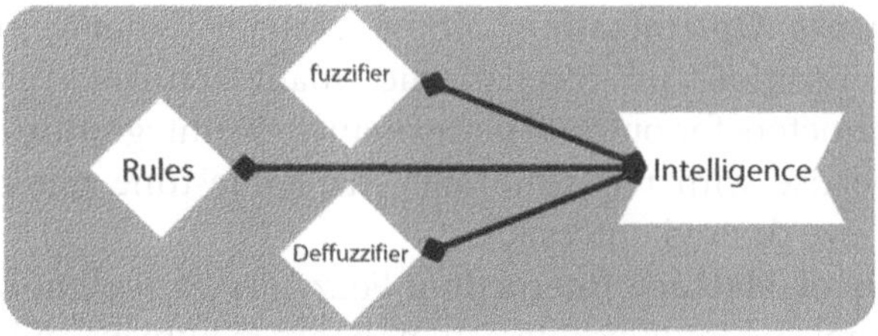

Figure 7.2 Fuzzy logic and machine learning in mechatronics.

and machine learning emerges as a powerful paradigm. This combined approach brings together the adaptability of fuzzy logic and the data-driven capabilities of machine learning to create intelligent [9] mechatronic systems that excel in control, decision-making, and optimization.

7.5.2 Fuzzy Logic: Handling Uncertainty and Complex Relationships

Fuzzy logic, rooted in the principles of human reasoning, addresses the challenges posed by uncertainty, imprecision, and nonlinear behaviors in mechatronic systems. By employing linguistic variables, fuzzy sets, and expert-defined rules, fuzzy logic captures the nuances of real-world conditions. Its ability to process imprecise data and make human-like decisions enables systems to navigate complex scenarios and adapt to dynamic environments.

7.5.3 Machine Learning: Data-Driven Adaptability

Machine learning, on the other hand, harnesses the power of data to derive patterns, insights, and predictive models. Through algorithms such as neural networks, decision trees, and support vector machines, machine learning algorithms learn from data to make informed decisions or predictions. In mechatronics, machine learning leverages historical and real-time data to optimize control parameters, detect anomalies, and enhance system performance are shown in Figure 7.3.

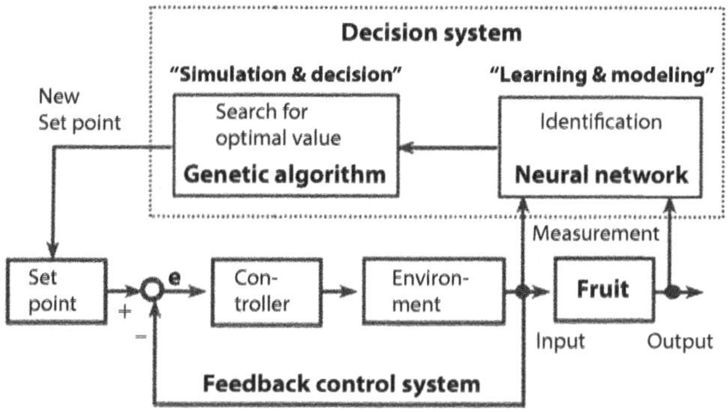

Figure 7.3 Handling uncertainty and complex relationships.

7.5.4 Applications in Mechatronics

1. **Predictive Maintenance:** The integration of fuzzy logic and machine learning facilitates predictive maintenance in mechatronic systems. Fuzzy logic captures expert knowledge to define rules for assessing system health, whereas machine learning algorithms analyze sensor data to predict maintenance needs and prevent unplanned downtime.
2. **Autonomous Systems:** Fuzzy logic's adaptive control and machine learning's pattern recognition enable autonomous mechatronic systems. Robots can combine fuzzy control for real-time decision-making with machine learning to learn and refine behaviors based on feedback and experience.
3. **Energy Efficiency:** Fuzzy logic optimizes energy usage by adjusting control parameters based on expert-defined rules. Machine learning models can identify optimal operating conditions from historical data, further enhancing energy efficiency in mechatronic systems.
4. **Human–Robot Interaction:** The fusion of fuzzy logic and machine learning enhances human–robot interaction. Fuzzy logic interprets human inputs, and machine learning

models adapt responses based on user preferences and past interactions.

5. **Process Optimization:** By integrating machine learning's data-driven insights and fuzzy logic's adaptability, mechatronic systems can optimize manufacturing processes based on both historical data and real-time conditions.
6. **Health Monitoring:** Fuzzy logic's ability to handle uncertain data pairs well with machine learning's predictive capabilities in health monitoring. This combination allows systems to assess the health of mechanical components and predict potential failures.

7.5.5 Benefits and Challenges

The combination of fuzzy logic and machine learning offers benefits that include enhanced adaptability, improved decision-making, and data-driven optimization are shown in Table 7.3. It empowers mechatronic systems to handle complex and uncertain scenarios while also learning from data to continuously improve performance. However, the successful integration of these approaches requires expertise in both fuzzy logic and machine learning techniques. The challenge lies in designing systems that effectively combine the strengths of both paradigms without introducing complexity that could hinder implementation and interpretability.

Table 7.3 Benefits and challenges of data-driven adaptability.

Challenge	Explanation
Rule Design Complexity	Crafting accurate and comprehensive fuzzy rule bases
Real-Time Execution	Ensuring efficient execution in time-sensitive systems
Rule Tuning and Optimization	Adjusting membership functions and weights for performance
Hybrid Approaches	Balancing fuzzy logic with other control paradigms
Interpretability and Transparency	Making fuzzy control decisions transparent and interpretable

Fuzzy logic and machine learning represent two complementary approaches that, when combined, create a holistic and intelligent framework for mechatronics. Their synergy facilitates systems that can handle uncertainty, adapt to changing conditions, and learn from data, enabling the creation of mechatronic systems that are both responsive and data-informed. As the field of mechatronics evolves, the integration of fuzzy logic and machine learning holds the promise of unlocking new levels of efficiency, autonomy, and intelligence in various applications.

7.6 Fuzzy Control in Multivariable Mechatronic Systems

7.6.1 Navigating Complexity with Adaptability

In the intricate domain of multivariable mechatronic systems, where numerous interconnected components interact within dynamic environments, achieving precise and efficient control becomes a formidable challenge. Fuzzy control emerges as a powerful strategy to address this complexity, offering adaptability, robustness, and human-like decision-making that align with the intricate nature of these systems.

7.6.2 Challenges in Multivariable Mechatronic Systems

Multivariable mechatronic systems involve the interplay of multiple inputs and outputs, creating intricate relationships that are difficult to model using traditional control approaches. These systems are subject to nonlinear behaviors, uncertainties, and cross-coupling effects, making it challenging to devise control strategies that yield optimal performance across all variables simultaneously.

7.6.3 Fuzzy Control: A Multivariable Solution

In Figure 7.4 fuzzy control shines in multivariable mechatronic systems by providing an approach that can manage the complexities inherent in such setups. Its ability to handle uncertainty, adapt to changing conditions, and accommodate nonlinear behaviors makes it a natural fit for controlling interconnected components with varying dynamics.

In multivariable systems, fuzzy control employs rules that capture the interactions between inputs and outputs using linguistic terms. By considering the interdependencies between variables, fuzzy control can adjust

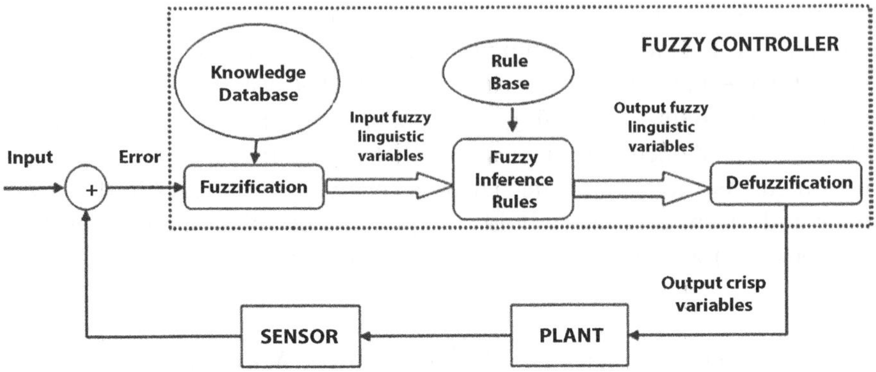

Figure 7.4 Knowledge database in fuzzy controller.

control actions in response to changing conditions. The adaptability of fuzzy control means that it can handle situations where conventional control methods struggle, such as situations with incomplete or noisy data.

7.6.4 Applications and Benefits

Fuzzy Logic applications in various fields are shown in Table 7.4.

1. **Robot Manipulators:** In multivariable robotic systems, fuzzy control optimizes the coordination of multiple joints, adapting to changing payloads and constraints. It addresses cross-coupling effects and nonlinearity, enabling precise and smooth movement.
2. **Aerospace Systems:** Fuzzy control manages the intricate interactions between various control surfaces and actuators in aircraft. It ensures stability, maneuverability, and response adaptability in varying flight conditions.
3. **Mechanical Systems:** Multivariable mechanical systems, such as complex machinery with interconnected components, benefit from fuzzy control's ability to balance forces, regulate speeds, and maintain stability.
4. **Process Industries:** Fuzzy control optimizes multivariable processes in industries like chemical manufacturing, where multiple variables influence product quality. It adapts to

Table 7.4 Fuzzy logic in various fields.

Industry	**Fuzzy logic applications**
Healthcare	Medical imaging enhancement, drug delivery optimization
Manufacturing	Process control in chemical plants, quality assurance
Automotive	Engine control for fuel efficiency, collision avoidance
Aerospace	Flight control, aircraft navigation
Robotics	Autonomous navigation, human–robot interaction
Consumer Electronics	Smart home automation, energy management

changing raw material properties, ambient conditions, and desired outcomes.

5. **Healthcare Devices:** In medical devices like surgical robots, fuzzy control adjusts multiple robotic arms to ensure precise movements and minimal tissue damage, even when interacting within a confined and sensitive environment.

7.6.5 Benefits in Specific Applications

The benefits of fuzzy logic in various fields are shown in Table 7.5.

Table 7.5 Benefits of fuzzy logic in various fields.

Application	**Benefits of fuzzy logic**
Robotic Manipulation	Precise, adaptive control for complex movements
Industrial Processes	Handling nonlinear behaviors, process optimization
Healthcare Devices	Patient-specific treatment, real-time adaptation
Aerospace Control	Stability, adaptability in varying flight conditions
Quality Control	Adaptation to varying product characteristics
Autonomous Systems	Responsive decision-making, learning from experience

7.6.6 Challenges and Considerations

Designing and implementing fuzzy control for multivariable systems require careful rule formulation, membership function design, and tuning. The challenge lies in capturing the complex interactions accurately while keeping the control system manageable and interpretable.

Fuzzy control stands as a valuable approach to manage the intricacies of multivariable mechatronic systems. Its adaptability, ability to handle uncertainty, and capture complex relationships make it an ideal candidate for controlling interconnected components with diverse dynamics. As mechatronic systems become increasingly sophisticated, the integration of fuzzy control offers a practical means of achieving efficient and responsive control in the face of intricate variables and evolving environments.

7.7 Industrial Automation and Fuzzy Logic

7.7.1 Enhancing Precision and Adaptability

Industrial automation [10], driven by the integration of mechanical, electrical, and computational systems, aims to improve efficiency, accuracy, and productivity in various manufacturing and production processes. Fuzzy logic plays a pivotal role in this domain by providing a flexible and robust approach to control and decision-making, thereby optimizing operations in complex and dynamic industrial environments.

7.7.2 Challenges in Industrial Automation

Industrial processes often involve intricate interactions between multiple variables, nonlinear behaviors, and uncertainties. Conventional control methods struggle to address these challenges due to the limitations of precise mathematical models and rigid control strategies. Industrial automation requires solutions that can adapt to changing conditions, handle uncertainty, and accommodate the complex relationships between variables.

7.7.3 Fuzzy Logic: A Solution for Industrial Automation

In Figure 7.5 fuzzy logic offers a solution that aligns seamlessly with the complexities of industrial automation [8]. It allows control systems to work

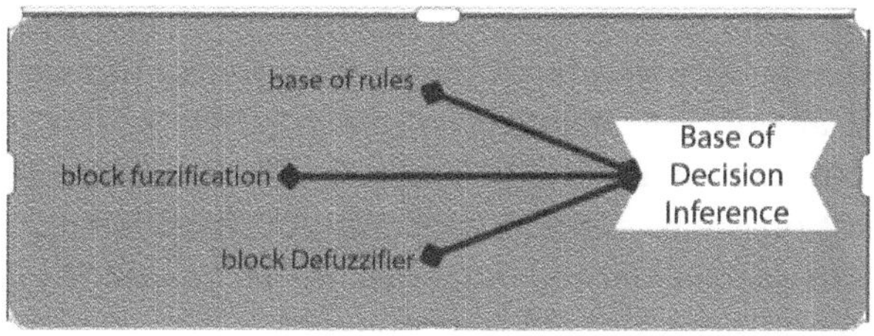

Figure 7.5 Fuzzy logic in industrial automation.

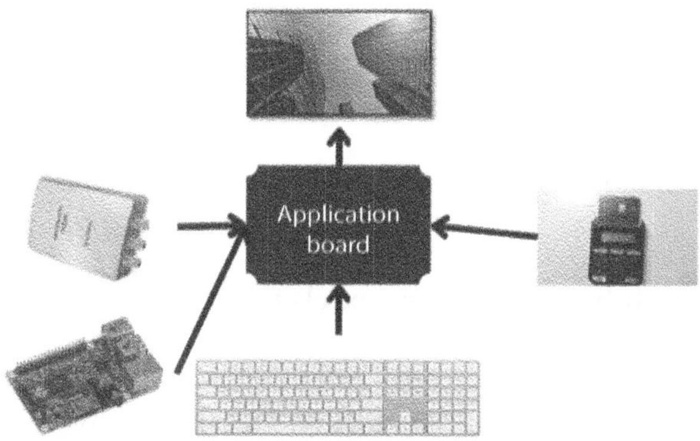

Figure 7.6 Fuzzy logic in process control.

with imprecise, vague, or uncertain data, making it well suited for applications where sensor readings might be noisy or where exact models are difficult to derive.

In industrial automation, fuzzy logic can be used for various tasks:

The simple illustration is the above-shown Rasperry Pi shown in Figure 7.6.

1. **Process Control:** Fuzzy control optimizes industrial processes by adjusting control parameters based on linguistic variables and expert-defined rules. It adapts to changes in raw materials, equipment conditions, and environmental factors, ensuring stable and efficient operations.
2. **Quality Control:** Fuzzy logic aids in quality control by assessing product characteristics based on linguistic terms.

It can make decisions about whether a product meets predefined quality standards, even when precise numerical values may not be available.

3. **Resource Management:** Fuzzy logic optimizes resource allocation, such as energy consumption, by adjusting parameters to achieve desired outcomes while considering varying load conditions and efficiency targets.
4. **Fault Detection:** Fuzzy logic–based fault detection systems [11] can identify anomalies and deviations from normal behavior in complex industrial machinery. It interprets sensor data to detect irregular patterns that may indicate impending failures.

7.7.4 Benefits and Considerations

The application of fuzzy logic in industrial automation yields several benefits:

- **Adaptability:** Fuzzy logic excels in adapting control strategies to changing conditions and variations in the production process.
- **Nonlinear Behavior Handling:** It can effectively manage nonlinearities and complex relationships between variables, ensuring stable and accurate control.
- **Expert Knowledge Integration:** Fuzzy logic allows experts to contribute their domain knowledge directly into control systems through linguistic rules.
- **Improved Productivity:** By optimizing control parameters, fuzzy logic enhances production efficiency and reduces waste in manufacturing processes.

However, designing a fuzzy control system requires careful consideration of linguistic variables, membership functions, and rule bases. Proper tuning and validation are crucial to ensuring effective and reliable performance.

In the realm of industrial automation [10], where the convergence of mechanical, electrical, and computational elements drives efficiency and productivity, fuzzy logic stands as a versatile and valuable tool. Its ability to handle uncertainty, adapt to changing conditions, and accommodate complex relationships between variables empowers industrial systems to achieve precision, flexibility, and optimal performance in the face of intricate production processes and dynamic environments.

7.8 Challenges and Future Directions

The integration of fuzzy logic into mechatronic systems [12] offers numerous benefits, but it also comes with its own set of challenges and opportunities for future advancements. Here are some key challenges and potential future directions in this field:

7.8.1 Challenges

1. **Complex Rule Design:** Crafting accurate and comprehensive fuzzy rule bases for complex mechatronic systems can be challenging. Designing rules that capture intricate relationships and interactions among variables requires deep domain knowledge and expertise.
2. **Rule Tuning and Optimization:** Fuzzy control systems often involve tuning membership functions and rule weights. This process can be time-consuming and requires a trial-and-error approach to achieve optimal performance.
3. **Interpretability and Transparency:** Fuzzy logic can sometimes yield complex rule sets that are difficult to interpret. Ensuring transparency and understanding of the decision-making process within fuzzy control systems is crucial, especially in safety-critical applications.
4. **Combining With Machine Learning:** While the integration of fuzzy logic and machine learning holds promise, finding the right balance and synergy between these two paradigms is a challenge. Developing hybrid systems that harness the strengths of both approaches requires careful design and validation.
5. **Real-Time Processing:** In mechatronic applications where real-time responses are essential, the computational requirements of fuzzy control systems might pose challenges.

Optimizing algorithms for efficient real-time execution is necessary.

7.8.2 Future Directions

1. **Hybrid Approaches:** Integrating fuzzy logic with other control methods, such as neural networks or reinforcement learning, can create hybrid systems that leverage the strengths of each approach. This can lead to more adaptable and intelligent control strategies.
2. **Machine Learning for Rule Generation:** Using machine learning techniques to automatically generate fuzzy rules based on data can streamline the rule design process and make fuzzy control more data-driven.
3. **Self-Adaptive Systems:** Developing mechatronic systems that can self-tune their fuzzy control parameters based on changing operating conditions can enhance adaptability and performance.
4. **Explainable AI:** As AI-driven systems become more prevalent, ensuring the explainability of fuzzy logic–based decisions is crucial. Researching techniques to provide transparent explanations for fuzzy control decisions can enhance trust and adoption.
5. **Internet-of-Things (IoT) Integration:** Integrating fuzzy control with IoT technologies can enable real-time data collection and decision-making, enhancing the responsiveness and efficiency of mechatronic systems.
6. **Education and Skill Development:** Fuzzy logic and mechatronics are specialized fields that require interdisciplinary expertise. Offering comprehensive education and training programs can cultivate a skilled workforce capable of designing and implementing effective control solutions.
7. **Standardization and Validation:** Establishing standards and guidelines for designing, implementing, and validating fuzzy control systems in mechatronics can ensure consistency and reliability across different applications.

In conclusion, the integration of fuzzy logic into mechatronics presents both challenges and exciting opportunities for advancement. Addressing these challenges while exploring innovative directions can lead to more

Table 7.6 Area of advancements and potential developments using fuzzy logic in various fields.

Area of advancement	Potential developments
Hybrid Systems	Combining fuzzy logic with AI, machine learning
Self-Adaptive Systems	Systems that tune their own parameters
Explainable AI	Transparent decision-making in fuzzy control
Internet-of-Things (IoT) Integration	Real-time data-driven decision-making
Standardization and Validation	Establishing guidelines for fuzzy control
Education and Skill Development	Training workforce in fuzzy logic and mechatronics

intelligent [13], adaptable, and efficient mechatronic systems that excel in diverse applications across industries are shown in Table 7.6.

7.9 Conclusion

The journey through the integration of fuzzy logic into the world of mechatronics reveals a transformative partnership that harmonizes intricate mechanical systems with adaptive decision-making. Fuzzy logic, a beacon of adaptability and human-like reasoning, enriches mechatronics by addressing challenges that traditional control methods struggle to conquer. Its ability to manage uncertainty, navigate nonlinear behaviors, and accommodate complex relationships has propelled mechatronic systems toward new realms of efficiency, precision, and intelligence.

In the realm of patient-centric healthcare, where personalized treatments and secure data management are paramount, fuzzy logic offers a resilient framework that aligns with the sensitive and dynamic nature of patient care. Its fusion with blockchain technology enhances patient privacy, consent management, and research integrity, revolutionizing the healthcare landscape.

The versatility of fuzzy logic extends to combatting counterfeit drugs, advancing medical research, and optimizing clinical trials. Its precision and adaptability empower industries to tackle pressing challenges while fostering a safer and more efficient environment for patients and practitioners alike.

As we gaze into the future, the integration of fuzzy logic into the intricate web of mechatronic systems continues to evolve. The fusion of fuzzy logic with machine learning propels us toward intelligent autonomy, enabling systems to learn, adapt, and optimize based on data-driven insights. The challenges ahead—such as rule design complexity and real-time execution—underscore the necessity of ongoing research, innovation, and collaboration to unlock the full potential of this partnership.

The tale of fuzzy logic in mechatronics is one of innovation, overcoming challenges, and harnessing the synergy of diverse disciplines. By integrating fuzzy logic into mechatronic systems, we pave the way for a future where complex systems respond intelligently, adapt to uncertainties, and emulate human-like reasoning—ushering in a new era of precision, efficiency, and progress across industries and applications.

References

1. Sharma, R., Gaur, P., Mittal, A.P., Design of two-layered fractional order fuzzy logic controllers applied to robotic manipulator with variable payload. *Appl. Soft Comput.*, 47, 565–576, 2016, [Google Scholar] [CrossRef].
2. Benli, E., Motai, Y., Rogers, J., Human behavior-based target tracking with an omni-directional thermal camera. *IEEE Trans. Cogn. Dev. Syst.*, 11, 36–50, 2019, [Google Scholar] [CrossRef].
3. Bhatia, V., Kalaichelvi, V., Karthikeyan, R., Application of a Novel Fuzzy Logic Controller for a 5-DOF Articulated Anthropomorphic Robot, in: *Proceedings of the 2015 IEEE International Conference on Research in Computational Intelligence and Communication Networks (ICRCICN)*, Kolkata, India, 20–22 November 2015, [Google Scholar].
4. Zavlangas, P.G., Tzafestas, S.G., Althoefer, K., Fuzzy Obstacle Avoidance and Navigation for Omnidirectional Mobile Robots. *European Symposium on Intelligent Techniques*, Aachen, Germany, 2000, [Google Scholar].
5. Zhang, Y., Wang, J., Han, D., Wu, H., Zhou, R., Fuzzy-logic based distributed energy-efficient clustering algorithm for wireless sensor networks. *Sensors*, 17, 1554, 2017, [Google Scholar] [CrossRef][Green Version].
6. Li, K., Zhao, X., Sun, S., Tan, M., Robust target tracking and following for a mobile robot. *Int. J. Robot. Autom.*, 33, 4, 2018, [Google Scholar] [CrossRef].

7. Mamdani, E.H., Application of fuzzy algorithms for control of simple dynamic plant. *Proc. Inst. Electr. Eng.*, 121, 1585–1588, 1974. [Google Scholar] [CrossRef].
8. Karras, G.C. and Fourlas, G.K., Model Predictive Fault Tolerant Control for Omni-directional Mobile Robots. *J. Intell. Robot. Syst.*, 97, 635–655, 2020, [Google Scholar] [CrossRef].
9. Nadour, M., Boumehraz, M., Cherroun, L., Puig Cayuela, V., Hybrid type-2 fuzzy logic obstacle avoidance system based on horn-schunck method. Electroteh. *Electron. Autom.*, 67, 45–51, 2019, [Google Scholar].
10. Mac, T., Copot, C., De Keyser, R., Tran, T., Vu, T., MIMO fuzzy control for autonomous mobile robot. *J. Autom. Control. Eng.*, 4, 65–70, 2016, [Google Scholar].
11. Xue, H., Zhang, Z., Wu, M., Chen, P., Fuzzy Controller for Autonomous Vehicle Based on Rough Sets. *IEEE Access*, 7, 147350–147361, 2019, [Google Scholar] [CrossRef].
12. Prakash, M. and Jajulwar, K., Design of adaptive fuzzy tracking controller for Autonomous navigation system. *Int. J. Recent Trend Eng. Res.*, 2, 268–275, 2016, [Google Scholar].
13. Aydogmus, Z., A real-time robust fuzzy-based level control using programmable logic controller. *Elektron. Elektrotech.*, 21, 1, 13–17, 2015, 10.5755/j01.eee.21.1.7812

Further Reading

Abdulla, A., Maria, J., Francisco, M., Arturo, E., Jose, M., An AppearanceBased Tracking Algorithm for Aerial Search and Rescue Purposes. *Sensors*, 19, 652, 2019, [Google Scholar] [CrossRef][Green Version].

Abror, B., Hyun, K., Myeong, C., Rymduck, O., Akmal, A., Heung, S., Application of fuzzy logic for problem of evaluating states of a computing system. *Appl. Sci.*, 9, 3021, 2019, [Google Scholar] [CrossRef][Green Version].

Ahmad, S., Ali, S., Tabasha, R., The design and implementation of a fuzzy gain-scheduled PID controller for the Festo MPS PA compact workstation liquid level control. *Eng. Sci. Technol.*, 23, 307–315, 2020, 10.1016/j.jestch.2019.05.014.

Baghli, F.Z., Bakkali, E.L., Lakhal, Y., Multi-input multi-output fuzzy logic controller for complex system: Application on two-links manipulator. *Proc. Technol.*, 19, 607–614, 2015, [Google Scholar] [CrossRef][Green Version].

Bandara, R.N. and Gaspe, S., Fuzzy logic controller design for an Unmanned Aerial Vehicle, in: *Proceedings of the IEEE International Conference on Information and Automation for Sustainability*, Galle, Sri Lanka, 16–19, pp. 1–5, December 2016.

Bobyr, M.V., Kulabukhov, S.A., Milostnaya, N.A., Fuzzy control system of robot angular attitude, in: *Proceedings of the 2nd International Conference on*

Industrial Engineering, Applications and Manufacturing, Chelyabinsk, Russia, 19–20 May 2016, pp. 1–6, [Google Scholar].

Chabni, F., Taleb, R., Benbouali, A., Bouthiba, M.A., The application of fuzzy control in water tank level using Arduino. *Int. J. Adv. Comput. Sci. Appl.*, 7, 4, 261–265, 2016, 10.14569/IJACSA.2016.070432.

Chatterjee, A. and Watanabe, K., An adaptive fuzzy strategy for motion control of robot manipulators. *Soft Comput.*, 9, 185–193, 2005, [Google Scholar] [CrossRef].

Chih, H. and Ya, F., Design of Takagi-Sugeno fuzzy control scheme for real word system control. *Sustainability*, 11, 3855, 2019, [Google Scholar] [CrossRef] [Green Version].

Edalati, L., Sedigh, A.K., Shooredeli, M.A., Moarefianpour, A., Adaptive fuzzy dynamic surface control of nonlinear systems with input saturation and time-varying output constraints. *Mech. Syst. Signal. Process*, 100, 311– 329, 2018, [Google Scholar] [CrossRef].

Experion overview, Honeywell Int. Inc, May 5, 2006, Release 300.1. Fuzzy logic toolbox: User's guide for use with MATLAB, TheMathWorks, Inc, Natick, MA, 1998.

Ghanooni, P., Yazdanib, A.M., Mahmoudi, A., MahmoudZadeh, S., Movaheded, M.A., Fathi, M., Robust precise trajectory tracking of hybrid stepper motor using adaptive critic-based neuro-fuzzy controller. *Comput. Electr. Eng.*, 81, 106535, 2020, [Google Scholar] [CrossRef].

Helbing, D., Farkas, S., Vicsek, T., Simulating Dynamic Features of Escape Panic. *Nature*, 407, 487–490, 2020, [Google Scholar] [CrossRef] [PubMed] [Green Version].

Herrera, F. and Herrera-Viedma, E., Linguistic Decision Analysis: Steps for Solving Decision Problems under Linguistic Information. *Fuzzy Sets Syst.*, 115, 67–82, 2000, [Google Scholar] [CrossRef].

Huang, C.-H., Wang, W.-J., Chiu, C.-H., Design and implementation of fuzzy control on a two-wheel inverted pendulum. *IEEE Trans. Ind. Electron.*, 58, 2988–3001, 2011, [Google Scholar] [CrossRef].

Hu, H., Wang, X., Chen, L., Impedance with Finite-Time Control Scheme for Robot-Environment Interaction. *Math. Probl. Eng.*, 2020, 2796590, 2020, [Google Scholar] [CrossRef].

Jahanshahi, H., Jafarzadeh, M., Sari, N.N., Pham, V.T., Huynh, V.V., Nguyen, X.Q., Robot motion planning in an unknown environment with danger space. *Electronics*, 8, 201, 2019, [Google Scholar] [CrossRef][Green Version].

Jantzen, J., Foundations of Fuzzy Control, John Wiley and Sons, New York, 2007, Online ISBN: 9780470061176, doi: 10.1002/9780470061176.

Jassbi, J.J., Serra, P.J.A., Ribeiro, R.A., Donati, A.A., Comparison of Mandani and Sugeno Inference Systems for a Space Fault Detection Application, in: *Proceedings of the 2006 World Automation Congress*, Budapest, Hungary, pp. 1–8, 24–26 July 2006, [Google Scholar].

Kahraman, C. and Kabak, O., Fuzzy Statistical Decisio-Making: Theory and Applications. *Studies in Fuzziness and Soft Computing*, Springer, Berlin/Heidelberg, Germany, 2018, ISBN 9783319817934, [Google Scholar].

Kanagasabai, N. and Jaya, N., Fuzzy gain scheduling of PID controller for a MIMO process. *Int. J. Comput. Appl.*, 91, 10, 13–20, 2014, 10.5120/15916-4803.

Kasabanov, N. and Kozma, R., Neuro-Fuzzy Techniques for Intelligent Information Systems, vol. 30, p. 449, Springer, Berlin/Heidelberg, Germany, 1999, ISBN1 -10 3790811874. ISBN2 -13 978-3790811872, [Google Scholar].

Kolbari, H., Sadeghnejad, S., Parizi, A.T., Rashidi, S., Baltes, J.H., Extended fuzzy logic controller for uncertain teleoperation system, in: *Proceedings of the 4th International Conference on Robotics and Mechatronics*, Tehran, Iran, 26–28 October 2016, pp. 78–83, [Google Scholar].

Lin, J., Zhou, J., Lu, M., Wang, H., Yi, A., Design of Robust Adaptive Fuzzy Controller for a Class of Single-Input Single Output (SISO) Uncertain Nonlinear System. *Math. Probl. Eng.*, 2020, 6178678, 2020, [Google Scholar] [CrossRef][Green Version].

Mahmoodabadi, M.J., Abedzadeh Maafi, R., Taherkhorsandi, M., An optimal adaptive robust PID controller subject to fuzzy rules and sliding modes for MIMO uncertain chaotic systems. *Appl. Soft Comput.*, 52, 1191–1199, 2017, [Google Scholar] [CrossRef].

Minea, M., Dumitrescu, C., Costea, I., Chiva, I., Semenescu, A., Developing a Solution for Mobility and Distribution Analysis Based on Bluetooth and Artificial Intelligence. *Sensors*, 20, 7327, 2020, [Google Scholar] [CrossRef] [PubMed].

Morari, M. and Zafiriou, B., Robust Process Control, Prentice Hall, New Jersey, 1989. Mousa, H.M., Koutb, M.A., El-Araby, S.M., Elsayed, H.M.A., Design of optimal fuzzy controller for water level of U-Tube steam generator in nuclear power station. *J. Am. Sci.*, 7, 4, 629–637, 2011.

Neshkov, T., Yordanova, S., Topalova, I., Process Control and Production Automation, Technical University of Sofia, Sofia, 2007.

Pandey, A., Pandey, S., Parhi, D., Mobile robot navigation and obstacle avoidance techniques: A review. *Int. Robot. Autom. J.*, 2, 3, 2017, [Google Scholar] [CrossRef][Green Version].

Precup, R.E., David, R.C., Petriu, E.M., Radac, M.B., Fuzzy logic-based adaptive gravitational search algorithm for optimal tuning of fuzzy-controlled servo systems. *IET Control Theory Appl.*, 7, 1, 99–107, 2013, 10.1049/iet-cta.2012.0343.

Roboaca, S., Dumitrescu, C., Mantan, I., Aircraft Trajectory Tracking using Radar Equipment with Fuzzy Logic Algorithm. *Mathematics*, 8, 207, 2020, [Google Scholar] [CrossRef][Green Version].

Rossomando, F., Serrano, E., Soria, C., Scaglia, G., Neural Dynamics Variations Observer Designed for Robot Manipulator Control Using a Novel Saturated

Control Technique. *Math. Probl. Eng.*, 2020, 3240210, 2020, [Google Scholar] [CrossRef].

Saleki, A. and Mehdi, M., Model-free control of electrically driven robot manipulators using an extended state observer. *Comput. Electr. Eng.*, 87, 106768, 2020, [Google Scholar] [CrossRef].

Shome, A. and Ashok, D., Fuzzy logic approach for boiler temperature and water level control. *Int. J. Sci. Eng. Res.*, 3, 6, 1–6, 2012.

Siegwart, R., Nourbakhsh, I.R., Scaramuzza, D., Introduction to Autonomous Mobile Robots, MIT Press, Cambridge, MA, USA, 2011, [Google Scholar].

Steels, L. and Brooks, R., The Artificial Life Route to Artificial Intelligence: Building Embodied, Situated Agents, Routledge, Oxford, UK, 2018, [Google Scholar].

Su, X., Wu, Y., Song, J., Yuan, P.A., Fuzzy Path Selection Strategy for Aircraft Landing on a Carrier. *Appl. Sci.*, 8, 779, 2018, [Google Scholar] [CrossRef] [Green Version].

Tan, W., Water level control for a nuclear steam generator. *Nucl. Eng. Des.*, 241, 5, 1873–1880, 2011, 10.1016/j.nucengdes.2010.12.010.

Tanaka, K. and Wang, H.O., Fuzzy Control Systems Design and Analysis: A Linear Matrix Inequality Approach, John Wiley and Sons, New York, 2001.

Terano, T., Asal, K., Sugeno, M., Fuzzy Systems Theory and Its Applications, 1st, Academic Press, San Diego, CA, USA, 1992, [Google Scholar].

Thieme, C., Sodium carbonates, in: Ullmann's Encyclopedia of Industrial Chemistry, pp. 299–316, Wiley-VCH Verlag GmbH & Co. KGaA, Weinheim, Germany, 2012, https://onlinelibrary.wiley.com/doi/10.1002/14356007.a24_299, 33.

Versaci, M., Calcagno, S., Cacciola, M., Morabito, F., Palamara, I., Pellicanò, D., Innovative Fuzzy Techniques for Characterizing Defects in Ultrasonic Nondestructive Evaluation, in: *Ultrasonic Nondestructive Evaluation Systems*, P. Burrascano, S. Callegari, A. Montisci, M. Ricci, M. Versaci, (Eds.), Springer: Cham, Switzerland, 2015, ISBN 978-3-319-10565-9, [Google Scholar] [CrossRef].

Xu, J.X., Guo, Z.Q., Lee, T.H., Design and implementation of a TakagiSugeno-Type fuzzy logic controller on a two-wheeled mobile robot. *IEEE Trans. Ind. Electron.*, 60, 5717–5728, 2013, [Google Scholar] [CrossRef].

Yordanova, S., Design of fuzzy supervisor-based adaptive process control systems, in: *New Approaches in Intelligent Control: Techniques, Methodologies and Applications*, K. Nakamatsu and R. Kountchev (Eds.), Springer Int. Publ., Switzerland, 2016, DOI: 10.1007/978-3-319-32168-4_1.

Yordanova, S., Intelligent approaches to real time level control. *Int. J. Intell. Syst. Appl.*, 7, 10, 19–27, 2015, 10.3233/IFS-141242.

Yordanova, S., Methods for Design of Fuzzy Logic Controllers for Robust Process Control, 2011, KING, Sofia, Bulgarian.

Yordanova, S., Slavov, M., Gueorguiev, B., Parallel distributed compensation for improvement of level control in carbonisation column for soda production.

Contr. Eng. Pract., 71, 53–60, 2018, 10.1016/j.conengprac.2017.10.003, https://www.aspentech.com/en/products/msc/aspen-infoplus21.

Yuste-Delgado, A.J., Cuevas-Martinez, J.C., Triviño-Cabrera, A., EUDFCEnhanced Unequal Distributed Type-2 Fuzzy Clustering Algorithm. *IEEE Sens. J.*, 19, 4705–4716, 2019, [Google Scholar] [CrossRef].

Zadeh, L., Fuzzy logic = computing with words. *IEEE Trans. Fuzzy Syst.*, 4, 103–111, 1996, [Google Scholar] [CrossRef][Green Version].

8

Drones and Autonomous Robotics Incorporating Computational Intelligence

R. Thangamani[1*], R. K. Suguna[2†] and G. K. Kamalam[3‡]

[1]*Department of Artificial Intelligence, Kongu Engineering College, Perundurai, Tamil Nadu, India*
[2]*Department of Computer Science and Engineering, KPR Institute of Technology, Coimbatore, Tamil Nadu, India*
[3]*Department of Information Technology, Kongu Engineering College, Perundurai, Tamil Nadu, India*

Abstract

Computational intelligence (CI) has emerged as a powerful paradigm in the field of autonomous robotics and drones, enabling the development of intelligent, adaptive, and self- learning systems. This chapter explores the application of CI techniques in autonomous robotics and drones, focusing on the utilization of artificial intelligence, machine learning, and swarm intelligence algorithms to enhance navigation, perception, decision-making, and collaboration capabilities. The journey starts with an introduction to autonomous robotics and drones, outlining the challenges and opportunities in developing self-sufficient systems. It then delves into the fundamentals of CI, explaining the concepts of artificial intelligence, machine learning, and swarm intelligence, and their significance in autonomous systems. The core aspects of autonomous robotics and drones are addressed, including navigation and path planning, object detection and recognition, and adaptive control. It highlights the use of CI algorithms for intelligent decision-making, real-time object detection using deep learning, and cooperative navigation with swarm robotics. Furthermore, the abstract explores the emerging field of autonomous drone delivery systems, discussing the challenges and regulatory considerations involved in implementing efficient and safe drone delivery services.

**Corresponding author*: thangamani.ai@kongu.edu
†*Corresponding author*: suguna.k@kpriet.ac.in
‡*Corresponding author*: kamalamparames@gmail.com

Kolla Bhanu Prakash, Satish Kumar Peddapelli, Ivan C. K. Tam, Wai Lok Woo and Vishal Jain (eds.) Computational Intelligent Techniques in Mechatronics, (243–296) © 2024 Scrivener Publishing LLC

Human–robot interaction and collaboration play a pivotal role in autonomous systems, and the abstract discusses how CI techniques facilitate natural language processing for seamless communication and the development of empathetic robots. The future trends and challenges in the field are also examined, such as AI at the edge, ethical implications, and the gap between research and real-world deployment. It emphasizes the transformative impact of CI in autonomous robotics and drones. By harnessing the power of AI, machine learning, and swarm intelligence, these intelligent systems are poised to revolutionize various industries and contribute to a more automated and efficient future. However, it also highlights the need for addressing ethical considerations and ensuring responsible deployment for the widespread acceptance and success of autonomous systems in the real world.

Keywords: Computational intelligence, autonomous systems, robotics and drones, ethical considerations

8.1 Introduction

Drones and autonomous robotics are transforming industries and regenerating the possibilities of automation. These devices are equipped with advanced computational intelligence (CI), exploiting the power of artificial intelligence (AI) algorithms and data-driven techniques. Drones and autonomous robots can make informed decisions, navigate through complex environments, and adapt to changing conditions, starting a new era of self-driven devices by incorporating CI [1, 2, 24, 25].

The incorporation of CI into drones and robots has opened a vast amount of applications across various sectors. These intelligent systems can perform tasks with enhanced effectiveness, accuracy, and autonomy; they can contribute to a greater height in the fields of aerial surveillance, logistics, agriculture, search and rescue, and industrial automation. Autonomous drones can sail through unknown territories to intelligent robots collaborating as a swarm to achieve complex objectives.

Computer vision algorithms enable drones to recognize and track objects, whereas simultaneous localization and mapping (SLAM) algorithms help robots build maps of unexplored environments. Reinforcement learning techniques allow these devices to learn from experience and enhance their actions to achieve specific goals. Predictive analytics and fault-tolerant algorithms further enhance the efficiency and reliability of these intelligent systems.

These advanced systems become increasingly fused into our daily lives, but ensuring ethical and safe use becomes a challenge. Balancing between innovation and ethics while addressing the potential challenges will be essential to make use of the full potential of drones and autonomous robotics incorporating CI [11].

The Role of Computational Intelligence in Advancing Autonomy
Computational intelligence empowers intelligent systems to operate independently and make informed decisions in complex and dynamic environments. The integration of CI techniques enables autonomous systems to learn from data, adapt to changing conditions, and optimize their performance, thus transforming industries and enhancing human experiences [27, 30].

- Decision-Making and Adaptive Behavior: Computational intelligence enables autonomous systems to make decisions based on real-time data and changing conditions. Techniques like fuzzy logic, neural networks, and evolutionary algorithms allow these systems to process complex information, weigh multiple factors, and make informed choices.
- Perception and Sensing: Autonomous systems need accurate perception and sensing capabilities to understand their environment. CI methods such as computer vision, sensor fusion, and signal processing help in interpreting sensory data from cameras, LiDAR (Light Detection and Ranging), RADAR (Radio Detection and Ranging), and other sensors. This information is crucial for navigation, object detection, and obstacle avoidance.
- Learning and Adaptation: One of the key features of CI is its ability to learn from data and adapt over time. Machine learning algorithms, including reinforcement learning, deep learning, and genetic algorithms (GAs), enable autonomous systems to improve their performance through experience. They can learn optimal strategies for different tasks, adapt to new scenarios, and handle uncertainties.
- Path Planning and Navigation: Autonomous systems need to plan optimal paths and navigate through complex environments. CI techniques help in generating paths that avoid obstacles, consider terrain characteristics, and optimize travel time. Algorithms like A* search, particle swarm optimization (PSO), and GAs contribute to efficient path planning.
- Uncertainty Management: Real-world environments are often uncertain and dynamic. Computational intelligence provides methods for handling uncertainty, including probabilistic modeling, Bayesian networks, and fuzzy logic. These techniques allow autonomous systems to make reasonable decisions even in the presence of incomplete or noisy data.

- Human–Robot Interaction (HRI): Advancing autonomy also involves better HRI. Natural language processing (NLP), sentiment analysis, and emotion recognition techniques help robots understand and respond to human commands and cues, making interactions more intuitive and effective [3].
- Swarm Intelligence: Computational intelligence principles are central to swarm intelligence, where groups of autonomous entities collaborate to achieve complex goals. Techniques like ant colony optimization (ACO), PSO, and artificial immune systems enable coordinated behavior and efficient problem-solving within a swarm.
- Cognitive Abilities: CI contributes to giving autonomous systems cognitive capabilities such as perception, reasoning, problem-solving, and learning. These abilities enable systems to perform tasks that require higher-order thinking, adapt to unforeseen situations, and learn from experience.
- Real-Time Adaptation: Autonomous systems often operate in dynamic environments where conditions can change rapidly. CI algorithms can process data in real time and make quick decisions to adapt to new circumstances, ensuring safe and effective operation.
- Continuous Improvement: Through data-driven insights and self-learning mechanisms, CI-driven autonomous systems can continually improve their performance, efficiency, and reliability over time, reducing the need for manual intervention.

Fundamentals of Computational Intelligence

Computational intelligence is a subdivision of AI that focuses on developing intelligent systems that are capable of learning, reasoning, and adapting to complex and unpredictable environments [1, 6].

It encloses a diverse set of AI techniques and algorithms that imitate human-like intelligence to solve complicated problems. The fundamentals of CI key components are shown in Figure 8.1.

- Machine Learning: It is a key aspect of CI. It involves training algorithms to learn patterns and make predictions from data without explicit programming. Common paradigms in CI are supervised learning, unsupervised learning, and reinforcement learning.

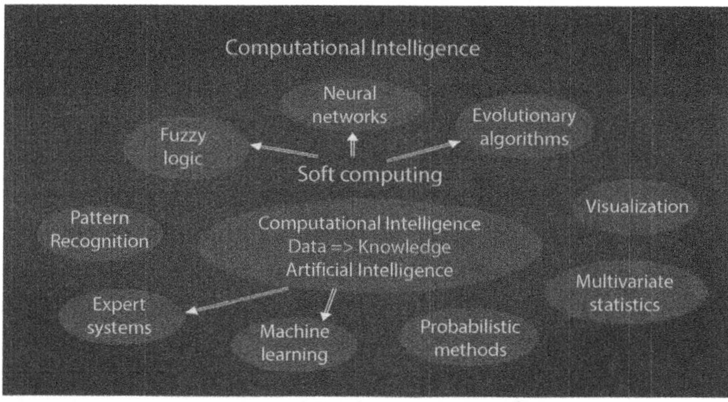

Figure 8.1 Computational intelligence includes key components.

- Neural Networks: They are inspired by the human brain's neural structure. It consists of interconnected nodes (neurons) organized in layers to process and learn patterns from the data. Deep learning comes under neural networks that can handle complex tasks and large datasets [28].
- Evolutionary Algorithms: They are inspired by the process of natural selection and evolution. Techniques like GAs, genetic programming, and evolutionary strategies are used to enhance solutions to complex problems by imitating the principles of survival of the fittest.
- Fuzzy Logic: Fuzzy logic deals with reasoning and decision-making in the presence of uncertainty or imprecision. It allows truth values, rather than strict binary values, enabling more human-like decision-making in ambiguous situations.
- Swarm Intelligence: Swarm intelligence is inspired by the interaction in the environment of social insects like ants, bees, and swarm animals where they coordinate and collaborate among multiple agents to solve complex problems causing the emergence of intelligent behavior.
- Probabilistic Reasoning: Probabilistic reasoning deals with uncertain information and probability distributions. Bayesian networks and Markov model techniques are used to model uncertainty and make probabilistic inferences.

- Expert Systems: Expert systems capture and imitate human expertise in specific domains by their decision-making capability. It uses rules and facts to make decisions and based on the accumulated knowledge of human experts, it provides solutions.
- Hybrid Approaches: Hybrid approaches combine multiple techniques to leverage their strengths and overcome their limitations. For example, neuro-fuzzy systems combine fuzzy logic and neural networks to improve decision-making in uncertain environments.
- Optimization Techniques: A wide range of optimization techniques, such as GAs and PSO are encompassed to find optimal solutions to complex optimization problems.

8.2 Literature Review

The convergence of drones, autonomous robotics, and CI has yielded a rich landscape of research and innovation that spans various domains. In recent literature, there is a growing emphasis on leveraging CI techniques to enhance the autonomy, adaptability, and intelligence of these unmanned systems.

Researchers have explored the integration of machine learning algorithms, such as convolutional neural networks (CNNs) and deep reinforcement learning, to enable drones and autonomous robots to learn from their environments and make data-driven decisions. These algorithms have been applied to tasks like image recognition, enabling drones to identify objects of interest or hazards, and enabling robots to autonomously navigate complex terrains.

Swarm intelligence techniques have also gained prominence, allowing groups of drones or robots to exhibit collective behaviors for tasks like exploration, surveillance, and environmental monitoring. These algorithms draw inspiration from natural systems like ant colonies or bird flocks, enabling coordinated actions and efficient information gathering.

Literature highlights the significance of real-time decision-making and adaptive control in the context of drones and autonomous robots. Researchers have developed predictive models and algorithms that account for uncertainties, enabling these systems to respond effectively to unforeseen changes in their environments. This is particularly crucial for applications such as disaster response, where autonomous robots need to navigate unpredictable and hazardous conditions.

Moreover, the literature underscores the importance of HRI and the role of CI in enabling seamless communication between autonomous systems and human operators. Natural language processing and affective computing techniques have been employed to enhance the interpretability of commands and enable robots to understand and respond to human emotions, thereby fostering more intuitive and effective collaboration.

Challenges and Opportunities in Autonomous Systems

Autonomous systems offer transformative opportunities in various industries like healthcare and manufacturing. However, they also face some challenges that must be addressed to explore their full potential [12, 34, 35]. Some of the key challenges and opportunities in autonomous systems are shown in Table 8.1.

Table 8.1 Challenges and opportunities in autonomous systems.

Challenges in autonomous systems	Opportunities in autonomous systems
1. Safety and Reliability: Ensuring the safety of autonomous systems, especially in critical applications like self-driving cars and medical robots.	1. Enhanced Safety: Autonomous systems can reduce human errors, potentially leading to safer operations in industries such as transportation and healthcare.
2. Uncertainty and Variability: Dealing with unpredictable environments, weather conditions, and unexpected events that can impact the performance of autonomous systems.	2. Adaptability: Autonomous systems equipped with learning algorithms can adapt to various scenarios, making them versatile and capable of handling dynamic environments.
3. Ethical and Legal Concerns: Addressing ethical dilemmas, liability issues, and regulatory frameworks as autonomous systems become more integrated into society.	3. Improved Efficiency: Autonomous systems can optimize resource utilization, energy consumption, and workflow efficiency, benefiting industries like manufacturing and logistics.
4. Human–Machine Interaction: Designing interfaces that facilitate effective communication and collaboration between humans and autonomous systems.	4. Human Augmentation: Autonomous systems can complement human capabilities in areas such as medical diagnosis, disaster response, and precision agriculture.

(*Continued*)

Table 8.1 Challenges and opportunities in autonomous systems. (*Continued*)

Challenges in autonomous systems	**Opportunities in autonomous systems**
5. Data Privacy and Security: Ensuring the protection of sensitive data collected and processed by autonomous systems, preventing unauthorized access and potential breaches.	5. Data-Driven Insights: Autonomous systems generate vast amounts of data that can be analyzed to gain insights and improve decision-making across industries.
6. Technical Limitations: Overcoming technological limitations, such as sensor accuracy, processing power, and connectivity issues, can impact the performance of autonomous systems.	6. Technological Advancements: Advances in sensor technology, computing power, and communication networks can drive the development of more capable and efficient autonomous systems.
7. Social Acceptance: Gaining public trust and acceptance of autonomous systems, especially in domains like transportation, where safety concerns can be significant.	7. Transportation Revolution: Self-driving cars and drones can revolutionize transportation, reducing traffic congestion, improving accessibility, and minimizing accidents.
8. Cost and Accessibility: Developing cost- effective autonomous systems that are accessible to a wide range of industries and applications.	8. Increased Accessibility: Affordable autonomous systems can democratize technology, making it accessible to small businesses and resource-limited regions.
9. Integration With Existing Infrastructure: Integrating autonomous systems into existing workflows and infrastructure without causing disruptions.	9. Process Optimization: Autonomous systems can optimize workflows, supply chains, and operations, leading to cost savings and improved productivity.
10. Public Perception and Education: Overcoming misconceptions and educating the public about the capabilities, benefits, and limitations of autonomous systems.	10. Education and Awareness: Opportunities exist to educate the public, policymakers, and businesses about the transformative potential of autonomous systems across sectors.

These challenges and opportunities highlight the multifaceted nature of the autonomous systems landscape, where technological advancements, ethical considerations, regulatory frameworks, and societal acceptance intersect to shape the future of automation.

Swarm Intelligence in Multi-Robot Systems

Swarm intelligence is inspired by the collaborative behavior of social insects such as ants, bees, and other swarm animals. In the case of multi-robot systems, the ability of a group of robots to work together collaboratively, reveals emergent intelligent behavior. This approach leverages the power of decentralized decision-making, communication, and cooperation, enabling the swarm of robots to accomplish complex tasks more efficiently and robustly than individual robots operating in isolation.

Here are some key aspects of swarm intelligence in multi-robot systems:

- Decentralization: Swarm intelligence focuses attention on decentralized control, where each robot makes decisions based on local information and simple rules. This lack of a central authority allows the swarm to exhibit self-organization and adaptability to changing environments.
- Cooperation and Coordination: In a swarm, robots cooperate and coordinate with each other to achieve common goals. Through communication and interactions, they share information and work collaboratively to accomplish tasks.
- Robustness and Fault Tolerance: Swarm intelligence endue multi-robot systems with robustness and fault tolerance. Even if individual robots fail or are removed from the swarm, the overall system can continue to function effectively, ensuring mission success in dynamic and unpredictable environments.
- Scalability: Swarm intelligence is highly scalable, allowing the number of robots in the swarm to vary without a significant impact on performance. This scalability makes swarm-based systems suitable for a wide range of applications, from small groups to large collectives of robots.
- Exploration and Coverage: Swarms of robots can efficiently explore unknown environments and achieve area coverage. By employing distributed algorithms and communication,

they can divide tasks and explore multiple regions simultaneously, reducing exploration time.

- Search and Rescue: Swarm intelligence is particularly valuable in search and rescue situations. Robots can collaboratively search for survivors in disaster-stricken areas, maximizing the chances of locating victims quickly.
- Collective Decision-Making: Swarms can collectively make decisions by aggregating the preferences or opinions of individual robots. This ability to perform collective decision-making helps in situations where multiple options need to be evaluated.
- Resource Efficiency: Swarm intelligence enables resource-efficient operations, as robots can distribute tasks, share responsibilities, and avoid duplication of efforts, optimizing energy consumption and prolonging the overall system's operation.
- Adaptive Behavior: Swarm-based systems can adapt to dynamic environments and changing conditions. The collective intelligence of the swarm allows it to adjust its behavior in response to new challenges and requirements.

8.3 Navigation and Path Planning

The essential components of autonomous systems are navigation and path planning, including drones, robots, and self-driving vehicles. It enables these intelligent devices to move safely and efficiently through their environments, avoiding obstacles and reaching their intended destinations. Let us explore navigation and path planning in more detail.

- Navigation: It refers to the ability of an autonomous system to decide its location, recognize its surroundings, and make decisions to reach a target destination. Various key aspects of navigation include the following:

    - Localization: It involves determining the exact position of the autonomous system within its environment. Several techniques like GPS (Global Positioning System) and SLAM are used to estimate the system's location accurately.

  - Perception: It gathers information about the environment by using various sensors, such as cameras, LiDAR, and RADAR. These data are processed to identify objects, obstacles, and other relevant information.
  - Mapping: It is the process of creating a representation of the environment that the autonomous system can use for navigation. It involves building and updating a map of the surroundings to plan safe and efficient paths.

- Path Planning: It is the process of determining a suitable path from the current location to the destination while avoiding obstacles and adhering to constraints. It involves analyzing the map and making decisions on the best route to take.

Path Planning

It enables autonomous systems to move efficiently while avoiding collisions and optimizing their trajectories. Key components of path planning include:

- Search Algorithms: Path planning algorithms use search techniques, such as A* (A- star) and Dijkstra's algorithm, to explore the possible paths from the current location to the destination. These algorithms find the shortest or most efficient path based on a cost function and constraints.
- Collision Avoidance: Path planning algorithms consider obstacles and dynamically update the route to avoid collisions with static or moving obstacles in the environment. Collision avoidance is important for ensuring safety and preventing damage to the autonomous system and its surroundings.
- Dynamic Path Planning: Path planning algorithms need to adapt in real-time to changing conditions, such as moving obstacles or dynamic road conditions. This technique uses predictive models and sensor data to adjust the path as the environment evolves.
- Optimization: Some path planning algorithms aim to minimize travel time and energy consumption or to focus on optimization. These algorithms are crucial for efficient and resource-conscious autonomous systems.
- Multi-Agent Path Planning: In scenarios with multiple autonomous systems, multi- agent path planning algorithms

coordinate the movement of these agents to avoid collisions and conflicts while achieving their objectives.

Navigation and path planning play significant roles in ensuring the safe and effective operation of autonomous systems in diverse applications, from delivery drones and self- driving cars to industrial robots and exploration rovers. As CI and AI algorithms continue to advance, navigation and path-planning capabilities will become even more sophisticated, contributing to the realization of a future with smart and autonomous devices.

AI-Based SLAM (Simultaneous Localization and Mapping) for Robotic Navigation

AI-based SLAM is an ultra-modern technology that combines AI algorithms with sensor data that enable autonomous robots and drones to navigate through unknown environments while simultaneously creating a map of their surroundings. SLAM is a crucial component of robotic navigation, as it allows robots to understand their location relative to the environment and make informed decisions to move safely and efficiently. Here is how AI-based SLAM works and its significance in robotic navigation:

- Sensor Data and Feature Extraction: In AI-based SLAM, the robot collects data from various sensors, such as cameras, LiDAR, and IMU (Inertial Measurement Unit). AI algorithms are used to extract key features, such as landmarks, edges, and corners, from the sensor data.
- Mapping: The robot uses AI techniques like computer vision and machine learning to process the extracted features and build a map of the environment. As the robot moves, it continuously updates the map with new sensor data to represent its current understanding of the surroundings.
- Localization: AI-based SLAM enables the robot to estimate its position (localization) within the map it created. This process involves aligning the sensor data with the map and calculating the robot's pose accurately, even in the presence of uncertainties and noise.
- Loop Closure: AI algorithms detect and close loops in the robot's path, in in-order to improve the precision of the map and localization. When the robot revisits a previously mapped location, loop closure algorithms recognize the connection and correct any accumulated errors in the map.

- Real-Time Adaptation: AI-based SLAM allows robots to adapt in real-time to changes in the environment, such as moving objects or dynamic obstacles. The AI algorithms continuously process sensor data and adjust the map and localization to handle dynamic scenarios.

Significance in Robotic Navigation

AI-based SLAM plays a major role in enabling robotic navigation in complex and unstructured environments:

- Autonomous Exploration: With SLAM, robots can autonomously explore and map unknown environments, making them ideal for search and rescue missions or mapping areas in hazardous conditions.
- Indoor Navigation: SLAM is essential for robots operating in indoor environments, where GPS signals are not available. Robots can create maps of buildings and navigate through offices, warehouses, and factories with high precision.
- Mobile Robotics: AI-based SLAM is essential for mobile robotics applications, such as delivery robots and service robots, allowing them to navigate safely and efficiently in dynamic urban environments.
- Mapping for Other Applications: The maps which are generated by SLAM can be used for other tasks like path planning, object detection, and augmented reality.
- Human–Robot Interaction: SLAM intensifies HRI by providing robots with situational awareness, allowing them to better understand and respond to human commands and gestures.
- AI-based SLAM continues to advance, leveraging the power of deep learning, probabilistic methods, and sensor fusion to improve the accuracy and efficiency of robotic navigation. As research progresses, AI-based SLAMs will remain at the forefront of enabling robots to operate autonomously and effectively in diverse and dynamic environments [28].

Swarm Robotics: Cooperative Navigation and Exploration

Swarm robotics draws inspiration from the collective behavior of social insects, such as ants and bees, to design multi-robot systems that exhibit intelligent and cooperative behaviors. The important applications of swarm robotics include cooperative navigation and exploration, where groups of robots collaborate to navigate through complex environments and explore unknown territories. Let us delve into these concepts.

Cooperative Navigation

In cooperative navigation, robots work together as a swarm to reach specific locations or destinations while avoiding obstacles and ensuring smooth movement. Key aspects of cooperative navigation in swarm robotics include the following:

- Decentralized Decision-Making: Each robot in the swarm operates independently and makes decisions based on local information and simple rules. This decentralized approach allows the swarm to exhibit emergent behavior and adapt to changes in the environment without relying on a central control unit.
- Communication and Coordination: Cooperative navigation relies on communication and coordination among the robots to share information about their positions, paths, and obstacles. This enables the swarm to avoid collisions and optimize its movement collectively.
- Path Planning and Exploration: Swarm robots use distributed algorithms for path planning, dividing the exploration area into sub-tasks that individual robots can accomplish. This collaborative approach enhances the efficiency of exploration in large and complex environments.
- Fault Tolerance: Cooperative navigation verifies robustness in the swarm. If a robot experiences a failure, then the other robots can adapt their paths and continue the mission.

Cooperative Exploration

Cooperative exploration involves swarm robots collaboratively exploring unknown and potentially hazardous environments, aiming to map the surroundings and gather valuable information. Key aspects of cooperative exploration in swarm robotics include the following:

- Coverage Optimization: Swarm robots use distributed algorithms to divide the exploration area into regions, ensuring that each robot covers a unique and unexplored section. This optimization enhances the overall exploration efficiency and minimizes redundant efforts.

- Exploration Strategy: Cooperative exploration employs different strategies for coordinated movement. Some robots may act as scouts, exploring distant regions, while others act as connectors, establishing communication paths between the scouts and the rest of the swarm.
- Mapping and Information Sharing: As swarm robots explore, they create maps of the environment and share this information among the swarm members. This collaborative mapping strengthens the robots' collective knowledge about the surroundings.
- Targeted Exploration: In targeted exploration, swarm robots may prioritize certain areas of interest based on specific criteria, such as a higher probability of finding resources or detecting potential hazards.

Applications

Cooperative navigation and exploration in swarm robotics applications refer to the collective behavior of a group of autonomous robots working together to navigate through an environment and explore it efficiently. This concept draws inspiration from natural swarms like bird flocks or ant colonies, where individual agents coordinate their actions to achieve common goals. Here is an overview of cooperative navigation and exploration in swarm robotics [33]:

- Search and Rescue: Swarm robots can be deployed in search and rescue missions to explore disaster-stricken areas and locate survivors efficiently.
- Environmental Monitoring: In ecological studies, swarm robots can cooperatively explore and monitor remote and challenging environments to gather data about wildlife and ecosystems.
- Precision Agriculture: Swarm robots can explore agricultural fields to improve resource allocation, monitor crop health, etc.
- Surveillance and Security: In surveillance and security applications, swarm robots can collaboratively vigil and monitor large areas, strengthening situational awareness.

The power of swarm robotics continues to grow, driven by advancements in AI, communication, and sensor technologies for cooperative navigation and exploration. As research progresses, swarm robots are expected to play an increasingly significant role in addressing complex real-world challenges and advancing exploration capabilities in various domains.

8.4 Perception and Object Detection

Perception and object detection are important components of autonomous systems, these processes involve using various sensors and algorithms to recognize and interpret objects, providing essential information for navigation, decision-making, and interaction. Let us explore perception and object detection in more detail:

- Perception and Object Detection: CI-powered computer vision techniques enhance perception capabilities, enabling autonomous systems to understand their surroundings better. Object detection algorithms help identify and track objects of interest, such as pedestrians, vehicles, and obstacles, critical for safe navigation.
- Sensor Fusion: It combines data from multiple sensors to create a comprehensive view of the environment. By integrating information from cameras, LiDAR, RADAR, and other sensors, autonomous systems gain comprehensive awareness and can handle uncertainty and sensor limitations.
- Path Planning and Navigation: CI-based path planning algorithms allow autonomous systems to find the most efficient and safe paths through complex environments. These algorithms optimize trajectories while considering constraints and avoiding collisions with obstacles.
- Reinforcement Learning for Adaptive Behavior: It empowers autonomous systems to interact with the environment and receive feedback to improve their behavior iteratively. This adaptive approach enables systems to optimize their actions to achieve specific objectives.
- Swarm Intelligence and Multi-Agent Systems (MASs): It allows multiple autonomous agents to coordinate and collaborate efficiently. These systems can collectively achieve complex tasks.
- Uncertainty Handling: CI techniques, including fuzzy logic and probabilistic reasoning, address uncertainty in the environment, enabling autonomous systems to make robust decisions even in vague situations.

- Real-Time Processing: CI supports real-time data processing, critical for time- sensitive applications like autonomous vehicles and drones. High-performance algorithms ensure rapid decision-making and response to dynamic changes in the environment.
- Anomaly Detection and Fault Tolerance: CI algorithms are utilized for anomaly detection and fault tolerance, enabling autonomous systems to identify and respond to unexpected events, ensuring safe and reliable operation.
- Explainable AI: As autonomy advances, the need for explainable AI becomes crucial, particularly in safety-critical applications. CI techniques help create interpretable models, providing insights into the decision-making process of autonomous systems.
- The fusion of CI with autonomous systems empowers them to navigate complex challenges and explore new possibilities. From self-driving cars and delivering drones to industrial robots and intelligent healthcare devices, CI's role in advancing autonomy is instrumental in creating safer, more efficient, and adaptive intelligent systems. As research and development in CI continue, the future holds even more exciting opportunities to leverage AI-driven autonomy for the benefit of society.

Perception

Perception refers to the ability of an autonomous system to recognize and understand its environment through sensory data. It involves collecting, processing, and interpreting information from sensors such as cameras, LiDAR, RADAR, and ultrasonic sensors. Key aspects of perception include the following:

- Sensor Fusion: Sensor fusion combines data from multiple sensors to create a comprehensive and accurate representation of the environment. By combining data from different sources, autonomous systems can obtain a more reliable perception of their surroundings.
- Feature Extraction: In perception, features relevant to object detection and recognition are extracted from raw sensor data. These features may include edges, corners, textures, and color information, which are used to identify objects.

- Localization: Perception also involves determining the precise location of objects in the environment relative to the autonomous system. Several techniques such as image registration and 3D point cloud processing are used for accurate localization.
- Object Segmentation: Object segmentation involves dividing the sensor data into meaningful segments corresponding to individual objects or regions of interest. This segmentation aids in identifying and tracking objects in the environment.

Object Detection

Object detection is a specific task within perception that focuses on identifying and locating specific objects of interest within the sensory data. Key aspects of object detection include the following:

- Object Recognition: Object detection algorithms use deep learning and computer vision techniques to recognize specific objects, such as pedestrians, vehicles, traffic signs, and obstacles.
- Real-Time Processing: In many autonomous applications, real-time object detection is crucial to enable quick decision-making and response. High-performance algorithms and hardware acceleration are employed to achieve real-time processing.
- Classification and Localization: Object detection algorithms not only identify objects but also provide information about their location within the sensor data. This includes both classifying the type of object and providing bounding boxes around the detected objects [20].
- Multiple Object Detection: Object detection algorithms must be capable of detecting multiple objects simultaneously, as autonomous systems encounter numerous objects in their surroundings.

Applications

Perception and object detection play pivotal roles in various autonomous systems and applications, are shown in Table 8.2.

These applications showcase how perception and object detection technologies play a crucial role across various industries, improving efficiency, safety, and decision-making in diverse contexts.

Table 8.2 Applications of perception and object detection.

Applications of perception and object detection	Description
Autonomous Vehicles	Perception and object detection are critical for self-driving cars to identify pedestrians, vehicles, traffic signs, and obstacles in their environment, enabling safe navigation and decision-making [21].
Surveillance and Security	Surveillance systems utilize object detection to monitor areas for unauthorized access, suspicious activities, or intruders.
Robotics	Robots use perception and object detection to navigate, interact with objects, and collaborate with humans in various environments.
Industrial Automation	In manufacturing, perception and object detection enable robots to perform tasks like sorting, quality control, and assembly on production lines [14, 15].
Healthcare	Object detection aids in medical imaging, assisting in identifying anomalies, tumors, or specific anatomical structures in diagnostic scans.
Agriculture	Drones and robots equipped with perception capabilities can monitor crops, identify diseases, and optimize planting and harvesting.
Retail	Object detection enhances inventory management, customer tracking, and checkout processes in retail environments.
Environmental Monitoring	In wildlife conservation, perception, and object detection help identify and track animals for research and conservation efforts.
Augmented Reality (AR)	AR applications overlay virtual objects onto the real world, relying on perception to understand the environment and context for accurate placement [19].
Smart Cities	Perception and object detection contribute to smart city initiatives by enabling the monitoring of traffic flow, waste management, and urban planning.

Computer Vision Techniques for Object Detection and Recognition

Computer vision techniques for object detection and recognition have significantly evolved over the years, thanks to advancements in deep learning and the availability of large- scale datasets. These techniques aim to automatically identify and localize objects within images or video frames. Here are some popular techniques used for object detection and recognition [28]:

- Haar Cascade Classifiers: Haar Cascades are simple and efficient feature-based classifiers that can be used for object detection tasks. They work by identifying features in an image using Haar-like features and then using a cascade of classifiers to detect objects.
- Histogram of Oriented Gradients (HOG): The HOG technique extracts local gradient information from an image and represents it as a histogram of gradient directions. This technique is often used for pedestrian detection and other similar tasks.
- Scale-Invariant Feature Transform (SIFT): SIFT is a method for detecting and describing local features in images. It is robust to scale, rotation, and illumination changes, making it useful for object recognition tasks.
- Speeded-Up Robust Features (SURF): SURF is a faster alternative to SIFT, offering similar performance for feature detection and matching.
- Faster R-CNN: Faster R-CNN is a deep learning–based object detection model that combines the advantages of region proposal methods (like R-CNN) and fast feature extraction using CNNs. It is one of the most widely used and accurate object detection techniques.
- You Only Look Once (YOLO): YOLO is a real-time object detection system that directly predicts bounding boxes and class probabilities from an entire image in one forward pass through a neural network.
- Single-Shot MultiBox Detector (SSD): Similar to YOLO, SSD is another real-time object detection algorithm that predicts multiple bounding boxes and class probabilities at different scales.
- Mask R-CNN: Mask R-CNN extends Faster R-CNN to also predict object masks in addition to bounding boxes and class probabilities. This is particularly useful for tasks that require pixel-level segmentation.

- MobileNet: MobileNet is a lightweight CNN architecture designed for efficient computation on mobile and embedded devices. It is often used for real-time object detection on resource-constrained platforms.
- EfficientDet: EfficientDet is an efficient and scalable object detection model that achieves state-of-the-art performance with fewer computational resources.
- Transfer Learning: Transfer learning is a technique where a pre-trained model on a large dataset (like ImageNet) is fine-tuned on a smaller dataset for object recognition tasks. This approach is useful when you have limited labeled data.
- Data Augmentation: This technique involves artificially increasing the size of the training dataset by applying transformations like rotation, scaling, flipping, and color jittering. This helps improve the generalization of the model.

Deep Learning for Real-time Object Detection in Drones

Deep learning has revolutionized real-time object detection in various domains. Deep learning models can efficiently process images and videos in real-time, with the arrival of powerful GPUs and dedicated hardware accelerators. Here are some key aspects and techniques involved in using deep learning for real-time object detection in drones [28]:

- Convolutional Neural Networks (CNNs): CNNs are the foundation of most modern deep learning–based object detection models. They are designed to automatically learn hierarchical features from images, making them highly effective for visual recognition tasks. CNN architectures like ResNet, MobileNet, and EfficientNet are commonly used in drone applications due to their efficiency and accuracy.
- Single-Shot Detection Models (SSDs) and You Only Look Once (YOLO): SSD and YOLO are popular real-time object detection models that offer high speed and accuracy. They directly predict bounding boxes and class probabilities in a single pass through the network, making them well suited for applications where real-time performance is critical, such as drones.
- EfficientNet and MobileNet: These lightweight CNN architectures are designed to achieve a good balance between accuracy and model size. They are often used in drone applications to reduce computational overhead while maintaining acceptable detection performance.

- Hardware Acceleration: For real-time object detection on drones, hardware acceleration is essential to perform inference efficiently. Dedicated hardware solutions like NVIDIA Jetson boards or custom field-programmable gate arrays (FPGAs)/application-specific integrated circuits (ASICs) implementations can significantly speed up the processing.
- Transfer Learning: It is a common technique used in drone applications, where pre- trained models on large image datasets (e.g., ImageNet) are fine-tuned on drone- specific datasets. This approach leverages the knowledge learned from general visual recognition tasks and adapts it to drone-specific object detection requirements, even when labeled drone data might be limited.
- Data Augmentation: Data augmentation is important to increase the diversity of the training dataset and improve the generalization of the object detection model. Techniques like rotation, scaling, flipping, and random cropping are often used to augment drone images.
- Object Tracking and Fusion: To ensure consistent and accurate object detection in real-time, object tracking algorithms can be used to follow objects from frame to frame, reducing the need for redundant processing on consecutive frames. Additionally, sensor fusion techniques can combine data from multiple sensors (e.g., cameras and LiDAR) on the drone to improve object detection accuracy and reliability.
- Onboard vs. Cloud-Based Processing: Depending on the drone's computational capabilities, object detection can be performed onboard the drone or offloaded to a cloud server for processing. Onboard processing allows for real-time response and can be more reliable when communication with the cloud is limited or unstable.
- Obstacle Detection and Avoidance: Beyond object recognition, drones may need to detect obstacles in their flight path to ensure safe navigation. Deep learning models can be trained to identify potential obstacles and trigger collision avoidance mechanisms.

Real-time object detection in drones is a challenging yet essential task for various applications, such as surveillance, search and rescue, precision agriculture, and environmental monitoring.

8.5 Adaptive Control and Decision-Making

Adaptive control and decision-making within the realm of CI involve the dynamic adjustment of control strategies and choices based on real-time data and changing circumstances. Drawing on machine learning algorithms, neural networks, and evolutionary computing, adaptive control systems continuously learn and optimize their responses to complex and uncertain environments. These systems autonomously assess and adapt their actions, enabling them to tackle intricate tasks, navigate unforeseen challenges, and fine-tune strategies over time. By seamlessly integrating data-driven insights and adaptability, adaptive control and decision-making harness the power of CI to enhance the autonomy and performance of autonomous systems across a wide spectrum of applications, from robotics and automation to finance and beyond [21, 29].

CI Algorithms for Intelligent Control in Robotics

Control algorithms play a crucial role in robotics, enabling intelligent decision-making and precise manipulation of robot movements. There are several CI techniques used for intelligent control in robotics. Some of the prominent CI algorithms for robotics are [6–8, 18] as follows:

- Fuzzy Logic Control (FLC): Fuzzy logic is widely used in robotics to deal with uncertainty and imprecision. FLC allows the representation of expert knowledge in the form of linguistic rules and enables the robot to make decisions based on fuzzy reasoning. It is effective for systems that have complex and non-linear relationships between inputs and outputs.
- Genetic Algorithms (GAs): GAs are evolutionary optimization techniques inspired by natural selection and genetics. In robotics, GA can be used to optimize control parameters or design robot behaviors to achieve specific objectives, such as path planning, motion control, or task allocation.
- Particle Swarm Optimization (PSO): PSO is a population-based optimization algorithm inspired by the social behavior of birds flocking or fish schooling. In robotics, PSO can be employed to find optimal control parameters or trajectories for robot motion planning and coordination tasks.
- Artificial Neural Networks (ANN): Neural networks have been extensively used for various robotic control tasks, including perception, path planning, motion control, and

reinforcement learning. They excel in learning complex mappings between inputs and outputs from data [22].
- Reinforcement Learning (RL): RL is an area of machine learning where agents learn from their interactions with the environment to maximize cumulative rewards. In robotics, RL can be employed for tasks like robot navigation, grasping, and task execution in uncertain and dynamic environments.
- Deep Reinforcement Learning (DRL): DRL combines deep learning and RL to handle high-dimensional input spaces and complex control problems. DRL has been successfully used for robotic tasks, such as robotic arm control, robotic locomotion, and autonomous vehicles.
- Neuro-Fuzzy Systems: Neuro-fuzzy systems combine fuzzy logic and neural networks to leverage the interpretability of fuzzy rules with the learning capabilities of neural networks. They can be applied in robotic systems for adaptive control and decision-making.
- Swarm Robotics: Swarm robotics involves coordinating multiple robots to accomplish tasks collectively, inspired by the collective behavior of social insects. CI techniques like PSO or ACO are employed to manage the swarm's behavior effectively.
- Hybrid Approaches: Many advanced robotic control systems use a combination of multiple CI techniques. For example, neuro-fuzzy systems, genetic-fuzzy systems, or fuzzy-PSO approaches may be used to design controllers that are both adaptive and efficient.
- Evolutionary Robotics: Evolutionary Robotics employs evolutionary algorithms like GAs to evolve robot control strategies and behaviors. It is useful when designing control systems for robots with complex morphologies or in environments with unknown characteristics.

These CI algorithms enable robots to adapt to changing environments, handle uncertainties, and optimize their control strategies for various tasks, making them valuable tools in the field of intelligent robotics.

8.6 Swarm Robotics and Multi-Agent Systems

Swarm robotics and MASs are interdisciplinary fields that focus on the study of groups of relatively simple agents or robots that collaborate to achieve complex tasks through decentralized coordination and interactions. Swarm robotics draws inspiration from natural swarms, such as insect colonies and bird flocks, to develop algorithms and principles for designing cooperative behaviors among robots. Multi-agent systems, on the other hand, encompass a broader range of scenarios where multiple autonomous agents, which can be robots, software agents, or even humans, interact to achieve shared goals.

Collective Decision-Making using Swarm Intelligence

Collective decision-making using swarm intelligence is a problem-solving approach inspired by the behavior of social swarms and colonies in nature. Swarm intelligence is a form of distributed and decentralized problem-solving, where a group of simple agents (also known as "swarm" or "colony" members) interact locally with one another and their environment to collectively achieve a common goal. This concept has been successfully applied in various domains, including robotics, optimization, and decision-making. Here is an overview of how swarm intelligence is used for collective decision-making [23, 26, 27]:

- Agent-Based Model: In collective decision-making using swarm intelligence, each agent in the swarm represents a potential solution or decision option. Agents may be simple entities or more complex, incorporating rule-based behavior or learning mechanisms.
- Local Interaction and Communication: Agents interact with one another and the environment through local interactions, meaning that they only have access to limited information about their neighboring agents. They communicate with each other by sharing information or signals through simple rules.
- Emergent Behavior: The collective decision-making emerges from the interactions between individual agents without any centralized control or global knowledge. It means that the system as a whole exhibits behavior that is not explicitly programmed but arises from the interactions of its components.
- Stigmergy: It is a key concept in swarm intelligence where agents modify their environment, leaving cues or signals that influence the behavior of other agents. For example, in ant colonies, ants leave pheromone trails, guiding others to resources like food.

- Swarm Algorithms: Various swarm algorithms have been developed to solve decision-making problems. Examples include ACO, PSO, and bee algorithm. Each algorithm is designed to exploit collective behavior to solve specific problems.
- Optimization Problems: Swarm intelligence is often used to find optimal solutions to complex optimization problems, such as the traveling salesman problem, routing problems, and resource allocation tasks. Agents explore the solution space through interactions and converge toward promising solutions over time.
- Multi-Agent Systems: In many cases, swarm intelligence is implemented using multi- agent systems (MAS), where multiple autonomous agents collaborate to achieve a common goal. These agents may have different capabilities, roles, and rules for interaction.
- Robotic Swarms: In robotics, swarm intelligence is applied to control groups of robots working collaboratively. Robotic swarms can perform tasks like exploration, surveillance, search and rescue, and environmental monitoring with increased efficiency and robustness.
- Consensus and Voting: Swarm intelligence can also be applied to collective decision- making in scenarios where agents need to reach a consensus or make a democratic vote. Agents share information and iteratively adjust their decisions to converge on a common choice.

Case Studies: Swarm Robotics Applications and Success Stories

Swarm robotics has shown great promise in various applications, with many success stories demonstrating the effectiveness of this approach. Here are some notable case studies of swarm robotics applications:

- Hazardous Environment Exploration: In situations where human presence could be dangerous, swarm robotics is employed for exploration tasks. For example, the EU- funded Swarmanoid project used a combination of ground robots, aerial drones, and climbing robots to explore and map complex environments, such as disaster-stricken buildings, for search and rescue missions.
- Agricultural Automation: Swarm robotics has been applied to precision agriculture to improve crop monitoring and

management. The BoniRob project by Deepfield Robotics involved a swarm of small robots equipped with cameras and sensors that cooperatively analyzed crops, identified weeds, and applied targeted herbicides, reducing the need for manual labor and minimizing chemical usage [14, 15].

- Construction and Infrastructure: Swarm robots can collaborate to build structures in environments where traditional construction methods are impractical. Harvard's Wyss Institute developed a termite-inspired swarm of robots that can autonomously build structures, such as walls and towers, by collectively assessing their surroundings and coordinating their actions.
- Search and Surveillance: Swarms of drones and ground robots can be deployed for surveillance and reconnaissance tasks. The US Navy's LOCUST (Low-Cost UAV Swarming Technology) program demonstrated the deployment of multiple drones that collaborate to perform surveillance and other tactical missions.
- Environmental Monitoring: Swarm robotics can be used for environmental monitoring in remote or hazardous locations. Researchers at Georgia Tech developed a swarm of underwater robots that can work collaboratively to monitor water quality and collect data on marine ecosystems.
- Goods Delivery and Warehouse Management: In logistics and warehousing, swarm robotics can optimize order fulfillment and inventory management. Companies like Amazon have explored the use of swarms of robots in their fulfillment centers to efficiently pick up and transport packages.
- Pollution Cleanup: Swarm robots can be deployed for environmental cleanup tasks. For instance, the European-funded project Dustbot employed a swarm of small robots to clean up dust and debris from city streets autonomously.
- Cooperative Transportation: Swarms of autonomous vehicles can collaborate to optimize traffic flow and improve transportation efficiency. Research initiatives, such as the European-funded project SMARTFLOWS, have explored swarm-based traffic management strategies.
- Underground Mining: Swarms of autonomous robots have been investigated for exploration and mapping in underground mining operations, where communication may be challenging.

- Aerospace and Space Exploration: Swarms of small satellites or rovers can be used for cooperative space exploration missions, such as asteroid mapping or planetary surface exploration.

These case studies demonstrate the versatility and potential of swarm robotics in solving real- world problems. The collective intelligence and cooperation among individual robots offer numerous advantages, including robustness, adaptability, and scalability, making swarm robotics an exciting field with a wide range of promising applications. As technology and research continue to advance, we can expect to see even more impactful and innovative uses of swarm robotics in the future.

8.7 Autonomous Drone Delivery Systems

Autonomous drone delivery systems are a type of unmanned aerial vehicle (UAV) application that aims to deliver packages, goods, or other items to specified locations without the need for direct human control. These systems have gained significant attention in recent years due to their potential to revolutionize last-mile logistics and provide faster, more efficient, and environmentally friendly delivery solutions. Here are key aspects and components of autonomous drone delivery systems:

- Autonomous Navigation: The drones in the delivery system are equipped with sophisticated navigation systems, including GPS, sensors (such as LiDAR and cameras), and onboard processors. These enable the drones to autonomously plan their routes, avoid obstacles, and make real-time adjustments during flight.
- Package Loading and Unloading: Automated systems are used to load packages onto the drones securely and retrieve them upon delivery. Some systems may involve automated docking stations or ground-based robots to assist in the loading and unloading process.
- Route Planning and Optimization: Efficient route planning is critical for optimizing delivery times and minimizing energy consumption. Algorithms consider factors such as distance, weather conditions, airspace regulations, and package weight to determine the most optimal routes.

- Air Traffic Management: To ensure the safe operation of autonomous drones, air traffic management systems are employed to monitor drone flights and prevent collisions. Integration with existing air traffic control systems is necessary to maintain a safe and organized airspace.
- Obstacle Avoidance: Drones are equipped with obstacle avoidance systems, such as LiDAR or depth cameras, to detect and avoid obstacles in their flight path, ensuring safe and collision-free navigation.
- Battery and Energy Management: Autonomous drones need to manage their battery usage efficiently to complete deliveries within the required range. Advanced battery management systems ensure that drones have enough power to complete their missions and return safely to their base.
- Regulatory Compliance: Drone delivery systems must comply with regulations set by aviation authorities, which include flight altitudes, restricted areas, and noise restrictions. Companies must work closely with regulators to ensure their systems adhere to the rules.
- Weather and Environmental Considerations: Weather conditions play a crucial role in drone delivery operations. Autonomous systems should be able to assess weather conditions and adjust flight plans accordingly to ensure safe and reliable deliveries.
- Communication and Connectivity: Drones require reliable communication links to receive real-time updates and transmit data about their flight status. This may involve satellite communication, 4G/5G networks, or other communication technologies.
- Security and Theft Prevention: Ensuring the security of the packages during delivery is essential. Drones may use secure lockboxes or other mechanisms to prevent theft or tampering.

Companies such as Amazon Prime Air, Google's Wing, UPS Flight Forward, and DHL have been actively developing and testing autonomous drone delivery systems. These systems have the potential to revolutionize the delivery industry by offering faster and more efficient delivery options, especially in remote or congested areas. However, there are still challenges to overcome, including regulatory hurdles, safety concerns, public acceptance, and integration with existing logistics networks. As technology and regulations progress, we can expect to see more widespread adoption of autonomous drone delivery systems in the future.

High-performance computing, vision systems, sensors, interconnect, memory, networking technologies, and increasingly advanced technologies like AI are working in concert to provide autonomous technologies with ever-higher levels of autonomy. When it comes to autonomous applications, devices such as drones, robots, and driverless vehicles come to mind. Typically, what they all have in common is the ability to gather information on their surroundings, process these data, interpret it, and act on it. The key difference between a simple remote-controlled or programmable device and an autonomous machine is, of course, its intelligent ability to do things on its own. This almost intuitive response to the data presented is called artificial intelligence or AI.

The push toward industrial digital transformation and automation is generating a growing demand for greater autonomy in robotics throughout the industrial sector. The level of autonomy in a robot or drone varies depending on the application. Some but not all machines will have an element of operator intervention or remote control. Greater autonomy comes from the use of AI for machine learning, to develop algorithms, apply them to new data, and act on the result. Called inference, the process needs significantly more computing power than previously programmable automated devices [1, 24, 37].

What Makes a Robot Smart?

Robots can be incredibly small, such as those used by surgeons, ranging in size up to vehicles such as the Mars Rover. In our mind's eye, we might see complex machines, static on production lines, or moving about in a warehouse, freeing up people from tedious or dangerous tasks. In operation, it is the repetition of the sense-process-act cycle, creating a feedback loop, that makes a machine smart. This smartness involves taking complex data and processing it with AI to enable machine learning.

This is what differentiates a truly autonomous robot from a simpler, pre-programmed piece of automation. The latest technology is having a dramatic effect on robot development. Machine vision, machine learning, and accessible computing power mean that robots can sense and process much more data much more efficiently. They are becoming smarter and more autonomous [14, 15].

The autonomous mobile robot (AMR) is the latest must-have. Seen as part of Industry 4.0, AMRs are ideal for applications such as automated material handling and in-house transportation, particularly not only in logistics operations with densely packed warehouses but also in industrial and automotive manufacturing. Figure 8.2 represents building a robot with a central controller, sensor hub, and manipulator section. The central controller is responsible for routing messages and maintaining task control information.

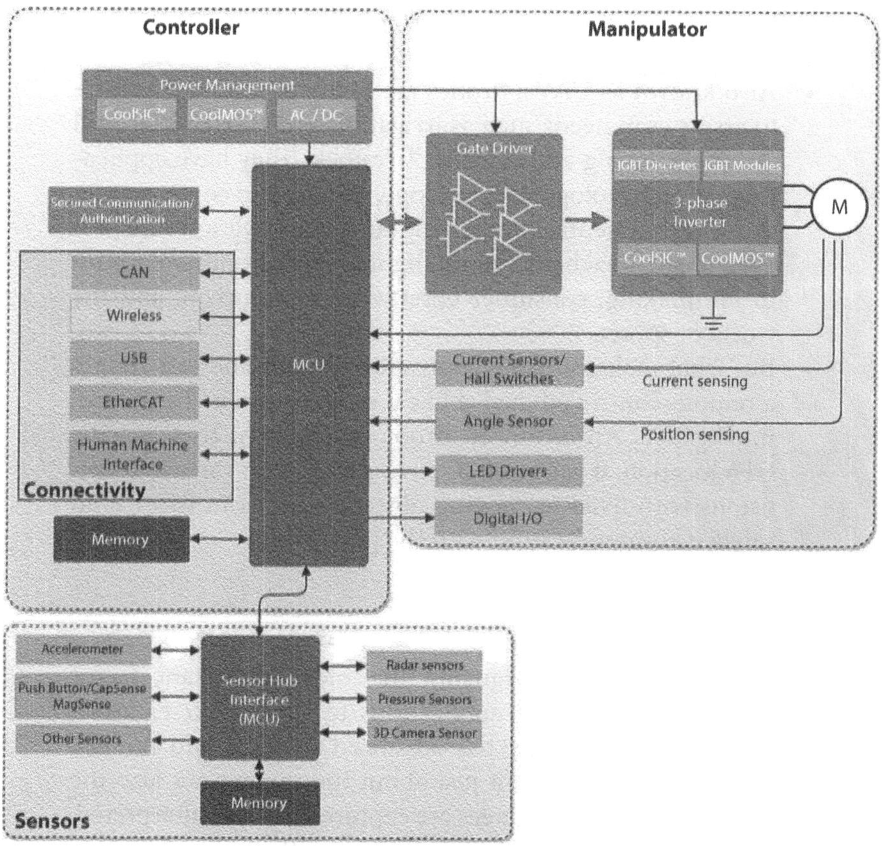

Figure 8.2 High-level block diagram of a typical robot showing central controller, sensor hub, and manipulator section (source: Infineon).

Can Drones Be Smart?

- Originally developed for military use at least 50 years ago for surveillance and more sinister applications, drones range in size from miniature machines that launch from the palm of your hand, to seriously large unmanned aircraft with fixed wings that require short runways to launch.
- Drones are often associated with dark and devious activity, but, today, they are used increasingly in humanitarian applications as well as for fun. They can save lives, survey disaster areas, deliver medication or equipment to people in trouble in remote locations, detect poisonous gases, or inspect potentially dangerous structures [13].
- Also known as UAVs, drones are being used in agriculture to assess crop needs such as irrigation and nutrients, as well as crop spraying and tracking livestock. They have applications in archaeology, meteorology, mining and construction, and environmental monitoring and conservation and can even locate poachers. On the lighter side, drones are useful in filmmaking, consumer entertainment, tourism, and the world of sports.
- But what makes a drone truly autonomous and not simply a remote-controlled device like a model aircraft? In a word: intelligence. An advanced drone can be sent to a specific GPS location, it can track a moving subject, it can avoid collisions with fixed or moving objects, and it can fly or land safely if communication with the pilot is lost.

Machine Insight

- Machine vision is arguably the most important factor enabling autonomous applications to sense their environment. The technology has advanced in leaps and bounds in recent years. It is not just about the camera but also the sensor and how the images are captured, stored, interpreted, and communicated. But let us start with the sensors that enable machines to see.
- There is a huge range of cameras available and selection depends on performance, as well as size, weight, and

cost constraints. Drone aerial photography and mapping, for example, might require the latest 4K video as well as 12-megapixel still shots. Inspecting the outside of structures might require Zoom facilities.

- It is the advent of 3D sensing that enables increased autonomy such that the machine can analyze the images and make more complex decisions. This is a fast-growing market comprising a range of advanced and still-developing technologies, including LiDAR, stereo vision, and structured light. Stereo vision is popular in drones and some automotive applications, giving long-range sensing above 10 m. Conversely, structured light has been adopted for short-range sensing, less than 1 m.
- LiDAR technology is already widely used in drones and unmanned vehicles as well as in driver assistance systems. Whether using mechanical, flash, or solid-state technology, it measures the distance from an object using laser light. A typical LiDAR includes lasers, photodetectors, and beam steering control circuitry.
- The advantage of mechanical LiDAR is its use of high-grade optics and a rotating assembly to create a wide field of view, up to 360°, with a superior signal-to-noise ratio. In operation, scanning LiDAR uses a focused pulsed laser beam for long-range, directed by a mechanical or MEMS mirror. MEMS (micro-electro-mechanical systems) sensor innovation is aiding the miniaturization of LiDAR mirrors. For detecting very small objects at longer ranges, mechanical LiDAR is popular in applications including AMRs and autonomous vehicles.
- Flash LiDAR, meanwhile, enables time-of-flight (ToF) sensing, using multiple beams of laser light simultaneously to measure distances to objects in a complete scene in a single shot. ToF depth-ranging camera sensors can be used for obstacle avoidance, object tracking, object scanning, 3D photography, and more, both indoors and outdoors. The field of view can be extended by mounting multiple side-looking sensors.
- Solid-state LiDAR, typically based on CMOS and/or VCSEL image sensing technology, has the advantages of being scalable in performance and small in size, with no moving parts

and low power consumption. It is critical in consumer products and finding favor in automotive and industrial systems. Even the smallest devices offer ToF technology and high performance and deliver VGA resolution.

- Laser and beam scanning technology avoids moving mirrors and includes optical phased arrays, laser diodes, edge-emitting lasers, and VCSELs. The latter are the new kids on the block, combining high-power density and simple packaging and featuring an infrared LED with the spectral width and speed of a laser. Although taking more space than edge-emitting lasers, VCSELs are targeting robotics in industrial applications and flash LiDAR.
- Solid-state image sensors, often based on CMOS technology, are becoming much smarter, incorporating on-chip AI processor units for compressing or pre-processing image data before shifting it to the CPU or image signal processor. This can significantly increase the speed of image recognition, for example, and simplify system design.

A Host of Sensors

- A host of other sensors can be mounted on drones or robotic systems. These include RADAR, ultrasonic, infrared, photogrammetry, and multispectral cameras. Sensors to detect CO_2, smoke, and other gases or particulates are useful in both indoor and outdoor applications for checking air quality, for example, or a potentially dangerous environment in a disaster area. Sound detection and voice recognition can be provided by audio sensors or MEMS microphones. For a noisy environment, active noise cancellation is vital.
- The application of RFID passive devices incorporating smart sensors is an emerging technology with wide potential in the industrial, machinery, and construction sectors. Parameters such as temperature, humidity, motion, ambient light, electrical continuity, and material characteristics can be measured passively. Drones equipped with RFID readers can gather relevant data at a range of 5 m to 10 m from buildings, bridges, open fields, greenhouses, and even livestock, fitted with these UHF RFID tags.

- Along with 3D, sensor fusion is revolutionizing data gathering for autonomous applications. The ability to link and interpret data from different sensors is making drones and robots even smarter, giving them the ability to better understand and act on their environment. Combining the data from a camera and a MEMS motion detector gives a more complete picture. Further, all sensors have a tolerance error.

  Sensor fusion improves integrity, reliability, and robustness and can mitigate to some extent against malfunction or tolerance errors.
- The key is to combine smart sensors with the necessary processing hardware software and computer algorithms for deep learning to analyze and interpret the data efficiently at the endpoint. Although initially under development for driverless vehicles, sensor fusion is already finding its way into industrial autonomous applications. Figure 8.3 represents sensors with smart devices and their components used in robots.

8.8 Human–Robot Interaction and Collaboration

Human–robot interaction (HRI) and collaboration refer to the study and design of interfaces, techniques, and systems that allow humans to interact with and work together with robots effectively and intuitively. This area of research focuses on creating robots that can understand human intentions,

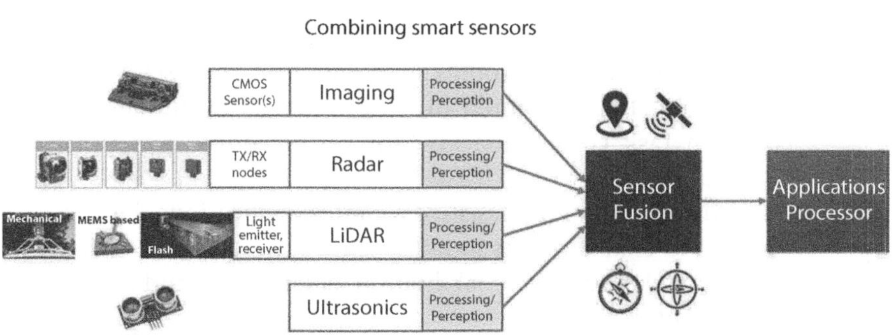

Figure 8.3 Robotic vehicles are often battery-powered, so reducing the total sensor count using smarter devices, plus the ability to merge the computation of multiple sensor data within a single processing unit, simplifies design, reduces power consumption, and lowers cost (Source: ARC Processors, Synopsys).

communicate with humans, and cooperate seamlessly in various tasks. Here are some key aspects of HRI and collaboration:

- Natural Interfaces: Human–robot interfaces aim to be natural and intuitive, allowing users to interact with robots using gestures, speech, touch, or other familiar means of communication. Natural interfaces enhance user experience and reduce the learning curve for interacting with robots.
- Gesture and Speech Recognition: Robots need to understand and interpret human gestures and speech to perceive human commands and intentions accurately. Advanced algorithms for gesture and speech recognition play a crucial role in achieving effective HRI.
- Social Cues and Expressions: To facilitate more natural interactions, robots can be designed to display social cues and expressions, such as eye contact, facial expressions, and body language. These cues help humans understand the robot's intentions and emotions better.
- Collaborative Task Planning: Human–robot collaboration often involves joint planning and coordination of tasks. Robots should be capable of understanding human instructions, preferences, and constraints to work collaboratively on shared tasks.
- Shared Autonomy: Shared autonomy is an approach in which both the human and the robot contribute to the decision-making process during collaborative tasks. The system can adjust the level of autonomy based on the user's expertise and preferences.
- Feedback and Communication: Effective communication between humans and robots is essential for successful collaboration. Robots should provide feedback to humans to confirm task completion, ask for clarifications, or report issues.
- Assistive Robotics: In assistive robotics applications, HRI is critical to enable robots to assist and support humans with various tasks, such as rehabilitation, elderly care, and medical procedures.
- Safety and Trust: Human–robot collaboration requires a high level of trust. Ensuring the safety of humans during

interactions and designing robots with transparent behavior are crucial factors for building trust between humans and robots.

- Learning From Demonstration: To facilitate HRI, robots can be trained to learn from demonstrations provided by humans. This enables robots to mimic human behavior and adapt to individual preferences.
- Ethical Considerations: As robots become more integrated into human lives, ethical considerations in HRI become crucial. Addressing issues related to privacy, consent, bias, and accountability is essential to ensure responsible and ethical robot design.
- Multi-Robot Collaboration: Human–robot collaboration can also involve multiple robots working together to achieve a common goal. Coordinating actions and interactions between multiple robots and humans adds complexity to the design of HRI systems.

Effective HRI and collaboration hold the key to developing robotics technologies that can seamlessly integrate into our daily lives, support us in various tasks, and improve overall efficiency and safety. Continued research and development in this area will further advance the capabilities and usability of robots, bringing us closer to a future where humans and robots can work together as trusted partners.

Natural Language Processing for Human–Robot Communication

Natural Language Processing (NLP) plays a vital role in enabling human–robot communication through natural and intuitive language interactions. NLP allows robots to understand and generate human language, making it easier for humans to interact with robots and give them commands or instructions in a way that resembles human-human communication. Here is how NLP is used in human–robot communication:

- Speech Recognition: NLP is used for speech recognition to convert spoken language into text, allowing robots to understand what humans are saying to them. Advanced speech recognition algorithms are employed to handle different accents, background noise, and various speech patterns.

- Natural Language Understanding (NLU): NLU is a crucial aspect of NLP that helps robots interpret and understand the meaning of human language. NLU algorithms process the textual input and extract the intent and entities to determine what the user wants the robot to do.
- Intent Recognition: Intent recognition involves identifying the user's intention or the desired action from their spoken or written input. For example, the robot should recognize whether the user is asking a question, giving a command, or making a request.
- Dialog Management: Dialog management is responsible for handling back-and-forth conversations between humans and robots. It allows the robot to maintain context, remember previous interactions, and ask for clarifications when necessary.
- Text-to-Speech (TTS): NLP is used for text-to-speech conversion, where the robot generates spoken responses in natural-sounding human-like voices. TTS systems use synthetic speech generation to communicate with users.
- Language Generation: NLP techniques also enable robots to generate text in response to user queries or as part of a natural conversation. This is particularly useful when the robot needs to provide information or explanations.
- Language Translation: NLP can be used for language translation to enable cross- language communication between humans and robots. Robots can understand and respond in different languages, breaking language barriers.
- Sentiment Analysis: Sentiment analysis is used to understand the emotions or sentiments expressed by users. This helps robots to respond appropriately and empathetically to emotional input.
- Chatbot Integration: NLP can be combined with chatbot technologies to create interactive and conversational robots capable of engaging users in more natural and human-like conversations.
- Personalization: NLP allows robots to personalize their interactions based on user preferences and historical data, creating more personalized and user-centric experiences.

By incorporating NLP into human–robot communication, robots become more accessible and user-friendly. They can understand and respond to natural language, making the interaction process more intuitive and effective. With ongoing research and advancements in NLP, human–robot communication will continue to improve, leading to more seamless and productive collaborations between humans and robots in various domains and applications.

Social Robotics: Developing Empathetic and User-Friendly Robots

Social robotics is a field of robotics that focuses on developing robots with the ability to interact with humans in a socially intelligent and empathetic manner. These robots are designed to recognize and respond to human emotions, engage in natural and intuitive communication, and build meaningful relationships with users. The goal of social robotics is to create robots that are user-friendly, approachable, and capable of providing emotional support and companionship. Here are key aspects of developing empathetic and user-friendly robots in social robotics:

- Emotion Recognition: Social robots are equipped with sensors and algorithms that allow them to recognize human emotions through facial expressions, body language, speech, and other cues. Emotion recognition enables the robot to adapt its responses and behavior based on the user's emotional state.
- Emotion Expression: Social robots can display emotions through facial expressions, gestures, and vocal intonations. By expressing emotions, robots can establish a more natural and engaging interaction with users.
- Natural Language Understanding and Generation: To communicate effectively with humans, social robots need to understand natural language input and generate responses that are contextually appropriate and linguistically natural.
- Context Awareness: Social robots should be context-aware, meaning they can understand the situational context and adjust their behavior accordingly. This includes recognizing the user's preferences, history of interactions, and the environment they are in.
- Personalization: Empathetic robots can be personalized to cater to individual users' preferences, interests, and needs.

Personalization helps to create a stronger emotional bond between the robot and the user.

- Proactive Behavior: Social robots can be programmed to initiate interactions and provide assistance proactively. For example, a robot may offer help or initiate conversation when it detects that the user is feeling sad or lonely.
- Ethical Considerations: Designing empathetic robots also involves addressing ethical considerations related to privacy, data security, and appropriate use of emotional data.
- User Studies and Feedback: User studies and feedback play a vital role in the development of social robots. Researchers and developers must involve users in the design process to understand their needs and preferences better.
- Companionship and Support: Social robots can provide companionship and emotional support to users, particularly in contexts where human interactions are limited, such as in eldercare or healthcare settings.
- Long-Term Interaction: Unlike conventional robots designed for specific tasks, social robots are intended for long-term interaction with users. This long-term interaction allows the robot to learn and adapt to the user's preferences over time.

Examples of empathetic and user-friendly social robots include Pepper by SoftBank Robotics, Miro by Consequential Robotics, and Paro, a therapeutic robotic seal developed by advanced industrial science and technology (AIST), Japan. These robots are designed to engage users socially, offer emotional support, and establish meaningful connections, making them valuable companions and assistants in various applications. As social robotics continues to advance, we can expect to see more empathetic and user-friendly robots that contribute positively to various domains, such as healthcare, education, customer service, and entertainment, enhancing HRIs and overall user experiences.

Collaborative Autonomous Systems in Industrial Settings

Collaborative autonomous systems in industrial settings refer to the integration of autonomous machines and robots working together with human operators to optimize and streamline industrial processes. These systems leverage advanced technologies such as robotics, AI, and the Internet of Things (IoT) to create efficient and productive collaborative environments.

Here are some key aspects of collaborative autonomous systems in industrial settings:

- Human–Robot Collaboration: In collaborative autonomous systems, robots work alongside human operators, complementing their skills and capabilities. Robots handle repetitive or dangerous tasks, whereas humans focus on more complex decision- making and supervisory roles.
- Robotic Co-Workers: Collaborative robots, also known as cobots, are designed to be safe and user-friendly, allowing them to work closely and directly with human workers without the need for safety barriers. This enables flexible and efficient collaboration in shared workspaces.
- Task Allocation and Scheduling: Autonomous systems use AI algorithms to allocate tasks among robots and human operators based on their respective strengths and expertise. Task scheduling is optimized to ensure smooth workflow and resource utilization.
- Real-Time Communication and Coordination: Industrial IoT and communication technologies enable seamless communication and coordination among autonomous machines and human operators. Real-time data exchange allows for efficient decision- making and task execution.
- Optimization and Efficiency: Collaborative autonomous systems focus on optimizing industrial processes to improve productivity, reduce downtime, and enhance overall efficiency. Robots and machines can work 24/7 and maintain consistent performance.
- Quality Control and Inspection: Autonomous robots equipped with sensors and computer vision capabilities can perform quality control and inspection tasks with high precision and consistency, ensuring product quality and safety.
- Autonomous Material Handling: Autonomous mobile robots (AMRs) and automated guided vehicles (AGVs) can handle material transportation within the industrial facility, reducing the need for manual labor and streamlining logistics.
- Predictive Maintenance: Collaborative autonomous systems can predict maintenance needs based on sensor data and machine learning algorithms. This proactive approach helps

minimize equipment downtime and optimize maintenance schedules [1, 26, 27].

- Safety and Risk Management: Safety is a paramount concern in collaborative autonomous systems. Advanced safety features, such as collision avoidance, emergency stop mechanisms, and human presence detection, are integrated into the robots to ensure safe collaboration with humans.
- Adaptability and Flexibility: Autonomous systems are designed to be adaptable to changing production demands and flexible enough to accommodate process variations and new tasks.

Examples of collaborative autonomous systems in industrial settings include automated assembly lines with human–robot collaboration, smart warehouses with robots working alongside warehouse personnel, and autonomous material handling systems in logistics and distribution centers.

Collaborative autonomous systems have the potential to transform industrial processes by increasing efficiency, productivity, and safety. As technology continues to evolve, we can expect further advancements and wider adoption of these systems in various industries, leading to a more optimized and agile industrial landscape.

8.9 Future Trends and Challenges

The future of drones and autonomous robotics with CI will witness heightened autonomy and collaboration, transforming industries. Challenges will emerge in terms of regulatory frameworks, ensuring ethical AI and maintaining data security.

Navigating these trends and challenges will shape the responsible and successful integration of these technologies [9, 31, 32, 36, 37].

Future Trends in Collaborative Autonomous Systems

- AI Advancements: Continued advancements in AI and machine learning will enable collaborative autonomous systems to become more intelligent, adaptive, and capable of learning from data and user interactions.
- Swarm Robotics: The use of swarm robotics, where large groups of robots collaborate to perform tasks, will become

more prevalent, leading to increased efficiency and scalability in various applications.

- Multi-Robot Collaboration: Future systems will involve more complex and diverse collaborations between multiple robots, human operators, and other autonomous machines, leading to enhanced capabilities and seamless integration.
- Human–Robot Teaming: The trend of developing robots as teammates rather than tools will lead to improved HRI and collaboration, creating more natural and effective partnerships.
- Edge Computing: Integrating edge computing with collaborative autonomous systems will reduce latency and enable faster decision-making, particularly in time-sensitive applications.
- 5G and Connectivity: The widespread adoption of 5G and improved connectivity will facilitate real-time communication and data exchange between autonomous systems and cloud-based platforms.
- Explainable AI: As collaborative autonomous systems become more complex, the need for explainable AI will increase to ensure transparency and build trust between humans and autonomous machines.
- Cross-Industry Integration: Collaborative autonomous systems will increasingly be applied across various industries, leading to shared knowledge and best practices between sectors.

Challenges in Collaborative Autonomous Systems

- Safety and Liability: Ensuring the safety of humans working with autonomous machines remains a significant challenge. Establishing liability frameworks for accidents involving collaborative autonomous systems is also a complex issue [16].
- Ethical Considerations: Ethical dilemmas, such as data privacy, job displacement, and robot decision-making in critical situations, need to be addressed to ensure responsible use of collaborative autonomous systems.
- Standardization and Interoperability: Developing standardized protocols and interfaces is crucial for seamless collaboration between different autonomous machines from various manufacturers.

- User Acceptance: Gaining user trust and acceptance of collaborative autonomous systems is vital for successful integration into industrial environments.
- Data Security: Protecting sensitive data and ensuring the integrity of communication between autonomous systems is a significant challenge, particularly in highly interconnected and networked environments.
- Regulatory Compliance: Adhering to diverse and evolving regulations across industries and regions requires continuous monitoring and compliance efforts.
- Cost and Implementation Complexity: The initial cost of implementing collaborative autonomous systems and integrating them into existing workflows can be significant.
- Environmental Impact: The environmental impact of increased automation and energy consumption by autonomous systems needs to be carefully considered [14, 15].
- Lack of Skilled Workforce: There may be a shortage of skilled workers capable of operating and maintaining collaborative autonomous systems, necessitating training and education initiatives.

Addressing these challenges will be critical for the successful adoption and realization of the full potential of collaborative autonomous systems. Through ongoing research, technological advancements, and responsible implementation, collaborative autonomous systems have the potential to revolutionize industries and enhance human–robot collaboration for a more efficient and sustainable future.

Shaping the Future of Robotics and Drones

Shaping the future of robotics and drones involves a collective effort from researchers, industry leaders, policymakers, and the public. Here are some key areas of focus to shape a positive and impactful future for these technologies [4, 5]:

Research and Innovation: Continued investment in research and innovation is crucial to advance the capabilities and applications of robotics and drones. Funding research projects, supporting academic institutions, and encouraging collaboration between researchers and industries can drive technological breakthroughs.

- Ethical and Responsible Development: Emphasizing ethical considerations in the design and development of robotics and drones is essential. Responsible AI practices, transparency, fairness, and privacy protection should be embedded in the technology's development process.

- Regulatory Frameworks: Governments and regulatory bodies need to establish clear and adaptive regulatory frameworks to govern the use of robotics and drones. These frameworks should address safety, privacy, security, liability, and other ethical concerns.
- Public Education and Awareness: Increasing public understanding and awareness of robotics and drones can foster acceptance and support. Educating the public about the benefits, potential risks, and responsible use of these technologies can lead to informed discussions and policies.
- Workforce Reskilling and Training: As robotics and automation impact the job market, investing in workforce reskilling and training programs is crucial to prepare individuals for the jobs of the future and bridge the skills gap.
- Interdisciplinary Collaboration: Robotics and drone development often require expertise from multiple disciplines. Encouraging interdisciplinary collaboration between engineers, computer scientists, ethicists, sociologists, and policymakers can lead to holistic solutions.
- Sustainable and Environmentally Friendly Applications: Emphasizing sustainable practices in the design and use of robotics and drones can minimize their environmental impact. Energy efficiency, waste reduction, and eco-friendly materials should be considered in their development.
- Accessibility and Inclusivity: Ensuring that robotics and drones are accessible and inclusive to people of all abilities is important. Technologies should be designed with consideration for different user needs and preferences.
- Collaborative Partnerships: Building collaborative partnerships between academia, industries, non-profit organizations, and governments can drive innovation and accelerate the deployment of robotics and drones for societal benefit.
- Addressing Global Challenges: Leveraging robotics and drones to address global challenges such as disaster response, environmental monitoring, healthcare access, and agricultural sustainability can have a transformative impact on society.

- Artificial General Intelligence (AGI): Engaging in responsible research and discussions around AGI and its potential implications for robotics is essential to ensure the safe and beneficial development of superintelligent systems.

By collectively focusing on these areas, we can shape a future where robotics and drones contribute positively to various sectors, improve human lives, protect the environment, and address pressing global challenges. Responsible and inclusive development is key to maximizing the benefits of these transformative technologies while mitigating potential risks.

Responsible AI: Ensuring Safe and Ethical Implementation

Responsible AI is an essential aspect of ensuring the safe and ethical implementation of AI technologies. It involves developing AI systems that are transparent, fair, secure, and aligned with human values. Here are some key principles and practices to ensure responsible AI implementation:

- Transparency and Explainability: AI systems should be designed to provide transparent and understandable explanations of their decisions and actions. Users and stakeholders should be able to understand how AI models arrive at their conclusions.
- Fairness and Bias Mitigation: Efforts should be made to identify and mitigate biases in AI algorithms that could lead to unfair outcomes. Fairness-aware AI models can be developed to ensure equitable treatment for all individuals and groups.
- Privacy and Data Protection: Protecting user privacy and securing sensitive data are crucial in AI systems. Implementing privacy-preserving techniques and adhering to data protection regulations are essential steps.
- Human Oversight and Control: While AI can automate tasks, there should be mechanisms for human oversight and intervention, particularly in critical decision- making processes.
- Accountability and Traceability: AI systems should be designed to be accountable for their actions. Traceability features should allow for auditing and understanding of how AI models were trained and how they make decisions.

- Continuous Monitoring and Evaluation: AI models should be continually monitored and evaluated to detect and address potential issues that may arise during deployment.
- Robustness and Security: AI models should be developed with robustness and security in mind to prevent exploitation and adversarial attacks.
- Education and Awareness: Educating AI developers, users, and the public about responsible AI practices is crucial in promoting ethical use and understanding the impact of AI technologies.
- Multidisciplinary Collaboration: Responsible AI implementation requires collaboration among technologists, ethicists, policymakers, and other stakeholders to ensure a comprehensive and holistic approach.
- Regulatory Compliance: Adherence to relevant regulations and ethical guidelines is essential to ensure compliance with legal and ethical standards.
- Ethical Frameworks and Codes of Conduct: Developing ethical frameworks and codes of conduct specific to AI implementation can guide developers and users.
- Bias Testing and Mitigation: Conducting thorough bias testing and implementing strategies to reduce bias in AI systems can help ensure equitable outcomes.
- Human-Centered Design: Prioritizing the needs and well-being of humans in AI development helps align technology with human values and requirements.

By integrating these principles into the design, development, and deployment of AI technologies, we can create AI systems that are beneficial, trustworthy, and respectful of human values, thus fostering responsible AI implementation. Responsible AI is essential to harness the full potential of AI technologies while mitigating potential risks and ensuring AI's positive impact on society.

8.10 Ethical Implications of Autonomous Robotics and Drones

The rapid advancement of autonomous robotics and drones raises several ethical implications that need careful consideration. As these technologies become more pervasive in various domains, it is essential to address the following ethical concerns [11, 17, 35]:

- Safety and Security: Ensuring the safety and security of autonomous robotics and drones is of utmost importance. Malfunctions, errors in AI algorithms, or cyberattacks could lead to accidents, injuries, or even loss of life. Implementing rigorous safety measures and cybersecurity protocols is crucial to prevent such incidents.
- Privacy and Data Protection: Autonomous systems, including drones, often collect and process large amounts of data. This raises concerns about privacy violations and data misuse. Striking a balance between data collection for legitimate purposes and safeguarding individual privacy rights is essential.
- Accountability and Liability: Determining liability in case of accidents involving autonomous robots or drones can be complex. Identifying who is responsible, whether it is the robot's manufacturer, the operator, or the AI system itself, raises legal and ethical challenges.
- Autonomous Weapon Systems: The development of autonomous weapon systems raises significant ethical concerns. Ensuring that robots and drones are not used to cause harm indiscriminately or without human oversight is a pressing issue in the field of military robotics.
- Job Displacement: The widespread adoption of autonomous systems could lead to job displacement and economic inequality. Addressing the potential negative impact on the workforce and finding ways to reskill and retrain affected workers is critical.
- Bias and Fairness: AI algorithms powering autonomous systems can be biased, leading to unfair outcomes and perpetuating existing social inequalities. Ensuring fairness and addressing algorithmic biases is crucial to avoid discrimination and injustice.
- Transparency and Explainability: As autonomous systems become more complex, they may be challenging to interpret or understand. Ensuring transparency and explainability of AI decision-making is essential for gaining user trust and confidence.

- Environmental Impact: The use of autonomous drones and robotics can have environmental consequences, such as increased energy consumption and waste generation. Considering the environmental impact and striving for sustainable practices is crucial.
- Public Acceptance and Perception: Public acceptance and perception of autonomous robots and drones can impact their adoption and integration. Educating the public about the benefits, risks, and ethical considerations is essential for fostering trust and support.
- Autonomy and Human Oversight: Striking a balance between autonomy and human oversight is critical. While autonomy can lead to efficiency and effectiveness, ensuring human control and decision-making in critical situations is essential to prevent unintended consequences.

Addressing these ethical implications requires collaboration between technologists, policymakers, ethicists, and the public. Robust regulations and ethical guidelines are necessary to ensure responsible development, deployment, and use of autonomous robotics and drones, ultimately leading to their beneficial and ethical integration into society.

Bridging the Gap Between Research and Real-world Deployment

Bridging the gap between research and real-world deployment is a critical challenge in the field of emerging technologies, including robotics and AI. Here are some strategies to facilitate a smoother transition from research to practical applications:

- Collaboration with Industry Partners: Collaboration between researchers, academia, and industry partners is essential. Industry partners can provide valuable insights into real-world challenges, requirements, and constraints, whereas researchers can contribute cutting-edge technologies and expertise.
- Industry-Driven Research: Engaging in industry-driven research ensures that academic efforts align with practical needs. Identifying real-world problems and actively seeking solutions in collaboration with industry partners helps researchers develop relevant and applicable technologies.
- Prototyping and Pilot Projects: Building prototypes and conducting pilot projects allow researchers to test their

technologies in real-world scenarios. Feedback from early adopters and end-users helps refine the solutions and address potential challenges.

- Validation and Evaluation: Robust validation and evaluation of research outcomes are crucial before real-world deployment. It is essential to demonstrate the effectiveness, reliability, and safety of technologies through rigorous testing and assessment.
- Technology Transfer Offices: Academic institutions can establish technology transfer offices to facilitate the commercialization and licensing of research outcomes. These offices can help researchers navigate the process of transitioning technology to the industry.
- Standardization and Certification: Establishing standards and certifications for emerging technologies is essential to ensure interoperability, safety, and compliance with regulations. Researchers should consider these aspects during the development phase.
- Public-Private Partnerships: Public-private partnerships can foster collaboration between research institutions, governments, and private companies. These partnerships can help fund research and encourage technology adoption in public sectors.
- Regulatory Compliance: Researchers must be aware of regulatory requirements and ensure that their technologies comply with relevant laws and guidelines. Engaging with regulatory bodies early in the development process can help identify potential regulatory hurdles.
- Education and Awareness: Educating the public, policymakers, and stakeholders about the benefits and risks of emerging technologies fosters acceptance and support. Raising awareness about research outcomes can lead to faster adoption and deployment.
- Long-Term Vision: Researchers should have a long-term vision for their technologies, considering scalability, sustainability, and future developments. A comprehensive roadmap for research-to-deployment helps maintain focus and ensures continuity.
- Technology Readiness Level (TRL) Assessment: Using TRL assessments can help researchers gauge the maturity of their

technologies and identify the necessary steps to reach higher readiness levels for deployment.

By adopting these strategies and fostering a collaborative and interdisciplinary approach, researchers can increase the likelihood of successfully bridging the gap between research and real-world deployment. The effective integration of cutting-edge technologies into practical applications can have a significant positive impact on society and industry.

8.11 Conclusion

The integration of drones and autonomous robotics with CI has ushered in a new era of technological advancement with profound implications across various industries. This convergence has resulted in a synergy that enhances the capabilities, efficiency, and potential applications of these technologies. Drones, equipped with CI algorithms, have become more than just remote-controlled devices. They now possess the ability to perceive their environment, make real-time decisions, and adapt to dynamic situations. This has led to significant progress in fields such as agriculture, environmental monitoring, disaster response, and logistics. Drones can autonomously navigate complex environments, optimize routes, and collect data with precision, leading to improved decision-making processes.

Autonomous robotics, empowered by CI, has similarly revolutionized industries such as manufacturing, healthcare, and transportation. These robots can perform tasks with high accuracy and reliability, even in intricate and unpredictable environments. Through machine learning and AI, they learn from their experiences and continuously refine their performance, making them versatile tools for various applications [10].

The integration of CI in drones and autonomous robotics also brings challenges that need to be addressed. These challenges include ensuring safety, privacy, and ethical considerations, as well as addressing technical limitations such as power constraints, sensor accuracy, and decision-making in complex scenarios.

Looking forward, the collaboration between drones, autonomous robotics, and CI is poised to advance even further. As technologies evolve, we can anticipate enhanced capabilities, reduced costs, and increased accessibility. These advancements will likely lead to the development of innovative solutions that tackle current global challenges, such as environmental conservation, healthcare accessibility, and disaster management.

In conclusion, the fusion of drones, autonomous robotics, and CI holds immense promise for reshaping industries and addressing societal needs. With ongoing research, development, and responsible deployment, these technologies will continue to push boundaries and contribute to a more efficient, safer, and technologically enriched world.

References

1. Woschank, M., Rauch, E., Zsifkovits, H., A review of further directions for artificial intelligence, machine learning, and deep learning in smart logistics. *Sustainability*, 12, 3760, 2020.
2. Fahle, S., Prinz, C., Kuhlenkötter, B., Systematic review on machine learning (ML) methods for manufacturing processes–Identifying artificial intelligence (AI) methods for field application. *Procedia CIRP*, 93, 413–418, 2020.
3. Vrontis, D., Christofi, M., Pereira, V., Tarba, S., Makrides, A., Trichina, E., Artificial intelligence, robotics, advanced technologies, and human resource management: a systematic review. *Int. J. Hum. Res. Manage*, 33, 1237–1266, 2022.
4. Howard, J., Artificial intelligence: implications for the future of work. *Am. J. Ind. Med.*, 62, 917–926, 2019.
5. Arinez, J.F., Chang, Q., Gao, R.X., Xu, C., Zhang, J., Artificial intelligence in advanced manufacturing: current status and future outlook. *J. Manuf. Sci. Eng.*, 142, pp. 142–146, 2020.
6. Andras, I., Mazzone, E., van Leeuwen, F.W., De Naeyer, G., van Oosterom, M.N., Beato, S., Buckle, T., O'Sullivan, S., van Leeuwen, P.J., Beulens, A., Artificial intelligence and robotics: a combination that is changing the operating room. *World J. Urol.*, 38, 2359–2366, 2020.
7. Bai, Q., Li, S., Yang, J., Song, Q., Li, Z., Zhang, X., Object detection recognition and robot grasping based on machine learning: a survey. *IEEE Access*, 8, 181855–181879, 2020.
8. Lee, W.J., Wu, H., Yun, H., Kim, H., Jun, M.B., Sutherland, J.W., Predictive maintenance of machine tool systems using artificial intelligence techniques applied to machine condition data. *Procedia CIRP*, 80, 506–511, 2019.
9. Panesar, S., Cagle, Y., Chander, D., Morey, J., Fernandez-Miranda, J., Kliot, M., Artificial intelligence and the future of surgical robotics. *Ann. Surg.*, 270, 223–226, 2019.
10. Sarker, S., Jamal, L., Ahmed, S.F., Irtisam, N., Robotics and artificial intelligence in healthcare during COVID-19 pandemic: a systematic review. *Rob. Auton. Syst.*, 146, 103902, 2021.
11. Wasilow, S. and Thorpe, J.B., Artificial intelligence, robotics, ethics, and the military: a Canadian perspective. *AI Magazine*, 40, 37–48, 2019.

12. Belk, R., Ethical issues in service robotics and artificial intelligence. *Serv. Ind. J.*, 41, 860–876, 2021.
13. Elallid, B.B., Benamar, N., Hafid, A.S., Rachidi, T., Mrani, N., A comprehensive survey on the application of deep and reinforcement learning approaches in autonomous driving. *J. King Saud Univ.-Comput. Inf. Sci.*, 34, 9, 7366–7390, 2022.
14. Ning, H., Yin, R., Ullah, A., Shi, F., A survey on hybrid human-artificial intelligence for autonomous driving. *IEEE Trans. Intell. Transp. Syst.*, 23, 6011–6026, 2021.
15. El-Shamouty, M., Kleeberger, K., Lämmle, A., Huber, M., Simulation-driven machine learning for robotics and automation. *TM. Tech. Mess.*, 86, 673–684, 2019.
16. Lammie, C. and Azghadi, M.R., Memtorch: a simulation framework for deep memristive cross-bar architectures, in: *2020 IEEE International Symposium on Circuits and Systems (ISCAS)*, IEEE, pp. 1–5, 2020.
17. Soori, M., Arezoo, B., Habibi, M., Accuracy analysis of tool deflection error modeling in prediction of milled surfaces by a virtual machining system. *Int. J. Comput. Appl. Technol.*, *55*, 4, 308–321, 2017.
18. Soori, M., Arezoo, B., Habibi, M., Virtual machining considering dimensional, geometrical and tool deflection errors in three-axis CNC milling machines. *J. Manuf. Syst.*, 33, 498–507, 2014.
19. Soori, M., Arezoo, B., Dastres, R., Machine learning and artificial intelligence in CNC machine tools, a review, Sustain. *Manuf. Serv. Econ.*, 2023, 100009, 2023.
20. Soori, M. and Ismael, M., Classification of research and applications of the computer aided process planning in manufacturing systems. *Indep. J. Manage. Prod.*, 12, 1250– 1281, 2021.
21. Dastres, R. and Soori, M., Advances in web-based decision support systems. *Int. J. Eng. Future Technol.*, 19, 1–15, 2021.
22. Dastres, R. and Soori, M., Artificial neural network systems. *Int. J. Imaging Robot. (IJIR)*, 21, 13–25, 2021.
23. Dastres, R. and Soori, M., The role of Information and Communication Technology (ICT) in environmental protection. *Int. J. Tomogr. Simul.*, 35, 24–37, 2021.
24. Baduge, S.K., Thilakarathna, S., Perera, J.S., Arashpour, M., Sharafi, P., Teodosio, B., Shringi, A., Mendis, P., Artificial intelligence and smart vision for building and construction 4.0: machine and deep learning methods and applications. *Autom. Constr.*, 141, 104440, 2022.
25. Sardar, P., Abbott, J.D., Kundu, A., Aronow, H.D., Granada, J.F., Giri, J., Impact of artificial intelligence on interventional cardiology: from decision-making aid to advanced interventional procedure assistance. *Cardiovasc. Intervent.*, 12, 1293–1303, 2019.
26. Dalzochio, J., Kunst, R., Pignaton, E., Binotto, A., Sanyal, S., Favilla, J., Barbosa, J., Machine learning and reasoning for predictive maintenance in Industry 4.0: current status and challenges. *Comput. Ind.*, 123, 103298, 2020.

27. Abdar, M., Khosravi, A., Islam, S.M.S., Acharya, U.R., Vasilakos, A.V., The need for quantification of uncertainty in artificial intelligence for clinical data analysis: increasing the level of trust in the decision-making process. *IEEE Syst. Man. Cybernet. Mag.*, 8, 28– 40, 2022.
28. Dimiduk, D.M., Holm, E.A., Niezgoda, S.R., Perspectives on the impact of machine learning, deep learning, and artificial intelligence on materials, processes, and structures engineering, Integr. *Mater. Manuf. Innov.*, 7, 157–172, 2018.
29. Shrestha, Y.R., Krishna, V., von Krogh, G., Augmenting organizational decision-making with deep learning algorithms: principles, promises, and challenges. *J. Bus. Res.*, 123, 588–603, 2021.
30. Károly, A.I., Galambos, P., Kuti, J., Rudas, I.J., Deep learning in robotics: a survey on model structures and training strategies. *IEEE Trans. Syst. Man. Cybernet.*, 51, 266– 279, 2020.
31. Tyagi, A.K. and Chahal, P., Artificial intelligence and machine learning algorithms, in: *Research Anthology on Machine Learning Techniques, Methods, and Applications*, pp. 421–446, IGI Global Publishing, 2022.
32. Arents, J. and Greitans, M., Smart industrial robot control trends, challenges and opportunities within manufacturing. *Appl. Sci.*, 12, 937, 2022.
33. Garg, P.K., Overview of artificial intelligence, in: *Artificial Intelligence*, pp. 3–18, Chapman and Hall/CRC, Taylor & Francis, 2021.
34. Kamruzzaman, M., New opportunities, challenges, and applications of edge-AI for connected healthcare in smart cities, in: *2021 IEEE Globecom Workshops (GC Wkshps)*, IEEE, pp. 1–6, 2021.
35. Kapeller, A., Felzmann, H., Fosch-Villaronga, E., Hughes, A.-M., A taxonomy of ethical, legal and social implications of wearable robots: an expert perspective. *Sci. Eng. Ethics*, 26, 3229–3247, 2020.
36. Milazzo, M. and Libonati, F., The synergistic role of additive manufacturing and artificial intelligence for the design of new advanced intelligent systems, *Adv. Intell. Syst.*, 4, 2100278, 2022.
37. Huang, Z., Shen, Y., Li, J., Fey, M., Brecher, C., A survey on AI-driven digital twins in industry 4.0: smart manufacturing and advanced robotics. *Sensors*, 21, 6340, 2021.

9

Exploring the Convergence of Artificial Intelligence and Mechatronics in Autonomous Driving

Ritika Wason[1]*, Parul Arora[1], Vishal Jain[2†], Devansh Arora[3] and M. N. Hoda[1]

[1]*Centre for Research and Innovation in Interdisciplinary Studies (CRIIS), Bharati Vidyapeeth's Institute of Computer Applications and Management (BVICAM), New Delhi, India*
[2]*Department of Computer Science and Engineering, Sharda School of Engineering and Technology Sharda University, Greater Noida, India*
[3]*Indraprastha Institute of Information Technology (IIIT), New Delhi, India*

Abstract
Vehicular automation, namely, autonomous driving, involves the application of mechatronics along with artificial intelligence (AI) to assist in vehicle control. This collaboration between AI and mechatronics (blend of mechanical engineering, electronics, and computer science) has replaced conventional means of vehicle production as well as control. This manuscript explores this intersection in light of the key components, challenges, and benefits of integrating AI in self-driving transportation systems. We also highlight the advancements made, current state of the art, and potential future directions.

Keywords: Mechatronics, artificial intelligence (AI), autonomous driving, self-driving, perception, sensing, transportation, vehicular automation

9.1 Introduction

The transport industry despite optimizations and advancements in its mechanical structures still relies heavily on human skills and experience [1].

**Corresponding author*: ritika.wason@bvicam.in
†*Corresponding author*: drvishaljain83@gmail.com

Kolla Bhanu Prakash, Satish Kumar Peddapelli, Ivan C. K. Tam, Wai Lok Woo and Vishal Jain (eds.) Computational Intelligent Techniques in Mechatronics, (297–316) © 2024 Scrivener Publishing LLC

As technology advances and customer demands alter, the need for advanced assistance systems (ADASs) is rising [2, 3]. Self-driving mechatronics implies the integration of mechatronics principles and artificial intelligence (AI) techniques in autonomous vehicles (AVs are also called as automated or driver-less vehicles) [4]. Competition in this industry is also intense, with several key players already involved in this novel territory. Such vehicles are tuned to sense their environment, make real-time decisions, and execute decisions without human intervention [5]. This ability demands effective collaboration between mechanical components, electronic systems, and AI algorithms and boasts of remarkable advancements in safety, efficiency, and convenience in these transportation systems [6]. Artificial Intelligence (AI) with self-driving mechatronics is an intricate marvel that combines multiple disciplines like robotics, computer vision, machine learning, and control systems [7]. If described simply, this involves the integration of mechanical and electronic components with intelligent AI algorithms to create AVs capable of navigating and interacting with their surroundings [8]. Autonomous driving, characterized by vehicles capable of operating without human intervention, has garnered significant attention in the recent years [9, 10]. Although self-driving vehicles have

Table 9.1 Levels of autonomous driving [2–4].

Level	Description
0	No automation
1	Driver assistance (independence of only one key control like self-parking, adaptive cruise control, and autonomous braking).
2	Partial automation (autonomy of two or more key jobs to discharge the driver of the same).
3	Conditional automation (limited self-driving; driver may yield control of all safety-critical jobs under certain environmental circumstances).
4	High automation (full self-driving where driver is not likely to be existing; example: local driverless taxi).
5	Full automation (same as level 4, can drive ubiquitously in all conditions).

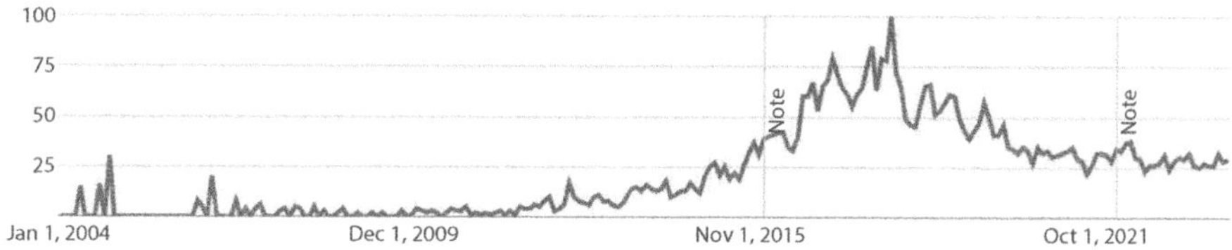

Figure 9.1 Autonomous car production worldwide statistics [6] (since 2004).

existed for ages, the technology development began as early as 1920 and has subsequently engulfed almost the entire world [11, 12]. To understand the operation of an AV, we should first understand that this functionality is included in vehicles at five levels [13], beginning from level 0 implying no automation to level 5 implying full automation [14, 15]. The present research is somewhere between levels 3 and 5 with level 5 systems majorly restricted to the labs [1]. These levels are detailed in Table 9.1.

Table 9.1 briefs the varied levels of autonomy in vehicles [5]. However, the statistics is soon expected to change. Figure 9.1 depicts the steep rise in autonomous car production worldwide since 2004 [6].

Implementation of AVs requires the seamless integration of numerous technologies as well as hardware and software components like AI algorithms and mechanical and electrical engineering [17]. This paper aims to explicate the various characteristics of this convergence and its consequence on the advancement of autonomous driving technology. We begin with presenting an overview of how this amalgamation of AI and mechatronic is delivering systems with the characteristics detailed below:

1. **Sensors and Perception:** Self-driving vehicles use multiple sensors, like LiDAR, Global Positioning System (GPS), cameras, Radio Detection and Ranging (RADAR), and ultrasonic sensors, to perceive their surroundings [18]. AI-empowered computer vision algorithms utilize these data to classify the detected surrounding objects like pedestrians, road signs, neighboring vehicles, and obstacles [19].
2. **Localization and Mapping:** AI algorithms are used to generate high-definition maps to precisely determine the vehicle's position in real time [20]. Simultaneous localization and mapping (SLAM) techniques combine sensor data and AI to generate a map of the surrounding and control the vehicle's position in it [21, 22].

3. **Path Planning and Decision Making:** Self-driving vehicles are empowered to control how to navigate through the traffic by themselves [7]. For this purpose, they use several path planning algorithms that receive sensor data as input and map the same to determine the optimal route, considering factors like traffic, road rules, and potential obstacles [8].
4. **Control Systems:** Reinforcement learning techniques are applied to regulate the vehicle's acceleration, steering, and braking to execute the planned path safely and efficiently [5]. Algorithms also help to fine-tune control algorithms by simulating various driving scenarios [10, 23, 24].
5. **Machine Learning and Deep Learning:** Vast amounts of sensor data are collected by each AV [25]. Convolutional neural networks (CNNs) are commonly used for image analysis, whereas recurrent neural networks (RNNs), and transformers can be used for processing sequential data and decision-making [24].
6. **Behavior Prediction and Interaction:** AI algorithms help the vehicle predict the behavior of other road traveler's and pedestrians [9]. This is crucial for safe navigation, as the self-driving vehicle needs to decide the actions of other agents on the road and adjust its driving behavior accordingly [10, 11].
7. **Safety and Redundancy:** To embed safety in self-driving vehicles requires certain degree of redundancy and fail-safe mechanisms [29, 30]. AI algorithms can help monitor the health of a vehicle's components and detect anomalies in real time [12]. In the incident of a system failure, the vehicle's AI system can take appropriate measures to ensure the safety of [13] the stakeholders involved [31].
8. **Continuous Learning and Adaptation:** AI can also help enable continuous learning and adaptation [32] in AVs. Such vehicles can improve themselves over time by analyzing real-world driving data and updating themselves to handle any novel scenarios, traffic conditions, and patterns [33].
9. **Regulations and Ethics:** The real-time placement of self-driving vehicles also increases concerns about regulations, liability, and ethical considerations [11] in case of an untoward incident. Algorithms must be programmed to follow traffic rules and make morally sound decisions in emergency situations [34].

10. **Testing and Simulation:** Virtual environments where the vehicle's performance can be tested in a wide range of scenarios aid in validating their performance and efficiency [25] in real-life scenarios.

AI-enabled AVs require collaboration of robotics, computer science, mechanical engineering, and many other relevant fields [5, 35]. The goal here is to create vehicles that are not only technically proficient but also safe, efficient, and capable of navigating real-world situations with a high degree of autonomy [5, 6, 22]. In this manuscript, we have analyzed the same. Section 9.1 introduces the application of AI-driven mechatronics in autonomous driving. Section 9.2 highlights the key components of autonomous driving. Section 9.3 elaborates the current state of affairs. Section 9.4 elaborates the mechatronic challenges of AVs. Section 9.5 enlists the advantages of the same. Section 9.6 delves into how self-driving affects environment sustainability. Section 9.7 discusses the legal and ethical issues in autonomous driving. Section 9.8 concludes the findings, whereas Section 9.9 highlights the future directions.

9.2 Key Components of Advanced Driver Systems

Autonomous vehicles comprise an intricate interplay of various enabling technologies, each contributing to the realization of safe and efficient autonomous operation [34–37]. Three key components enable autonomous driving, namely, sensors, semiconductors, and software [4]. Table 9.2 lists some common components used in AV technology [14–16].

Autonomous vehicle operation requires actuation in terms of three major controls [17]:

Table 9.2 Component technologies of autonomous vehicles.

Feature	Sensors	Semiconductors	Software
See/Sense	LiDAR, RADAR, Cameras, Global Navigational Satellite System (GNSS), Stereo Cameras, and Ultrasonic Range Sensors	Analog, Optoelectronics, Discrete Logic, and memory.	YOLO and Computational Neural Network (CNN)

Table 9.3 Actuation in autonomous vehicles.

S. no.	Actuation type	Achieved through
1	Throttle Actuation	Electronic Power Steering
2	Steering Actuation	Servo Motor
3	Brake Actuation	Electronic Stability Control

i. Throttle Actuation (Acceleration Control)
ii. Steering Actuation (Direction Control)
iii. Brake Actuation (Movement Control)

These actuations are already built in present vehicles in the following forms as outlined in Table 9.3.

Common components in AVs include perception systems (LiDAR, GPS, cameras, RADAR, etc.), sensor fusion algorithms, decision-making systems, control systems, and actuation mechanisms [18, 19, 21, 31, 38]. AI techniques enable the interpretation of sensor data, whereas mechatronics provide the physical means to execute decisions made by AI algorithms [18]. We discuss below a few considerable ones.

9.2.1 LiDAR (Light Detection and Ranging)

A laser-based sensor is a remote sensing framework that estimates space by illuminating a goal with a light beam and estimating the reflected light [19]. It is generally riding on the roof of the vehicle inside a cylindrical inclusion that can rotate 360° [20]. A LiDAR is generally made of a emitter, mirror, and receiver. LiDAR maps a 3D structure of the surrounding environment and the road integrating the data read from incoming reflected light [21]. The emitter releases a laser beam that bounces off a mirror rotating at 10 revolutions per minute [22]. After bouncing from an object, the beam returns to the mirror from where it is bounced to the receiver which interprets it appropriately [23].

9.2.2 RADAR (Radio Detection and Ranging)

This is an important sensor of AV systems. RADAR sensor is used to help the vehicle estimate objects and calculate their speed, distance, and direction [24]. It is capable of estimating the mutual velocity of the object and the vehicle, through electro-magnetic waves [22]. It achieves this by releasing a signal and waiting for it to comeback [24]. The frequency of the received signal is further estimated. In case of slight change, it is likely to immediately approximate the velocity of the driving vehicle ahead. As compared to LiDAR, the RADAR uses a larger wave length and lower signal energy. However, it does not describe the form of the scanned space. RADARs operate approximately in the 77-GHz range with a relatively directional scanning beam. It can scan only a distance of up to approximately 200-m distance [24].

9.2.3 Ultrasonic Sensors

Can be installed on several edges of the vehicle to sense objects nearby the vehicle or measure the spot of neighbouring vehicles during parking [22, 24]. These sensors enable parking assistance, collision warning, lane departure, among others.

9.2.4 Video Cameras

This enables real-time 3D images of the street ahead [22]. These can help sense traffic lights, traffic signs, unexpected things like animals or pedestrians, etc. [25]. They also perceive diverse road signs like "STOP" signs, zebra crossings, and sign boards. Video cameras also help in identifying certain signs, which other sensors cannot comprehend like hand waving or traffic cones.

9.2.5 GPS (Global Positioning System)

This space-based satellite navigation system enables existing location and time information anyplace on earth where there is an unhindered line of sight to four or more satellites [22, 24]. It forms the foundation of all maps that vehicle employ while on the road. All satellites broadcast using two similar frequencies, namely, 1.57542 GHz and 1.2276 GHz.

9.3 Current State of AI-Enabled Self-Driving Mechatronics

Major vehicle manufacturers and research institutions are at the forefront of driving automation development. Companies like Waymo, Tesla, and Cruise are actively testing and deploying AVs equipped with advanced AI algorithms and mechatronic systems [26, 27]. IHS automotive projects level 3 functionality by 2020, level 4 by 2025, and level5 by 2030, with AVs reaching 9 % of sales in 2035 and 90% of the vehicle convoy by 2055 [4]. Here are a few notable examples:

1. **Waymo (Alphabet/Google):** Waymo has been a pioneer in autonomous driving technology [28]. They have developed a self-driving system called "Waymo Driver" that combines AI, advanced sensors (LiDAR, RADAR, and cameras), and detailed maps for safe and reliable autonomous driving [28]. They employ several high-resolution cameras to enable a 360° view of the vehicle at long range in daylight as well as low-light conditions [4]. It also applies a 360° RADAR system to track the speed of vehicles around the vehicle. Waymo has been testing their technology in various cities and offering a limited ride-hailing service with AVs in certain areas.
2. **Tesla (Nvidia):** This is deeply involved in the AV market with goods like AI-enabled chips and virtual reality driving simulation. Tesla's Autopilot and Full Self-Driving (FSD) systems use AI and a mixture of cameras and sensors to empower features like adaptive cruise control, lane centering, and automatic lane changes [29]. Tesla's graphics processing units (GPUs) style relies heavily on neural networks and machine learning for perception and control. However, their system is not completely autonomous and requires driver supervision. Nvidia also provides the DRIVE technology, a unified AI computing architecture that trains deep neural networks within a vehicle.
3. **Cruise [General Motors (GM)]:** GM is working on completely AVs. Cruise [30] is working on an AV technology named "Cruise Origin." It integrates AI algorithms, LiDAR, and RADAR to harness the speed of moving objects. Further, LiDAR and cameras are applied to classify and trail objects [4]. It uses an assortment of 16 cameras to detect different objects including pedestrians, traffic lights, and free space [4].
4. **Aurora:** Aurora is building a self-driving platform that applies AI, LiDAR, RADAR, and cameras to create a comprehensive perception and control system [31]. They have partnerships with numerous industries, including trucking and logistics, aiming to bring autonomous technology to commercial applications.
5. **Auronomous (Uber ATG):** Uber's Advanced Technologies Group (ATG) was working on developing AV technology for ride-sharing services [32]. While their efforts were ongoing, the status and direction of their self-driving program may

have changed due to various factors, including regulatory challenges and shifts in corporate strategy [32].

6. **Nuro:** Nuro specializes in developing AVs for goods delivery [33]. Their compact vehicles are designed to carry goods in urban environments. Nuro's approach emphasizes safety and efficiency in autonomous delivery services.
7. **Baidu Apollo:** Baidu's Apollo platform provides open-source software and hardware for autonomous driving [34]. It aims to accelerate the expansion of AVs by offering a range of tools and components that developers can use to build their self-driving systems.
8. **Zoox (Amazon):** Zoox, acquired by Amazon, was working on creating a purpose-built, fully AV from the ground up [7]. Their focus was on designing a vehicle that reimagines transportation with passengers facing each other, which would be made possible by the absence of a traditional driver's seat.
9. **Carmera:** They are working on developing AVs mapping technology for AVs [16].

Please note that the self-driving industry is dynamic, and the status of companies and their technologies can change rapidly. It is recommended to check the latest news and updates from these companies for the most current information on AI-enabled self-driving systems.

9.4 Challenges in Self-Driving Mechatronics

Developing self-driving mechatronics, which involves integrating mechanical, electrical, and software components for AVs, presents a range of complex challenges [9, 20, 35, 36]. These challenges span technical, regulatory, and societal aspects. Here are some key challenges associated with self-driving mechatronics:

i. **Ownership Issues:** An ADAS is driverless. Such a scenario shall expand public travel options thus reducing private vehicle ownership [37].
ii. **Ethical and Legal Issues:** The choices made by self-driving vehicles in circumstances that include potential harm may elevate ethical and legal questions. Determining how

vehicles prioritize different safety considerations will be a complex problem [38].

iii. **Safety and Reliability:** Ensuring the security and reliability of self-driving mechatronics is paramount [39]. The technology must perform flawlessly in a wide range of conditions, and failures can have severe consequences. Rigorous testing, redundancy, fail-safe mechanisms, and fault tolerance are essential to address safety concerns.

iv. **Common Sense:** This may be uncommon in AVs. There can be specific real-world scenarios where AI may struggle. Example is detection of a bird on road or comprehending a cyclist hand gesture. Although AI continues to improve, its performance on above concerns requires a much deeper understanding [3].

v. **Real-Time Processing:** Autonomous vehicles need to process massive amounts of data in real time to make split-second decisions [40]. This requires powerful hardware, efficient algorithms, and optimization to handle sensor data, perception, path planning, and control in real time.

vi. **Environmental Variability:** Performance of self-driving automotives in complex scenarios under extreme environmental conditions is still questionable [41, 42]. Developing algorithms that can handle these variations while maintaining safety is challenging.

vii. **Regulatory and Legal Hurdles:** The legal and regulatory foundation for autonomous automotives is still sprouting in many jurisdictions [30]. Steering these regulations, liability issues, and safety morals is anintricate challenge that moves the development, testing, and deployment of self-driving mechatronics.

viii. **Cybersecurity:** Autonomous vehicles are also liable to cybersecurity threats that could possibly negotiate their safety and functionality [43]. Protecting the vehicle's software, communication networks, and data is crucial to avoid hacking and unauthorized access.

ix. **Data Privacy:** Autonomous vehicles collect and transmit significant volumes of data, including location and sensor data [44]. Protecting user privacy and ensuring secure data transmission and storage is essential.

x. **Cost and Scalability:** Developing and deploying self-driving mechatronics is a resource-intensive endeavor [20]. Ensuring that the technology becomes cost-effective and scalable for widespread implementation remains a challenge.

Addressing above challenges requires collaboration among experts in various fields, including robotics, AI, computer science, engineering, and law [35]. As technology advances and the industry mature, these challenges will likely continue to be a focal point of research, development, and innovation.

9.5 Advantages of Self-Driving Mechatronics

The integration of AI in self-driving mechatronics offers numerous benefits. Understanding it is especially important as human life is at stake.

i. **Safety:** AI-powered systems shall eliminate human error, a noteworthy aspect of many accidents. Self-driving vehicles can react faster and make decisions based on a broader range of data [18]. This may lead to safer travel as the number of road accidents shall reduce by a huge margin.
ii. **Efficiency:** Self-driving vehicles shall further optimize routes, vehicle speeds, and acceleration patterns to improve fuel efficiency and reduce traffic congestion [20]. This shall impact driving and overall transportation efficiency, reducing the overall travel time.
iii. **Accessibility:** Autonomous driver assistance systems (ADASs) shall enable mobility for individuals who are unable to drive due to debility, age, etc. Further, it shall make transportation available 24*7 (on-demand mobility), a remarkable breakthrough that may eliminate the concept of rush hour [35].
iv. **Traffic Management:** AI-equipped vehicles are empowered toconnect with each other as well as traffic management systems to further optimize traffic flow and overcome crowding [9].
v. **Energy Use and Environmental Impact:** Autonomous vehicles shall be a savior for the climate as well as the environment by saving fuel and energy consumption along with positive environmental impact [9]. They may further help create a sustainable environment for humanity.

The convergence of AI and mechatronics in autonomous driving presents several challenges [42]. Guaranteeing the security and reliability of AVs remains a paramount concern. The complex interactions between AI algorithms and mechatronic components demand rigorous testing and validation. Developing algorithms that can handle unpredictable scenarios and edge cases is also a challenge. Furthermore, the real-time nature of autonomous driving requires low-latency processing, necessitating the optimization of AI algorithms and mechatronic systems.

To address these challenges, researchers and engineers are exploring techniques such as simulation-based testing, sensor redundancy, fail-safe mechanisms, and advanced control strategies [45]. Additionally, ongoing advancements in hardware, such as powerful GPUs and dedicated AI accelerators, contribute to the efficient execution of AI algorithms in real-time applications.

9.6 Self-Driving and Environmental Sustainability

In an era, where we are struggling to ensure survival of the mankind by saving the environment [3] through sustainable practices. The environmental impact of AVs also needs to be negotiated before rolling them out in mass production [46]. These include many factors like impact on air quality, water consumption, energy, and greenhouse gas emissions [3]. Urban land use may be affected by decreased parking requirements. It would also impact human life by reducing accidents. Self-driving systems have the potential to contribute significantly to a sustainable environment in several ways.

1. **Fuel Efficiency and Traffic Optimization:** Autonomous vehicles can be programmed to drive more efficiently, optimizing speed and route choices to minimize fuel consumption [46]. They can also connect with each other and with traffic structure, reducing congestion and idling, which, in turn, leads to lower fuel consumption and discharges.
2. **Platooning and Drafting:** Self-driving technology enables vehicles to drive in close formations, known as platooning, and take advantage of drafting (driving closely behind another vehicle to reduce air resistance) [3]. These practices can improve aerodynamic efficiency and fuel economy, particularly for long-haul trucks.
3. **Electric and Alternative Fuel Adoption:** Autonomous technology can accelerate the adoption of electric and alternative

fuel vehicles [46]. Self-driving systems can be integrated into electric fleets, leading to higher utilization rates and faster adoption of clean energy options.

4. **Reduced Emissions:** With optimized driving patterns and reduced traffic congestion, AVs can contribute to lower emissions of greenhouse gases and other pollutants [43]. This is especially important in urban areas where traffic congestion is a significant contributor to air pollution.
5. **Shared Mobility:** Self-driving cars could facilitate the expansion of ride-sharing and car-sharing services [46]. This means fewer vehicles on the road, as people share rides and utilize vehicles more efficiently, leading to reduced manufacturing and fewer vehicles contributing to pollution.
6. **Efficient Urban Planning:** Self-driving vehicles could lead to changes in urban planning and design [42]. With reduced need for parking spaces, cities could allocate more space to green areas, public transportation, and pedestrian-friendly zones.
7. **Conservation of Resources:** Vehicles empowered with self-driving may perform better in terms of resource conservation as well as reducing the accident count [42]. This would result in non-renewable resource conservation as well as reduced environmental impact.
8. **Transportation Accessibility:** Autonomous vehicles offer a new ray of hope to the one's incapable of driving like the elderly or those with incapabilities [41]. The easy, anytime transportation option would also lead to lower number of personal vehicles and thus vehicles on the road.
9. **Behavioral Changes:** With humans getting used to autonomous driving, significant changes in transportation behavioral patterns may also be observed. For example, people may prefer public autonomous traveling options rather than their own vehicles, thus, in turn, reducing congestion as well as pollution.

However, it is vital to note that these profits are not guaranteed and depend on how self-driving technology is implemented, adopted, and regulated. Ensuring that self-driving vehicles are electric or use clean fuels, integrating them into sustainable transportation ecosystems, and addressing potential negative externalities such as increased vehicle miles traveled due to reduced perceived driving effort are all critical considerations to fully realize the environmental benefits of self-driving systems.

9.7 Legal and Safety Issues in Autonomous Driving

The road to completely autonomous driving is quite long [43]. A major hurdle in realizing autonomous technology remains the legal and safety issues involved. A completely AV implies a computer on four wheels. In such a scenario, the computer would be responsible for all tasks like steering, acceleration, deceleration, as well as perceiving the surrounding environment. Complete AVs shall completely reshape the automotive industry worth several trillion of dollars. However, if simply put, a completely AV could be easily equated to an array of microprocessors. Before widespread use and adoption of this technology, several legal and security concerns need to be addresses like the ones listed in Table 9.4.

Addressing the concerns listed in Table 9.2 requires real-time collaboration between technology companies, regulatory bodies, legal experts, and cybersecurity professions. It is high time when humanity needs to strike a balance between innovation and safety, to ensure successful integration of the self-driving mechatronics into our transportation systems. As per Intel research [35], a self-driving vehicle can be proven to be safe if it fulfills two crucial parameters:

i. Standardization of safety assurance: Minimal requirements are met by the vehicle.
ii. Scalability: Solutions that result in exorbitant costs will not materialize into mass production.

Thus, safety of AVs is a vital apprehension that needs to be ingrained in the vehicles before we leave them on roads.

9.8 Conclusion

The amalgamation AI with the self-driving mechatronics has surely transformed the transportation landscape. The collaboration between mechatronics and AI disciplines has enabled the development of vehicles capable of perceiving their environment, making informed decisions, and navigating safely. Although challenges remain, the benefits of self-driving mechatronics include improved safety, efficiency, and accessibility. They hold the potential to reshape how we move and interact with the world around us. The convergence of AI and Mechatronics in autonomous driving also exemplifies the symbiotic relationship between software and hardware.

Table 9.4 Legal and safety issues in autonomous driving.

Legal concerns [35]		
S. no.	**Concern**	**Remark**
1.	Liability and Responsibility	Determining accountability in accidents involving self-driving cars can be multifaceted. Who owns the responsibility—the vehicle manufacturer, the software developer, the vehicle owner, or the human "driver" (if there is one)? Clear guidelines must be established to allocate responsibility and ensure fair compensation for victims in case of accidents.
2.	Regulation and Legislation	Existing traffic laws and regulations were designed with human drivers in mind. Adapting these regulations to autonomous vehicles requires addressing various scenarios, such as traffic violations, insurance requirements, and licensing procedures. Developing new laws and regulations that govern self-driving technology while ensuring safety is a major challenge.
3.	Data Privacy	Self-driving cars gather vast amounts of data, including real-time position, routes taken, and potentially even passengers' activities. This raises concerns about data privacy, storage, and potential misuse of personal information.
4.	Cross-Border Issues	Autonomous vehicles could operate across different jurisdictions with varying laws and regulations. Harmonizing regulations internationally to ensure a unified experience for autonomous vehicles traveling across the borders is a challenge at the time of this writing.
Security Concerns [9, 20, 35, 46]		
1.	Cybersecurity	Advanced driver systems' dependency on complex software systems, sensors, and communication networks makes them vulnerable to hacking, malware, and cyberattacks that could negotiate the safety of the vehicle, passengers, and other pedestrians.

(*Continued*)

Table 9.4 Legal and safety issues in autonomous driving. (*Continued*)

Legal concerns [35]		
S. no.	**Concern**	**Remark**
2.	Sensor Manipulation	Autonomous vehicles depend on sensors to perceive their environment. Adversaries could exploit vulnerabilities in these sensors, such as using fake signals to confuse the vehicle's perception system and cause accidents.
3.	Data Integrity	Accurate and reliable data are crucial for safe autonomous driving. If the data from sensors is compromised or tampered with, then it could lead to incorrect decisions by the vehicle's control system.
4.	Remote Takeover	Some autonomous vehicles provide the capability for remote control or intervention. This feature can be useful for safety reasons, but it also opens up the possibility of unauthorized access, cyber-attacks and unauthorized control, potentially leading to malicious actions.
5.	Ethical Decisions	In real-time road driving, there may arise situations where a decision is required in split of seconds based on humanitarian grounds. For example, changing lane to save an animal on road. Programming autonomous vehicles to take these decisions raises ethical dilemmas and public debate.
6.	Software Complexity	The complexity of self-driving software makes it challenging to ensure its complete reliability. Bugs, glitches, and unexpected interactions could lead to system failures.

This convergence is propelling the development of AVs by leveraging AI's cognitive capabilities and mechatronics' physical execution.

In conclusion, the convergence of AI and mechatronics holds immense potential to transform the way we distinguish, interrelate with, and experience transportation through the development of safe, efficient, and intelligent AVs.

9.9 Future Directions in Self-Driving Mechatronics

AI-empowered mechatronics in self-driving systems is a dynamic field. With the increasing sophistication and ability among AI algorithms, AV's ability to observe and steer through complex surroundings shall only improve. However, the future still holds room for crucial research in the following areas:

- Full Autonomy: Level 5 autonomy is still a distant dream.
- Human Interaction: Autonomous systems should not perform all decisions by themselves. They should also be able to effectively communicate with all stakeholders for their seamless integration into the society.
- Regulatory Framework: Sound regulations and standards need to be redefined for AVs to ensure clarity of norms.

AI coupled with mechatronic systems shall lead to a shift toward Vehicle-to-Everything communication and shall further enhance capabilities with integration with other technologies like the edge and cloud computing.

References

1. Dakota, N., Navigating The Future of AI in Self-Driving Cars, Udacity, Accessed: Dec. 19, 2023. [Online], Available: https://www.udacity.com/blog/2023/11/ai-in-self-driving-cars.html.
2. Ignatious, H.A., El Sayed, H., Khan, M., An overview of sensors in Autonomous Vehicles, in: *Procedia Compt. Sci.*, pp. 736–741 , Elsevier B.V., Netherlands, 2021, doi: 10.1016/j.procs.2021.12.315.
3. Silva, Ó., Cordera, R., González-González, E., Nogués, S., Environmental impacts of autonomous vehicles: A review of the scientific literature. *Sci. Total Environ.*, 830, 154615, Jul. 2022, doi: 10.1016/J.SCITOTENV.2022.154615.
4. Coffin, D., Oliver, S., Verwey, J., Building Vehicle Autonomy: Sensors, Semiconductors, Software and U.S. Competitiveness, 2019. Office of Industries Working Paper ID-063, November 2019 , Available at SSRN: https://ssrn.com/abstract=3492778 or http://dx.doi.org/10.2139/ssrn.3492778.
5. Liu, H., Jiang, R., Tian, J., Zhu, K., Traffic flow of connected and automated vehicles at lane drop on two-lane highway: An optimization-based control algorithm versus a heuristic rules-based algorithm. *Chin. Phys. B*, 32, 1, 014501-1–014501-12, Jan. 2023, doi: 10.1088/1674-1056/AC9369.
6. Autonomous Vehicles Factsheet., Center for Sustainable Systems, University of Michigan. 2023, Pub. No. CSS16-18., Accessed: Dec.

22, 2023. [Online], Available: https://trends.google.com/trends/explore?date=all&q=autonomous%20vehicles.
7. Kuutti, S., Fallah, S., Katsaros, K., Dianati, M., Mccullough, F., Mouzakitis, A., A Survey of the State-of-the-Art Localization Techniques and Their Potentials for Autonomous Vehicle Applications. *IEEE Internet Things J.*, 5, 2, 829–846, Apr. 2018, doi: 10.1109/JIOT.2018.2812300.
8. Yu, L., Kong, D., Shao, X., Yan, X., A path planning and navigation control system design for driverless electric bus. *IEEE Access*, 6, 53960–53975, 2018. doi: 10.1109/ACCESS.2018.2868339.
9. Kesting, A., Treiber, M., Helbing, D., Enhanced intelligent driver model to access the impact of driving strategies on traffic capacity. *Philos. Trans. R. Soc. Lond. A*, 368, 1928, 4585–4605, Oct. 2010, doi: 10.1098/rsta.2010.0084.
10. Jermakian, J.S., Crash avoidance potential of four passenger vehicle technologies. *Accid. Anal. Prev.*, 43, 3, 732–740, May 2011, doi: 10.1016/j.aap.2010.10.020.
11. Fernandes, P. and Nunes, U., Platooning with IVC-enabled autonomous vehicles: strategies to mitigate communication delays, improve safety and traffic flow. *IEEE Trans. Intell. Transp. Syst.*, 13, 1, 91–106, Mar. 2012, doi: 10.1109/tits.2011.2179936.
12. Hang, P., Lv, C., Xing, Y., Huang, C., Hu, Z., Human-like decision making for autonomous driving: A noncooperative game theoretic approach. *IEEE Trans. Intell. Transp. Syst.*, 22, 4, 2076–87, 2020 Nov 17.
13. Donghao, X., Zhezhang, D., Xu, H., Huijing, Z., Mathieu, M., François, A., Franck, G., Learning From Naturalistic Driving Data for Human-Like Autonomous Highway Driving. *IEEE Trans. Intell. Transp. Syst.*, 1–14, 2020. 10.1109/TITS.2020.3001131.
14. Wenzel, T.A., Burnham, K.J., Blundell, M.V., Williams, R.A., Dual extended Kalman filter for vehicle state and parameter estimation. *Veh. Syst. Dyn.*, 44, 2, 153–171, Feb. 2006, doi: 10.1080/00423110500385949.
15. Wischnewski, A., Stahl, T., Betz, J., Lohmann, B., Vehicle Dynamics State Estimation and Localization for High Performance Race Cars. *IFAC-PapersOnLine*, 52, 8, 154–161, Jan. 2019, doi: 10.1016/J.IFACOL.2019.08.064.
16. Bresson, G., Rahal, M.C., Gruyer, D., Revilloud, M., Alsayed, Z., "A Cooperative fusion architecture for robust localization: Application to autonomous driving. *IEEE Conference on Intelligent Transportation Systems, Proceedings, ITSC*, pp. 859–866, Dec. 2016, doi: 10.1109/ITSC.2016.7795656.
17. Heng, L., Choi, B., Cui, Z., Geppert, M., Hu, S., Kuan, B., Liu, P., Nguyen, R., Yeo, Y.C., Geiger, A., Lee, G.H., Pollefeys, M., Sattler. T., Project AutoVision: Localization and 3D Scene Perception for an Autonomous Vehicle with a Multi-Camera System. 4695–4702, 2019. 10.1109/ICRA.2019.8793949.
18. Jeong, J., Cho, Y., Kim, A., The Road is Enough! Extrinsic Calibration of Non-overlapping Stereo Camera and LiDAR using Road Information. *IEEE Robot. Autom. Lett.*, 1-1, 2019. 10.1109/LRA.2019.2921648.

19. Mishra, S., Pandey, G., & Saripalli, S., Extrinsic Calibration of a 3D-LIDAR and a Camera. *2020 IEEE Intelligent Vehicles Symposium (IV)*, 1765-1770, 2020.
20. Yeong, D.J., Velasco-hernandez, G., Barry, J., Walsh, J., Sensor and Sensor Fusion Technology in Autonomous Vehicles: A Review. *Sensors, 21*, 6, 2140, Mar. 2021, doi: 10.3390/S21062140.
21. Piao, J. and McDonald, M., Advanced driver assistance systems from autonomous to cooperative approach. *Transp. Rev.*, 28, 5, 659–684, 2008, doi: 10.1080/01441640801987825.
22. Malik, S., Khan, M.A., El-Sayed, H., Khan, J., Ullah, O., How Do Autonomous Vehicles Decide? *Sensors*, 23, 317, 2023. https://doi.org/10.3390/s23010317.
23. Carballo, A., LIBRE: The Multiple 3D LiDAR Dataset. *2020 IEEE Intelligent Vehicles Symposium (IV)*, 1-9, 2020.
24. Silva, D., Földes, D., Csiszár, C., Autonomous vehicle use and urban space transformation: A scenario building and analysing method. Sustainability, 13, 6, 3008, 2021 Mar 10.
25. Roe, M., Who's Driving That Car: An Analysis of Regulatory and Potential Liability Frameworks for Driverless Cars. 60, 317–348, 2019.
26. Yanyan, D., Suk, L., Perception, planning and control for self-driving system based on on-board sensors. *Adv. Mech. Eng.*, 12. 1--3, 2020. 168781402095649. 10.1177/1687814020956494.
27. Li, L., Ota, K., Dong, M., Humanlike Driving: Empirical Decision-Making System for Autonomous Vehicles. *IEEE Trans. Veh. Technol.*, 67, 8, 6814–6823, Aug. 2018, doi: 10.1109/TVT.2018.2822762.
28. Rosenband, D.L., "Inside Waymo's self-driving car: My favorite transistors," *2017 Symposium on VLSI Circuits, Kyoto, Japan*, 2017, pp. C20–C22, doi: 10.23919/VLSIC.2017.8008500.
29. Maradin, D., Malnar, A., Kaštelan, A., Sustainable and clean energy: the case of tesla company. *Journal of Economics, Finance and Management Studies*, 5, 3531–42, 2022.
30. Hanlon ML. Self-Driving cars: Autonomous technology that needs designated duty passenger. *Barry L. Rev.*, 22, 1, 2016.
31. Zhang, T., Toward automated vehicle teleoperation: Vision, opportunities, and challenges. *IEEE Internet Things J.*, 7, 11347–54, 2020 Oct 6.
32. Wong, K., Zhang, Q., Liang, M., Yang, B., Liao, R., Sadat, A., Urtasun, R.. Testing the safety of self-driving vehicles by simulating perception and prediction. In *Computer Vision–ECCV 2020: 16th European Conference, Proceedings*, Part XXVI 16 2020, pp. 312-329, Springer International Publishing, Glasgow, UK, August 23–28, 2020.
33. García, S., Pelliccione, P., Menghi, C., Berger, T., Bures, T., High-level mission specification for multiple robots. In: *Proceedings of the 12th ACM SIGPLAN international conference on software language engineering*, pp. 127-140, 2019 Oct 20.

34. Xu, K., Xiao, X., Miao, J., Luo, Q., Data driven prediction architecture for autonomous driving and its application on apollo platform. In :*2020 IEEE Intelligent Vehicles Symposium (IV),* pp. 175-181, IEEE, 2020 Oct 19.
35. Shalev-Shwartz, S., Shammah, S., Shashua, A., On a Formal Model of Safe and Scalable Self-driving Cars, Aug. 2017, [Online], Available: http://arxiv.org/abs/1708.06374.
36. Marsden, G., McDonald, M., Brackstone, M., Towards an understanding of adaptive cruise control. *Transp. Res. Part C,* 9, 1, 33–51, 2021 doi: 10.1016/s0968-090x(00)00022-x.
37. Glassner, Y., Gispan, L., Ayash, A., Shohet, T.F., Closing the gap towards end-to-end autonomous vehicle system, Jan. 2019, Accessed: Aug. 18, 2023. [Online], Available: http://arxiv.org/abs/1901.00114.
38. Hassan, A., Ali, M., Durrani, N.M., Tahir, M.A., An Empirical Analysis of Deep Learning Architectures for Vehicle Make and Model Recognition. *IEEE Access,* 9, 91487–91499, 2021, doi: 10.1109/ACCESS.2021.3090766.
39. Bojarski, M., Del Testa, D., Dworakowski, D., Firner, B., Flepp, B., Goyal, P., Jackel, L.D., Monfort, M., Muller, U., Zhang, J., Zhang, X., End to End Learning for Self-Driving Cars, Apr. 2016, Accessed: Aug. 24, 2023. [Online], Available: http://arxiv.org/abs/1604.07316.
40. Jeppsson, H., Östling, M., Lubbe, N., Real life safety benefits of increasing brake deceleration in car-to-pedestrian accidents: Simulation of Vacuum Emergency Braking. *Accid. Anal. Prev.*, 111, 311–320, Feb. 2018, doi: 10.1016/J.AAP.2017.12.001.
41. Urmson, C. *et al.*, Autonomous driving in urban environments: Boss and the urban challenge. *J. Field Robot.*, 25, 8, 425–466, Aug. 2008, doi: 10.1002/ROB.20255.
42. Folsom, T.C., Energy and autonomous urban land vehicles. *IEEE Technol. Soc. Mag.*, 2, 2, 28–38, 2012, doi: 10.1109/mts.2012.2196339.
43. Schellekens, M., Human–machine interaction in self-driving vehicles: a perspective on product liability. *Int. J. Law Inf. Technol.*, 30, 2, 233–248, Nov. 2022, doi: 10.1093/IJLIT/EAAC010.
44. Yogamani, S., Hughes, C., Horgan, J., Sistu, G., Varley, P., O'Dea, D., Uricár, M., Milz, S., Simon, M., Amende, K., Witt, C., WoodScape: A multi-task, multi-camera fisheye dataset for autonomous driving In *Proceedings of the IEEE/CVF International Conference on Computer Vision*, pp. 9308–9318. May 2019, Accessed: Aug. 31, 2023. [Online], Available: http://arxiv.org/abs/1905.01489.
45. Fajardo, D., Au, T.C., Waller, S., Stone, P., Yang, D., Automated intersection control: Performance of future innovation versus current traffic signal control. *Transp. Res. Rec.*, 1, 2259, 223–232, Dec. 2011, doi: 10.3141/2259-21.
46. Igini, M., Environmental Pros and Cons of Self-Driving Cars. *Earth. Org.* 2022 May 16.

10

Improving Power Quality for Industry Control Using Mechatronics Devices

Pulkit Kumar, Harpreet Kaur Channi* and Surbhi Gupta

Department of Electrical Engineering, Chandigarh University, Mohali, Punjab, India

Abstract

Modern organizations are using mechatronics technologies to increase industrial output. Industrial operations are more susceptible to power quality issues, requiring novel dependability methods. This presentation discusses employing mechatronics devices to avoid commercial power quality issues. "Power quality" refers to producing electrical dependability and consistency. Harmonic distortions, voltage sags, surges, and interruptions may diminish productivity, prolong downtime, and raise maintenance costs. Traditional power conditioning has efficiency constraints. Mechatronics integrates mechanical engineering, electronics, computer science, and control systems for industrial regulation. Mechatronics devices in the power distribution network may help companies address power quality issues. Active power filters, voltage regulators, and energy storage devices are examples. Advanced control algorithms and real-time monitoring allow active power filters to regulate load voltage and harmonics. Active voltage regulators' improved control algorithms also keep output voltage constant regardless of input voltage. Energy storage might protect vital equipment from transient voltage changes. Integrating mechatronics devices into industrial power systems may have advantages and drawbacks. It also illustrates how mechatronics-based power quality solutions have increased industrial productivity. Mechatronic power quality improvement technologies may create smart and strong industrial infrastructures. More advanced and cheap mechatronics devices will enable dependable and environmentally friendly industrial processes as science and technology advance. Companies adapting to this new paradigm will help protect and prosper in the future and gain a competitive advantage.

**Corresponding author*: harpreet.ee@cumail.in

Kolla Bhanu Prakash, Satish Kumar Peddapelli, Ivan C. K. Tam, Wai Lok Woo and Vishal Jain (eds.) Computational Intelligent Techniques in Mechatronics, (317–346) © 2024 Scrivener Publishing LLC

Keywords: Automation, controllers and devices, Industry 4.0, microcontroller, power drives, semiconductor

10.1 Introduction

"The term 'power quality' describes the reliability and tidiness of the electricity delivered to control systems in manufacturing facilities. All control systems, machines, and processes can run smoothly and precisely provided the power supply is up to par."

The ability of electrical equipment and systems to function correctly is directly related to power quality, defined as the qualities of the electrical supply. Power quality is crucial when it comes to the efficiency, dependability, and life span of machinery and processes in industrial control systems. Inadequate power quality may result in broken machinery, halted output, higher repair bills, and potential dangers. To keep industrial control systems running smoothly and reliably, it is essential to have a firm grasp of and strategy for power quality [1]. Voltage stability is a major issue in the field of power quality. The utility system, electrical equipment, and significant load variations are all potential causes of voltage fluctuations like dips, spikes, and interruptions. Sensitive industrial control equipment is vulnerable to breakdowns, data loss, and production halts due to these voltage fluctuations. For instance, voltage sags are temporary drops in voltage that may be triggered by the beginning of powerful motors or power system failures. These dips may malfunction the control system or equipment resetting, disrupting manufacturing. Harmonic distortion is another factor in the power quality puzzle. Unwanted harmonics impair the power supply's sinusoidal waveform and should be avoided. They are caused by electronic devices, rectifiers, and variable frequency drives that are nonlinear loads. Overheating from harmonics may degrade the efficiency and shorten the life span of transformers, motors, and other machinery. Harmonics may also cause instability or system breakdowns when they spread across an electrical grid and disrupt delicate control signals [2]. Harmonic distortion may be reduced by carefully planning and implementing filtering and cancellation strategies. Another essential component of power quality is the power factor. The efficiency with which electrical power is being used is represented by this metric, which is the ratio of actual power to perceived control. A decrease in power factor may increase energy usage and hence utility costs. In addition, utilities often assess fines for inadequate power factor, which may add significantly to the already high operation costs for manufacturing operations. To enhance power factor, reduce reactive power

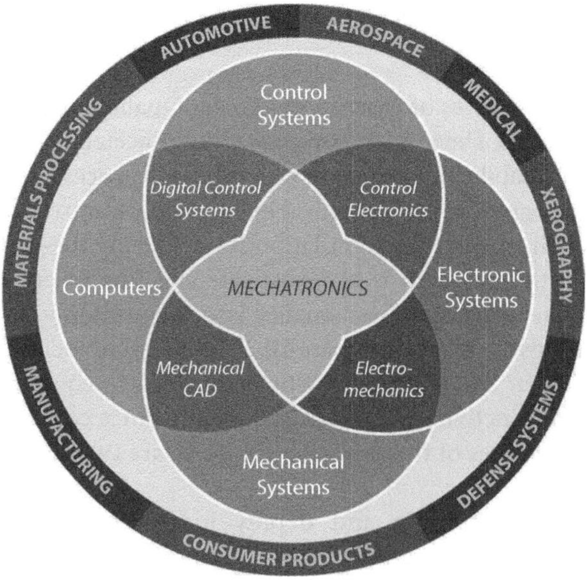

Figure 10.1 Various fields in which mechatronics deals with.

demand, and optimize energy efficiency through the implementation of power factor correction solutions such as capacitor banks can be used [3].

Figure 10.1 depicts diverse fields benefiting from mechatronics: Robotics, Automotive Systems, Industrial Automation, Medical Devices, Aerospace Systems, Consumer Electronics, Renewable Energy, Biomechanics, Smart Homes, Virtual Reality, and Augmented Reality. Another factor impacting power quality in industrial control systems is voltage imbalance. It happens in a three-phase power system when the three phases' voltage magnitudes and phase angles are unequal. Overheating, torque pulsations, higher losses, and reduced motor efficiency are all possible outcomes of voltage imbalance in a motor. In addition, industrial control systems' functionality and dependability may be negatively impacted by imbalanced voltages and currents. Voltage imbalance must be monitored and corrected so electrical devices function properly and last as long as possible [4]. Short-lived, high-amplitude disruptions in the power supply are known as transients, sometimes called voltage spikes or surges. Lightning strikes, switching processes, and power system problems are all potential sources of transients. Control circuits, processors, and communication devices are all susceptible to harm from these high-energy occurrences. Protecting industrial control systems against transient overvoltages calls for installing surge protection devices and using correct grounding procedures [5].

10.1.1 Scope and Objectives of Power Quality in Industrial Control

The scope and objectives of "Improving Power Quality for Industry Control Using Mechatronics Devices" encompass a multifaceted approach aimed at enhancing the reliability and efficiency of industrial processes. Mechatronics, a synergistic integration of mechanical engineering, electronics, and control engineering, will serve as the foundation for achieving these goals. The primary scope of this research is to investigate and implement mechatronics devices that can mitigate power quality issues commonly encountered in industrial settings as shown in Figure 10.2. These issues include voltage sags, harmonics, flicker, and transients, which can lead to equipment malfunction, reduced productivity, and increased maintenance costs.

The core objectives of this research initiative are twofold.

- Firstly, to develop and deploy innovative mechatronics devices, such as active power filters, voltage regulators, and energy storage systems, which will enable the mitigation of power quality disturbances in real-time. These devices will not only safeguard critical industrial equipment but also improve energy efficiency.
- Secondly, the study aims to establish advanced control strategies that integrate data analytics and sensor technologies for intelligent decision-making in power quality management. This involves the creation of adaptive control algorithms and predictive maintenance protocols to ensure long-term system performance and sustainability.
- Ultimately, this research seeks to provide industry with cost-effective and cutting-edge solutions for maintaining optimal power quality, leading to enhanced productivity and reduced operational downtime [6].

10.1.2 Literature Review

Kotenev *et al.* (2020) have designed and researched a reactive power compensation control system to improve the quality of electricity which proposes a solution to the actual problem of reactive power compensation based on the proposed power quality control algorithm. The rapid growth and development of manufacturing technology for controlling processors and high-power semiconductor devices has led to the enhancement of flexible AC transmission system (FACTS) controllers, thus improving

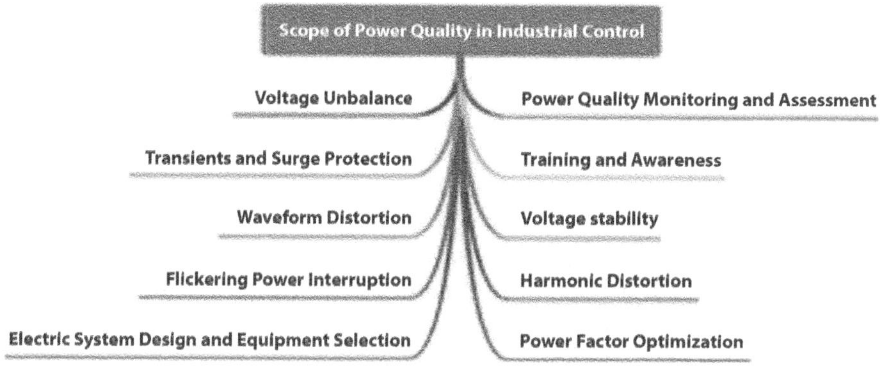

Figure 10.2 Scope of power quality in industrial control.

the performance of power systems [7]. Abasi *et al.* (2020) presented the accurate modeling of different parts of the control system and power electronics of the 72-pulse VSC-based generalized unified power flow controller (GUPFC) in detail to satisfy operation constraints and improve power quality in terms of harmonic level reduction. The wide use of these fast switching devices in the utility system leads to an increase in disturbances associated with harmonics and reactive power. Many researchers are working on the enhancement of power quality (PQ) using different custom power devices (CPDs) [8].

Das *et al.* (2021) highlighted the significance of the PQ in the utility network, its effect, and its solution, using different CPDs, such as passive, active, and hybrid filters. The power quality and energy conversion efficiency of the power conditioning system of the superconducting magnetic energy storage (SMES) and renewable energy systems depend on the control algorithms of the voltage source inverter. Jahan *et al.* (2021) proposed an advanced control technique named PI + LLC controller is proposed, which is based on the proportional integral (PI) controller and lead-lag compensator (LLC) [9]. Yue *et al.* (2021) proposed a comprehensive modeling framework is newly developed to quantify the connections of electricity savings, coal-based electricity systems, air pollutant emissions, control investments in China, a country exposed to poor air quality and find that industrial efficiency improvements can significantly decrease the dependence on coal-fired power generation, particularly the most polluting power fleet [10]. Takabayashi *et al.* (2021) provided a study on a smart body area network (SmartBAN) physical layer (PHY), as an of the Internet of medical things technology, for an advanced human monitoring system in industrial use. It is possible to develop more advanced human monitoring systems for industrial use by combining the SmartBAN with multimedia devices. Recent clinical trials have not demonstrated clear advantages of these devices over traditional passive devices [11]. Huang *et al.* (2021) advocated for new holistic approaches in which human motor control and intelligent prosthesis control function as one system (defined as human-prosthesis symbiosis). An excessive use of non-linear devices in industry results in current harmonics that degrades the power quality with an unfavorable effect on power system performance [12]. Rafique *et al.* (2022) proposed a soft-computing technique based on particle swarm optimization and adaptive fuzzy technique to avoid the phase delays caused by conventional control methods [13].

10.2 Power Quality in Industrial Settings

The efficiency and longevity of electrical devices and systems are compromised when poor power quality is present in manufacturing environments. Voltage stability, harmonic distortion, power factor, voltage imbalance, transients, and other properties of the power supply are all things that fall under this umbrella term. To keep industrial control systems running smoothly and securely, reliable electricity is a need. In the business world, stable voltage is of paramount importance. Causes of voltage sags, surges, and interruptions include utility grid problems and significant load shifts [14]. These changes can potentially corrupt data and cause faults in industrial control devices that rely on it. For instance, voltage dips might trigger mistakes in the control system or force a reset of the equipment, both of which disrupt the manufacturing process as a whole.

Harmonic distortion is one of the most serious problems with power quality in manufacturing. Harmonics are unwelcome frequencies to alter the sinewave nature of the power supply. Variable frequency drives and rectifiers are examples of nonlinear loads contributing to this phenomenon [15]. Harmful harmonics may lead to overheating, decreased efficiency, and increased wear and tear in machinery. In addition, they might cause unpredictable behavior or even system failures by disrupting delicate control signals. To keep the power quality at its best, it is crucial to reduce harmonic distortion via correct design, filtering, and cancellation [16].

10.2.1 Importance of Power Quality in Industrial Control

Power quality assumes a paramount role in the effectiveness, dependability, and productivity of industrial control systems, making it a matter of utmost importance. Several compelling reasons underscore the critical significance of power quality in industrial control: Firstly, equipment reliability hinges on the quality of the power supply. Clean and consistent power minimizes the risk of equipment failures, downtime, and damage to sensitive control components, leading to prolonged equipment life spans and reduced maintenance requirements. Secondly, product quality is intrinsically tied to power quality, as it is essential for maintaining tight and precise control over production processes [17]. Voltage fluctuations, harmonic distortion, and faulty wiring, all products of power quality issues, can introduce inconsistencies and flaws in the final output. A reliable power source ensures that industrial control systems can consistently produce goods that meet industry standards and customer expectations.

Thirdly, operational efficiency benefits greatly from power quality improvements. Enhancing power quality maximizes electrical power consumption, resulting in cost savings and heightened productivity by reducing power losses, optimizing power factors, and eliminating energy wastage. Fourthly, system performance is intricately linked to power quality, as it directly impacts the generation of precise control signals, accurate measurements, and dependable inter-component communication. Maintaining a consistent and clean power source is crucial in preventing signal degradation, noise interference, and communication breakdowns, all of which can compromise the efficiency of the system. Fifthly, safety in manufacturing plants is contingent upon the reliability of the power supply.

Decreasing the likelihood of power outages, voltage fluctuations, and other power quality-related issues plays a pivotal role in safeguarding the well-being of control system operators. Lastly, regulatory compliance is a non-negotiable aspect for industrial facilities, as they must adhere to stringent power quality regulations. Doing so not only ensures legal compliance but also fosters positive relationships with utility providers and regulatory bodies while protecting the welfare of employees [18].

10.2.2 Challenges and Issues in Power Quality for Industrial Control

Power quality issues present a host of formidable challenges within industrial control systems, potentially resulting in operational disruptions, equipment damage, and amplified maintenance expenses. These issues encompass voltage sags and swells, which can disrupt industrial processes and sensitive equipment due to sudden voltage fluctuations. Harmonics introduced by non-linear loads, like variable speed drives and arc furnaces, can distort waveforms and trigger equipment overheating and control system malfunctions. Voltage transients, including spikes and surges, pose a threat to electronic components. Unbalanced voltages in three-phase systems can impair motor efficiency and even lead to overheating. Voltage flicker, often induced by variable load consumption, can discomfort operators and affect control system and sensor performance. Industrial processes necessitate stable power frequencies, and variations can disrupt control and product quality [19].

Electromagnetic interference stemming from poor power quality can disrupt control systems and communications. Ensuring equipment compatibility with the power supply, controlling costs, and addressing diagnostic complexities challenge industrial operators. Compliance with stringent power quality standards and the selection of appropriate mitigation strategies require careful consideration. Adequate education and training are essential to empower personnel to recognize and respond to power quality issues effectively. In response, industrial facilities commonly implement power quality monitoring and mitigation solutions, including voltage stabilizers, harmonic filters, surge suppressors, and voltage regulators. Regular maintenance, equipment checks, and adherence to best practices are pivotal in upholding the consistent and dependable performance of industrial control systems in the face of power quality challenges [20].

10.3 Mechatronics Devices for Power Quality Improvement

By solving a wide range of problems related to power quality and enhancing the efficiency of power grids, mechatronics devices are crucial to making strides in this direction. Mechanical engineering, electronics, computer science, and control systems are all combined in mechatronics. Power quality monitoring, analysis, and mitigation are all tasks that may be

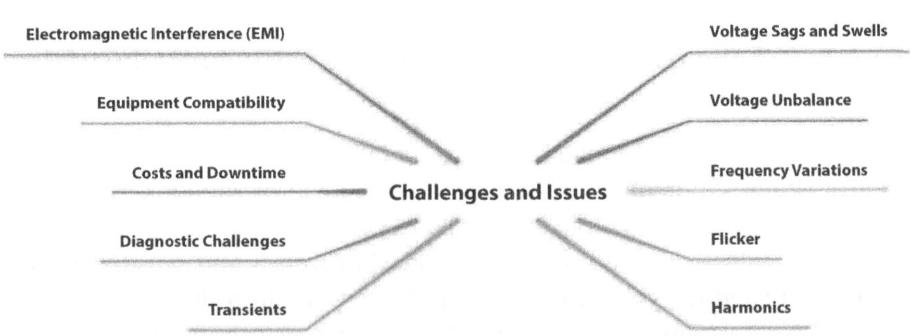

Figure 10.3 Challenges and issues in power quality for industrial control.

performed with the help of mechatronics devices as shown in Figure 10.3. These gadgets may fix voltage fluctuations, lower harmonic distortion, boost power factor, and lessen other power quality problems. A voltage regulator or compensator is a piece of mechatronics that maintains a constant voltage in an electrical system. To prevent voltage drops, spikes, and fluctuations, they use control algorithms to monitor the power supply in real time and make necessary adjustments [4].

Harmonic and active power filters are harmonic distortion attenuation devices built using mechatronics. By detecting harmonics and canceling them using control methods, they provide a clean power supply and protect against device failure. To further enhance power factor and voltage management in electrical grids, mechatronics devices like static VAR compensators (SVCs) and FACTSs are implemented [21]. These devices use sophisticated control algorithms to improve system efficiency by controlling power flow, reactive power demand, and other factorsas shown in Figure 10.4. In conclusion, mechatronics equipment provides novel approaches to enhancing power quality. These systems monitor, evaluate, and mitigate power quality problems by combining sensors, actuators, and control algorithms, guaranteeing a clean, efficient, and reliable power supply for commercial and industrial use.

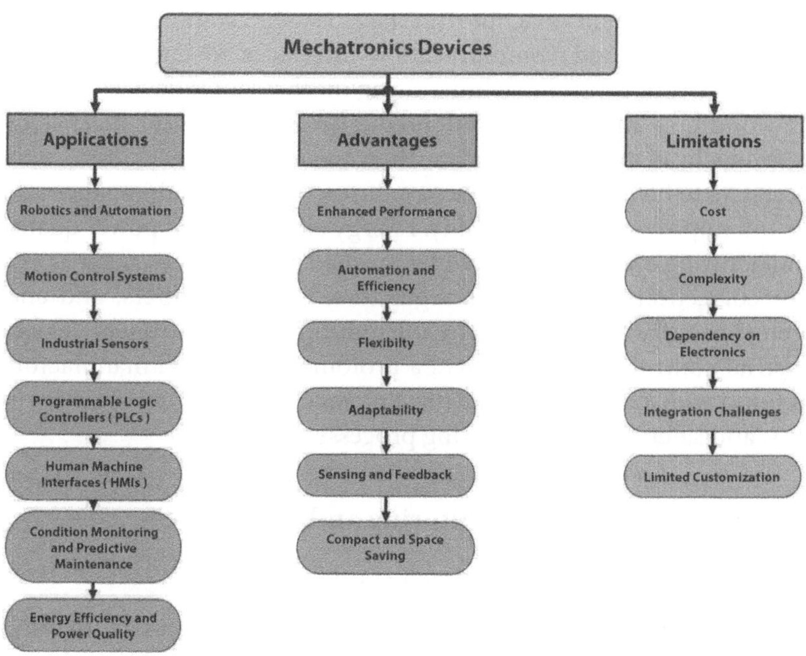

Figure 10.4 Mechatronics devices and its applications, advantages, and limitations.

10.3.1 Applications of Mechatronics Devices in Industrial Control

Various processes may benefit from integrating mechanical, electrical, and control components found in mechatronics devices widely used in industrial control systems. Mechatronics devices have revolutionized industrial control systems by providing a plethora of benefits, including enhanced efficiency, accuracy, reliability, and safety. These devices integrate mechanical, electrical, and control components to provide precise motion control, force control, object identification, and other critical functions. In robotics and automation, mechatronics devices enable robotic arms, grippers, and manipulators to perform complex tasks with high precision and repeatability [22]. Motion control systems rely on mechatronics devices to ensure accurate positioning, speed control, and trajectory tracking for CNC machines, conveyor systems, and robotic systems. Industrial sensors, such as proximity detectors, temperature monitors, pressure gauges, and vision systems, gather critical information about manufacturing processes, enabling continuous monitoring, feedback, and safety measures. Programmable logic controllers (PLCs) are digital devices that use mechatronics devices to automate manufacturing processes by monitoring sensor inputs, executing control algorithms, and generating outputs to actuate different devices. Human–machine interfaces (HMIs) combine touchscreens, buttons, displays, and communication interfaces to enable humans to interact with and control machines using mechatronics devices [23].

In condition monitoring and predictive maintenance, mechatronics devices such as vibration sensors, thermocouples, and acoustic sensors collect data to help avoid equipment failures, optimize maintenance schedules, and improve system reliability. Energy efficiency and power quality in manufacturing are also enhanced by mechatronics devices, such as power factor correction units, voltage regulators, harmonic filters, and energy monitoring software. Overall, the integration of mechatronics devices in industrial control systems has had a profound impact on manufacturing, enabling a wide range of benefits that enhance the efficiency, accuracy, reliability, and safety of manufacturing processes [24].

10.3.2 Power Quality Monitoring and Analysis

10.3.2.1 Power Quality Parameters and Standards

Power quality parameters are the specific characteristics of electrical power that must be closely monitored to ensure a stable and reliable power supply. These parameters include voltage magnitude (ensuring it stays within acceptable limits), voltage imbalances (detecting disparities between phases in a three-phase system), voltage sags and swells (temporary decreases and increases in voltage that can affect sensitive equipment), voltage harmonics (identifying harmonic frequencies that distort the power signal), current harmonics (assessing distortions in the current waveform due to non-linear loads), flicker (measuring voltage fluctuations that can affect lighting quality and human comfort), and transient disturbances (short-duration, high-amplitude voltage disruptions that can potentially damage equipment). Standards and regulations for power quality vary by region, and compliance with these standards is vital for ensuring the reliability and safety of the electrical supply. Common power quality standards and guidelines include the IEC 61000 series, IEEE 519, EN 50160, and ANSI C84.1, each addressing specific aspects of power quality [25]. Power quality standards and regulations vary by region, but some common standards and guidelines include the following:

- **IEC 61000 series:** International standards for electromagnetic compatibility and power quality.
- **IEEE 519:** Guidelines for harmonics in electrical power systems.
- **EN 50160:** European standard for voltage characteristics in public distribution systems.
- **ANSI C84.1:** American standard for voltage ratings (in the United States) [26].

10.3.2.2 Power Quality Monitoring Techniques

Power quality monitoring techniques encompass the use of specialized equipment and methods designed to measure and record various power quality parameters. Among these techniques, power quality analyzers stand at the forefront, as they are directly connected to the electrical system, continuously recording a wide range of parameters in real time, providing valuable data logging capabilities, and enabling comprehensive analysis. Oscilloscopes, versatile instruments with high-resolution waveform analysis capabilities, play a crucial role in capturing transient disturbances and waveform distortions, offering a visual representation of power quality issues [14]. Current clamps, specialized sensors, are employed to measure current harmonics and imbalances, particularly valuable for assessing current quality and often used in conjunction with other monitoring equipment. Data loggers are utilized for long-term monitoring, collecting data over extended periods to identify trends and recurring power quality issues. These loggers store the collected data for later analysis. Voltage recorders, on the other hand, are specifically designed for capturing voltage quality data over time, making them particularly useful in tracking voltage sags, swells, and transients and providing a detailed history of voltage variations. Power quality meters offer a comprehensive solution, combining monitoring and analysis capabilities, with a particular focus on harmonics, waveform analysis, and data logging, making them versatile tools for assessing power quality in a variety of applications. [27].

10.3.2.3 Data Analysis and Interpretation for Power Quality Assessment

Data analysis and interpretation constitute the heart of the power quality assessment process, comprising several essential steps. Initially, data collection involves gathering information from power quality monitoring equipment, encompassing vital aspects like voltage and current waveforms, harmonic content, transient event records, and other pertinent parameters. Following this, data preprocessing takes center stage, wherein collected data undergoes rigorous cleaning and filtering to ensure accuracy and suitability for analysis. This step often involves the removal of noise and outliers. The subsequent phase employs data visualization techniques, such as plots, graphs, and charts, to visually represent the power quality data, facilitating the identification of trends and anomalies [28].

Furthermore, frequency analysis delves into the frequency components of the data, identifying harmonics, assessing their amplitudes, frequencies, and phase relationships. This is indispensable in comprehending the harmonic content within the power system. Simultaneously, time-domain analysis scrutinizes voltage and current waveforms to detect and analyze transients, sags, swells, and other disturbances, providing insights into the temporal aspects of power quality issues. Event correlation is a pivotal step that associates power quality data with other events within the electrical system, including equipment start-ups or shutdowns. This correlation aids in identifying potential causes of power quality issues that may be linked to specific events. Subsequently, the process leads to the generation of comprehensive reports summarizing the power quality assessment. These reports highlight identified power quality issues, furnish recommendations for corrective actions, and offer insights for improving power quality in the future.

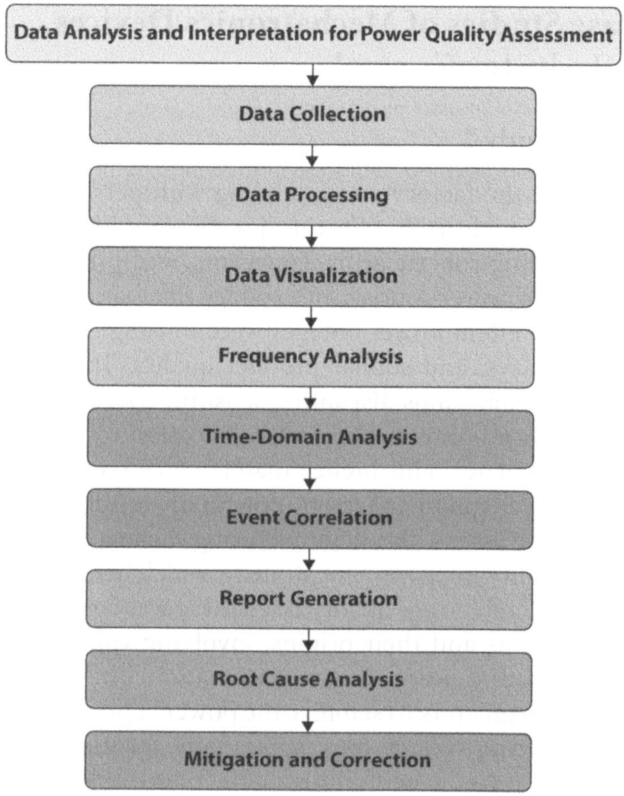

Figure 10.5 The process of assessing power quality.

The investigation then progresses to root cause analysis, entailing further testing, equipment analysis, or an examination of the entire electrical system to pinpoint the sources of power quality problems. Finally, the mitigation and correction stage is reached, where corrective measures are implemented based on the analysis findings. These measures may include the installation of filters, voltage stabilizers, or modifications to operational procedures, addressing the identified issues and enhancing power quality as discussed in Figure 10.5. Power quality analysis, through this systematic process, plays a critical role in ensuring the reliable operation of electrical systems, compliance with regulatory standards, and safeguarding sensitive equipment. Its implementation reduces the risk of equipment damage and downtime, establishing it as a vital component in maintaining a stable and efficient electrical supply [29].

10.4 Case Studies of Mechatronics Devices in Industry Control

10.4.1 Case Study 1

In this case study, the factory was grappling with persistent power quality issues that were adversely affecting production. The facility's critical equipment, including robotic arms, conveyors, and precision gear, relied on clean and stable power sources, but voltage fluctuations, harmonic distortion, and intermittent power outages were causing equipment failures, manufacturing delays, and subpar product quality. The challenges were substantial, with voltage dips disrupting sensitive control systems, leading to delays and reduced output. Harmonic distortion resulted in overheating, decreased efficiency, and higher maintenance costs, whereas power outages necessitated equipment resets, potentially resulting in data loss.

To address these issues, the manufacturing facility devised a comprehensive power quality improvement strategy, which involved various measures. First, a thorough power quality assessment was conducted to identify the specific problems and their origins, involving voltage measurement, harmonic analysis, and power factor examination. The solution included installing voltage regulators to stabilize the power supply to industrial control systems, reducing voltage dips, spikes, and fluctuations. Active harmonic filters were introduced to mitigate the negative effects of harmonic distortion by monitoring and canceling out harmonics, thus keeping the power supply cleaner and lowering the risk of equipment overheating. Surge protection measures, such as surge protectors and transient voltage surge suppressors, were implemented to safeguard against transient overvoltages induced by lightning strikes or switching activities, preventing damage to delicate control components. Additionally, capacitor banks were installed for power factor correction, optimizing energy use, reducing utility expenses, and avoiding fines.

The results of this power quality enhancement strategy were significant. Voltage stability greatly improved, leading to fewer disruptions and equipment failures. Lower levels of harmonic distortion translated into improved equipment efficiency and reduced maintenance costs. The surge protection measures ensured the safety of expensive machinery from short circuits. Power factor adjustment led to cost savings and a reduction in energy consumption. Overall, the manufacturing facility witnessed increased productivity, reduced downtime, enhanced product quality, and maximized energy efficiency, thanks to the power quality improvement strategy.

The industrial control systems operated more smoothly and reliably, benefiting from the clean and consistent electricity they now received.

10.4.2 Case Study 2

A data center faced severe harmonic distortion issues attributed to various nonlinear loads, including computers, uninterruptible power supply (UPS) systems, and cooling equipment, which resulted in equipment overheating, decreased efficiency, and operational disruptions. To address these challenges, a mechatronics-based harmonic mitigation solution was introduced. Harmonic distortion presented multiple problems in the data center, causing equipment malfunctions, shorter equipment life spans, and electromagnetic interference affecting critical machinery and communication systems. The data center required an efficient approach to mitigate harmonic distortion and ensure the proper functioning of its essential systems. The data center was equipped with mechatronics-based harmonic mitigation technology, featuring several essential components. Active harmonic filters were implemented to minimize the impact of harmonic distortion, effectively cleaning and stabilizing the power supply using sophisticated management algorithms and real-time monitoring. Strategically positioned sensors monitored the harmonic content of the power supply in real-time, enabling the evaluation of the mitigation strategy's efficiency. An intricate control system tracked and managed the activity of the harmonic filters, dynamically adjusting filter parameters in response to the evolving harmonic environment. Adaptive filtering was incorporated, allowing the control system to assess harmonic signatures and adjust filter properties for optimal harmonic cancellation under varying operating conditions.

The results were notable, with substantial improvements in power quality and data center efficiency. The mitigation of excessive harmonics led to reduced voltage distortion and enhanced equipment performance, decreasing the risk of equipment failures and downtime for cooling systems, servers, and UPS units. Additionally, the technology reduced energy losses caused by harmonic distortion, further enhancing the data center's efficiency. The adaptive filtering capabilities of the solution ensured ongoing optimization, even in situations of high load variability, guaranteeing efficient harmonic mitigation. Real-time monitoring and management systems improved the understanding of power quality and allowed for more proactive maintenance, ultimately enhancing system reliability. In summary, the data center experienced improved power quality, increased equipment performance, reduced downtime, and maximized energy efficiency thanks to a mechatronics-based harmonic mitigation system.

By leveraging cutting-edge mechatronics technology, the data center effectively managed harmonic issues and ensured the efficient and reliable operation of critical data center equipment.

10.4.3 Case Study 3

At a petrochemical facility, inconsistent voltage in the electrical supply had been causing equipment failures, operational disruptions, and safety concerns. Voltage fluctuations were undermining the efficiency and reliability of critical control systems and machinery. To address these challenges, a voltage control system based on mechatronics components was devised. The petrochemical factory faced significant issues due to power instability, leading to reduced output and increased equipment downtime resulting from control system disturbances caused by voltage dips, spikes, and fluctuations. Furthermore, voltage fluctuations posed safety risks to employees working with potentially hazardous machinery. A stable and reliable power source was crucial for the facility's seamless operations and the protection of its equipment and personnel. The solution to the petrochemical plant's voltage stability problems was the implementation of mechatronics-based voltage regulating technology, which comprised several key elements.

Cutting-edge voltage regulators were installed to stabilize the electrical power supply, automatically detecting voltage changes and adjusting the output to maintain the voltage within the specified range. This was made possible through advanced control algorithms and real-time monitoring. Sensors were strategically placed to monitor voltage levels at critical points in the power grid, providing continuous data to the control system for precise voltage regulation in response to dynamic conditions. A central control and monitoring system oversaw the voltage regulation process, processing sensor data and instructing the voltage regulators to make necessary adjustments based on the analysis results. To address voltage fluctuations, alarm systems and safety measures were put in place to alert operators and personnel to potentially hazardous voltage variations, allowing for swift responses in cases of equipment failure or safety concerns. The implementation of the mechatronics-based voltage control technology significantly improved the petrochemical plant's operations. Interruptions and downtime caused by voltage sags, swells, and fluctuations were reduced, resulting in improved voltage stability. This, in turn, led to increased output and enhanced product quality, as both equipment and control systems became more reliable.

The workplace also became safer due to improved voltage control, as the risk of equipment failures and safety hazards was significantly reduced.

The deployment of centralized management and monitoring systems ensured that the status of voltage was consistently visible, enabling preventative maintenance and rapid responses to voltage fluctuations. Operators' heightened awareness facilitated quicker issue diagnosis and more effective problem mitigation. Ultimately, the mechatronics-based voltage control system guaranteed a dependable power supply for the petrochemical plant, increasing productivity, reducing downtime, improving product quality, and enhancing worker safety. The advanced control and monitoring capabilities of the solution played a crucial role in enhancing the facility's operations and maintaining smooth production in a highly sensitive industrial environment.

10.5 Integration of Mechatronics Devices in Industrial Control Systems

Overcoming obstacles, including compatibility, system complexity, multidisciplinary cooperation, scalability, maintenance, dependability, safety, cost, and skill development, is necessary to integrate mechatronics devices in industrial control systems. Mechanical, electrical, and control engineers must work together on this. For smooth integration, it is essential to guarantee compatibility across devices, control systems, and preexisting infrastructure. Future growth and adjustments need planning for scalability and adaptability. Staff must be well-trained and supported; reliability, safety, and efficiency must be top priorities. To boost productivity, efficiency, and system performance, mechatronics devices must be properly integrated into industrial control systems [28].

10.5.1 Challenges and Considerations for Integration

Integration of mechatronics devices in industrial control systems presents several challenges and considerations that must be carefully addressed, as shown in Figure 10.6.

The integration of mechatronics devices into industrial control systems offers significant advantages, but it also presents a set of challenges that must be carefully addressed. These challenges encompass compatibility, system complexity, interdisciplinary collaboration, scalability, maintenance and support, reliability and safety, cost considerations, and the need for training and skill development. Ensuring compatibility across a diverse range of mechatronics devices, control systems, and existing infrastructure

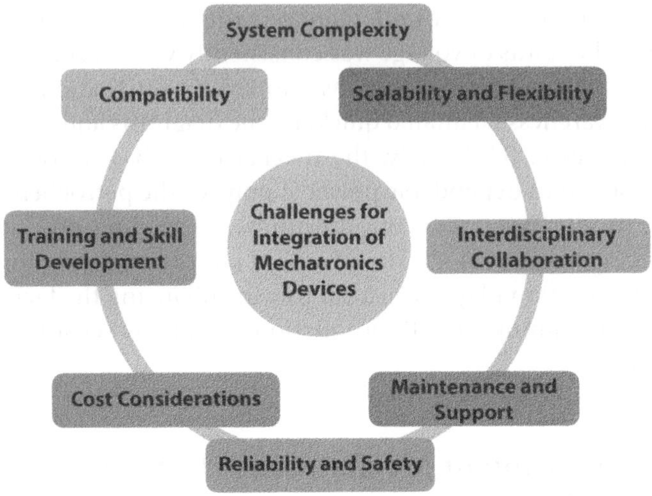

Figure 10.6 Challenges for the integration of mechatronics devices.

can be a complex task, necessitating adherence to communication protocols, interfaces, and software standards [30].

The increased system complexity resulting from the integration of mechatronics devices demands effective coordination among various teams or suppliers involved in the process, as well as a comprehensive understanding of mechanical, electrical, and control engineering. Successful integration requires robust interdisciplinary collaboration between mechanical engineers, electrical engineers, software developers, and control system experts, emphasizing communication and coordination. Scalability and flexibility are crucial to accommodate future growth and adapt to changing needs, enabling the system to incorporate new devices and features with ease [31]. The availability of maintenance and support, including replacement parts and technical assistance, is essential to keep mechatronics-equipped systems running smoothly and uninterrupted. Prioritizing reliability and safety through robust design, redundancy, fail-safe systems, and compliance with safety standards is imperative to mitigate the risk of system failures, equipment malfunctions, and safety concerns.

While the initial investment in integrating mechatronics devices may be substantial, potential advantages such as increased productivity, energy savings, and improved product quality should be considered. Additionally, adequate training and skill development are essential to ensure that personnel can effectively operate and maintain mechatronics-equipped systems. The compatibility problems, system complexity, multidisciplinary cooperation, scalability, maintenance, reliability, safety, cost, and skill development must be met before mechatronics devices can be successfully integrated into industrial control systems. The advantages of mechatronics devices may be used by industrial facilities to improve control systems and optimize operations if these problems are properly considered and addressed [32].

10.5.2 Communication and Control Interfaces

Industrial control systems rely heavily on communication and control interfaces because they facilitate interaction between disparate parts and pieces. With the help of these connectors, industrial operations may be monitored, controlled, and coordinated more effectively. Various communication protocols and interfaces guarantee interoperability and compatibility between equipment and systems in an industrial setting. Ethernet, Modbus, Profibus, CAN bus, and OPC (OLE for Process Control) are only a few examples of these interfaces' wired and wireless communication technologies [33].

Ethernet's fast data transfer rates, solid durability, and adaptability have made it the *de facto* standard for network connections in today's factory management systems. It is a standardized environment for connecting diverse components, including PLCs, HMIs, sensors, and actuators. Transmission Control Protocol/Internet Protocol and User Datagram Protocol are two Ethernet-based protocols that make remote monitoring and control possible. In addition to being utilized for communication between PLCs, SCADA systems, and other industrial equipment, Modbus is also a popular protocol for doing so. Its ease of use, adaptability, and compatibility make it a favorite in industrial automation settings [34].

Distributed control systems and industrial networks often use Profibus and CAN bus. Regarding real-time control applications, the CAN bus offers a stable and predictable communication technology, whereas Profibus facilitates an efficient connection between field devices and control systems. In industrial control systems, OPC is a standard that ensures compatibility between various pieces of software and hardware. With its standardized data-exchange protocol, it is simpler to combine products from different manufacturers [35]. Wireless technologies like Wi-Fi, Bluetooth, and Zigbee are becoming more commonplace in industrial control systems alongside conventional communication interfaces. Wireless connectivity paves the way for transportable and adaptable devices, facilitating remote sensing, control, and data collection. Good communication and control interfaces are essential for industrial control systems to function in a dependable, efficient, and secure manner. They help operators and automated systems make educated judgments and react quickly to situations by enabling real-time monitoring, data collecting, and control. Industrial control systems may achieve seamless integration, increased efficiency, and enhanced performance via carefully selecting communication protocols and interfaces [36].

10.5.3 Safety and Reliability Aspects

Industrial control systems must adhere to the highest standards of safety and dependability to safeguard workers, machines, and the environment. It is crucial to emphasize safety measures and keep the system reliable in industrial control systems because of the potentially dangerous situations in which they operate.

- Identifying and eliminating risks is an important part of industrial control system safety. Potential dangers are identified via risk assessments and hazard evaluations so that preventative actions may be taken. Safety interlocks, emergency stop systems, and protection barriers are installed to avoid mishaps and guarantee secure operation. They are establishing and teaching employees to follow safety regulations and procedures.
- Maintaining continuous operation and avoiding system breakdowns necessitates high levels of reliability in industrial control systems. Backup power sources, controllers, and communication networks are all examples of redundancy mechanisms put in place to guarantee system availability in the event of a failed component. To prevent unexpected breakdowns, it is important to undertake preventative maintenance such as inspections, tests, and calibrations regularly [37].

The consequences of malfunctions are lessened by using fault-tolerant layouts. Methods like failure modes and effects analysis and reliability-centered maintenance are used to zero in on the most important parts of a system and create upkeep plans that keep dependability as a top priority. Remote monitoring and diagnostic technologies are also utilized to improve further Detection of abnormalities and early resolution of possible difficulties. The safety and dependability of industrial control systems also depend on their cyber security. Protection against cyber-attacks is essential in this age of increased interconnection and digitalization of control systems. Networks are separated, users are given certain permissions, data are encrypted, and frequent security upgrades keep hackers out and system failures at bay [38].

In addition, industrial control systems must adhere to safety standards and regulations, such as the International Electrotechnical Commission's (IEC) 61508 and IEC 61111. Safety instrumented systems may be better planned, implemented, and maintained with the help of these regulations. If safety and reliability concerns are considered, then industrial control systems may run with less risk, less downtime, and safer working conditions. When safety and dependability are prioritized, people and property are protected, and output, efficiency, and overall system performance are improved [39].

10.6 Future Trends and Innovations in Mechatronics for Power Quality Improvement

Power quality may be greatly improved with the help of smart grid technology and the incorporation of renewable energy sources. Proactive and predictive maintenance will become possible using cutting-edge sensors and constant monitoring systems. Thanks to energy storage options and large-scale energy management systems, power grids will be more adaptable and stable. In addition, real-time data analysis and decision-making made possible by combining IoT and edge computing technologies will lead to more efficient and effective approaches to enhancing power quality [40].

10.6.1 Emerging Technologies in Mechatronics

Future electrical power systems will be influenced by new mechatronics technologies aimed at enhancing power quality. The future of electrical power systems is set to undergo a profound transformation driven by a host of innovative mechatronics technologies. These advancements offer numerous opportunities to enhance power quality, ensuring a more efficient, sustainable, and robust electrical grid [41]. Energy storage systems, including lithium-ion batteries, supercapacitors, and flywheels, are poised to play a pivotal role by storing excess energy during periods of low demand and releasing it when needed, thereby mitigating voltage and power fluctuations. Grid-connected renewable energy sources such as solar photovoltaics and wind turbines are becoming integral to the nation's electrical infrastructure, bolstered by smart grid technologies that enable greater integration and control, reducing dependence on fossil fuel–based production and enhancing grid stability [42].

Advanced power electronics, with voltage source converters, active power filters, and STATCOM devices, are ushering in a new era of nuanced power quality regulation, addressing issues like harmonic distortion and power factor in real-time. The convergence of the Internet of Things (IoT) and artificial intelligence (AI) is revolutionizing power system management, allowing for real-time monitoring, analysis, and control. Smart sensors and IoT-enabled devices provide precise measurements, whereas AI algorithms optimize power quality management techniques based on real-time data. Wide-area monitoring and control systems, utilizing synchronized phasor measuring units and advanced communication networks, offer the ability to monitor power quality characteristics over large areas, facilitating improved coordination and reducing voltage fluctuations across the network [43].

Microgrids and local energy management systems, composed of distributed energy resources and local control, contribute to better power quality, enhancing grid stability, reducing energy losses in transmission, and guaranteeing a consistent electricity supply in specific regions. In an increasingly interconnected and digitalized grid, cybersecurity and resilience have gained paramount importance. Technologies like blockchain, encrypted communication protocols, and intrusion detection systems safeguard the integrity and reliability of power quality management systems against cyber threats. These transformative mechatronics technologies are paving the way for a future of power distribution that ensures a constant supply of high-quality electricity, meeting the demands of an evolving and electrified world [44–46].

10.6.2 Potential Applications in Industry Control

Mechatronics technologies are proving to be a versatile and essential solution for power quality improvement in industrial control systems, offering a spectrum of key applications. These applications encompass a comprehensive approach to optimizing power quality, reliability, and efficiency within industrial settings [47]. Within industrial control systems, mechatronics devices excel in precise voltage regulation, eliminating or significantly reducing voltage fluctuations to ensure the uninterrupted and reliable operation of critical control equipment. Harmonic filtering, another crucial application, leverages mechatronics to mitigate harmonic

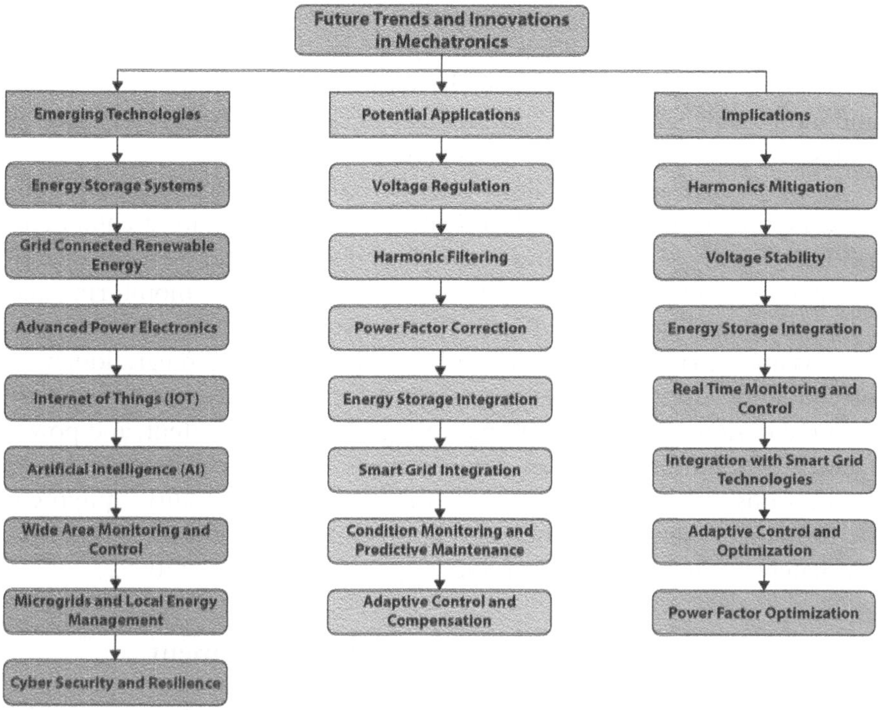

Figure 10.7 Futuristic trends and innovations in mechatronics.

distortion in industrial power systems. Harmonics can be detrimental, causing device failures, overheating, and efficiency losses. Mechatronic filters effectively detect and counteract these distortions, safeguarding equipment and enhancing overall system performance [48].

Furthermore, mechatronics devices like SVCs and active power filters play a pivotal role in power factor correction, actively regulating reactive power in industrial control systems. This enhancement results in increased power system efficiency by reducing power losses, improving energy utilization, and ultimately saving costs. Energy storage integration, made possible by mechatronics technology, is instrumental in ensuring voltage and frequency stability within industrial control systems. These systems store excess energy during periods of low demand and deliver it during high demand, providing backup power during outages and contributing to power quality improvement [49]. The integration of mechatronics devices into smart grid systems is yet another facet of their application. Smart grid systems, underpinned by mechatronics technology, enable intelligent load control, peak shaving, and demand response, utilizing real-time data and communication technologies to enhance power quality and efficiency in industrial settings. Condition monitoring and predictive maintenance systems within industrial control, driven by mechatronics-based sensors and monitoring technology, continuously track crucial parameters such as voltage levels, harmonics, and temperature.

This proactive approach allows early detection of potential power quality concerns, facilitating timely maintenance and preventing equipment failures. Finally, mechatronics technology facilitates active control and compensation of power quality factors through real-time monitoring and adaptive control algorithms. Figure 10.7 depicts futuristic trends and innovations in mechatronics, including advanced robotics, AI integration, nanotechnology, sustainable energy systems, smart materials, and biologically inspired designs. It ensures optimal voltage, harmonic content, and power factor with minimal human intervention. These mechatronics-driven applications collectively contribute to improving power quality, equipment performance, energy efficiency, and uptime in industrial control systems, ultimately leading to greater productivity and cost savings [50–52].

10.6.3 Implications for Power Quality Enhancement

Power quality enhancement through the application of mechatronics technologies represents a multifaceted approach with far-reaching implications for improving the performance and reliability of power systems. These implications encompass a comprehensive range of benefits: First and foremost, mechatronics devices enable precise voltage control, maintaining voltage levels within defined parameters [53]. This enhancement significantly contributes to power quality improvement by reducing the frequency and magnitude of voltage spikes and dips, ultimately preventing faults in industrial control systems. Voltage stability is a cornerstone of reliable and efficient power distribution. Harmonic mitigation is another critical implication of mechatronics in power quality enhancement. Mechatronics-based active power and harmonic filters prove highly effective in reducing harmonic distortion in electrical systems. By doing so, these devices not only enhance power quality but also protect electrical equipment from efficiency losses and overheating attributed to harmonics. The reduction of harmonics ensures the reliable operation of sensitive equipment and reduces the risk of damage and downtime [54].

Power factor optimization is a paramount consideration in the quest for better power quality, and mechatronics technology plays a vital role in achieving it. Through mechatronics-enabled power factor correction, the use of reactive power is minimized, resulting in more efficient power utilization in industrial control systems. This, in turn, leads to enhanced power quality, reduced power losses, and improved energy consumption, translating into cost savings and greater efficiency. Energy storage integration within the framework of mechatronics is a transformative implication that contributes to consistent voltage and frequency. Mechatronics-driven energy storage systems serve as a crucial safety net, ensuring uninterrupted power supply during periods of high demand or low voltage. This resilience in the face of power fluctuations and outages bolsters the overall reliability of the electrical grid, maintaining power quality even under adverse conditions [55].

Real-time monitoring and control, facilitated by mechatronics-based devices, offer a proactive approach to power quality management. Metrics such as voltage, harmonics, and power factor can be continuously tracked in real-time, allowing for the prompt detection of any discrepancies or anomalies. This real-time monitoring empowers operators to take immediate action, swiftly addressing any issues that may arise and preventing potential disruptions, thus ensuring consistent power quality and system reliability.

Mechatronics-based adaptive control and optimization further elevate power quality management. These systems continually assess data and dynamically adjust control settings in real time to optimize performance. This adaptability ensures that power quality is maintained even as operating conditions change, contributing to system stability and reliability. The integration of mechatronics with smart grid technologies represents a pivotal connection that augments power quality improvement. This synergy enables advanced monitoring, communication, and control capabilities. The incorporation of intelligent load control, demand response mechanisms, and the seamless integration of renewable energy sources into the power grid are all made possible through this integration. Smart grid technologies enhance power quality by facilitating real-time decision-making and enhancing grid stability, making power distribution more efficient and sustainable [56].

In conclusion, mechatronics technologies offer a holistic approach to power quality enhancement, addressing critical aspects such as voltage stability, harmonic mitigation, power factor optimization, energy storage integration, real-time monitoring and control, adaptive control and optimization, and integration with smart grid technologies. These implications collectively contribute to the improved performance, reliability, and efficiency of power systems, ensuring a consistent and high-quality power supply for industrial and commercial applications. As the integration of mechatronics continues to advance, it holds the promise of revolutionizing the way we manage and maintain power quality in the ever-evolving landscape of electrical distribution. Power quality improvement via mechatronics has several practical applications in industrial control systems. System uptime increases, equipment performance is boosted, energy consumption minimizes, and renewable energy sources may be more easily integrated. Mechatronics technologies allow industrial facilities to operate more efficiently, save money, and reduce their carbon footprint [57].

10.7 Conclusion

Enhancing industrial processes' stability, reliability, efficiency, and power quality for industry control by employing mechatronics devices is a crucial and successful technique. Voltage fluctuations, harmonics, and power factor imbalances are all examples of poor power quality that may negatively impact sensitive industrial equipment, resulting in lower output, more downtime, and greater maintenance costs. Mechanical, electrical, and computer technology integration results in mechatronics devices that provide novel approaches to these problems. Better voltage control is possible in industrial settings via mechatronics-based power quality improvement systems, lowering equipment failure probability. Sophisticated control algorithms and real-time monitoring capabilities enable quick reactions to power interruptions. In addition to improving energy use and lowering electricity costs, mechatronics devices may reduce harmonics and boost power factor.

In addition, mechatronics tools allow for a simple interface with pre-existing industrial control systems, lowering the entry barrier and the cost of upgrading to better power quality. Mechatronics devices play an increasingly important role in the smart factory idea, helping create a more sustainable and competitive industrial environment as sectors adopt automation and digitalization. Using mechatronics equipment to improve power quality in manufacturing is, thus, a sensible and innovative approach. It protects vital machinery, increases output, decreases overhead expenses, and fortifies businesses. By adopting this technology, companies can ensure the longevity and effectiveness of industrial control and automation far into the future.

References

1. Valtierra-Rodriguez, M. *et al.*, Detection and classification of single and combined power quality disturbances using neural networks. *IEEE Trans. Ind. Electron.*, 61, 5, 2473–2482, 2013.

2. Hossain, E. *et al.*, Analysis and mitigation of power quality issues in distributed generation systems using custom power devices. *IEEE Access*, 6, 16816–16833, 2018.
3. Martinenas, S., Knezović, K., Marinelli, M., Management of power quality issues in low voltage networks using electric vehicles: Experimental validation. *IEEE Trans. Power Deliv.*, 32, 2, 971–979, 2016.
4. Naderi, Y. *et al.*, An overview of power quality enhancement techniques applied to distributed generation in electrical distribution networks. *Renew. Sustain. Energy Rev.*, 93, 201–214, 2018.
5. Nadweh, S., Hayek, G., Atyeh, B., Power quality improvement in variable speed drive systems VSDS with 12-pulse rectifier topology for industrial applications. *Majlesi J. Mechatron. Syst.*, 8, 2, 1–6, 2019.
6. Karthik, T. *et al.*, Power quality improvement using DSTATCOM, in: *2019 Innovations in Power and Advanced Computing Technologies (i-PACT)*, IEEE, 2019.
7. Kotenev, V., Kochetkov, V., Elkin, D., The reactive power control of the power system load node at the voltage instability of the power supply, in: *2017 International Siberian Conference on Control and Communications (SIBCON)*, IEEE, 2017.
8. Abasi, M. *et al.*, Accurate simulation and modeling of the control system and the power electronics of a 72-pulse VSC-based generalized unified power flow controller (GUPFC). *Electr. Eng.*, 102, 1795–1819, 2020.
9. Jahan, S. *et al.*, An advanced control scheme for voltage source inverter based grid-tied PV systems. *IEEE Trans. Appl. Supercond.*, 31, 8, 1–5, 2021.
10. Yue, H. *et al.*, The potential of industrial electricity savings to reduce air pollution from coal-fired power generation in China. *J. Clean. Prod.*, 301, 126978, 2021.
11. Takabayashi, K., Tanaka, H., Sakakibara, K., Toward an advanced human monitoring system based on a smart body area network for industry use. *Electronics*, 10, 6, 688, 2021.
12. Huang, H.H. *et al.*, Taking both sides: seeking symbiosis between intelligent prostheses and human motor control during locomotion. *Curr. Opin. Biomed. Eng.*, 20, 100314, 2021.
13. Rafique, W. *et* al., Adaptive Fuzzy Logic Controller for Harmonics Mitigation Using Particle Swarm Optimization. *Comput. Mater. Contin.*, 71, 3, 4275-4293 2022.
14. Hassan, S.R. *et al.*, Comparative analysis of power quality monitoring systems. *NFC IEFR J. Eng. Sci. Res.*, 7, 1, 19–23, 2020.
15. Gao, Z. *et al.*, The discussion on power quality technology, in: *2016 IEEE International Conference on Mechatronics and Automation*, IEEE, 2016.
16. Riaz, M.T. *et al.*, Analysis and evaluating the effect of harmonic distortion levels in industry, in: *2021 4th International Conference on Energy Conservation and Efficiency (ICECE)*, IEEE, 2021.

17. Yaghoobi, J. *et al.*, Power quality issues of distorted and weak distribution networks in mining industry: A review. *IEEE Access*, 7, 162500–162518, 2019.
18. Tarasiuk, T. *et al.*, Review of power quality issues in maritime microgrids. *IEEE Access*, 9, 81798–81817, 2021.
19. Bollen, M.H. *et al.*, Trends, challenges and opportunities in power quality research. *Eur. Trans. Electr. Power*, 20, 1, 3–18, 2010.
20. Xie, H. *et al.*, IntelliSense technology in the new power systems. *Renew. Sustain. Energy Rev.*, 177, 113229, 2023.
21. Keddar, M. *et al.*, Power quality improvement for 20 MW PEM water electrolysis system. *Int. J. Hydrogen Energy*, 47, 95, 40184–40195, 2022.
22. Haghshenas, M., Ebadian, M., Shariatinasab, R., Autonomous control of inverter-interfaced distributed generation units for power quality enhancement in islanded microgrids. *Int. J. Mechatron. Electr. Comput. Technol.*, 4, 10, 2305–0543, 2014.
23. Purcaru, D., Purcaru, A., Gordan, C., Interface with simultaneous sampling of analog inputs for power quality measurement: Testing programs and experimental results. *Ann. Univ. Craiova Ser.: Autom. Comput. Electron. Mechatron.*, 9, 2-36, 33–42, 2012.
24. Moloi, K. *et al.*, Power Quality Analysis For A Solar-Grid Integrated System Using Support Vector Machine, in: *2021 Southern African Universities Power Engineering Conference/Robotics and Mechatronics/Pattern Recognition Association of South Africa (SAUPEC/RobMech/PRASA)*, IEEE, 2021.
25. Zhou, X. *et al.*, A review of reactive power compensation devices, in: *2018 IEEE International Conference on Mechatronics and Automation (ICMA)*, IEEE, 2018.
26. Van den Broeck, G., Stuyts, J., Driesen, J., A critical review of power quality standards and definitions applied to DC microgrids. *Appl. Energy*, 229, 281–288, 2018.
27. Kopacek, P. and Bula, I., Development Trends of Mechatronics, 2021,UBT International Conference. 291, pp. 1-6, UBT Kampus, Lipjan, 10.33107/ubt-ic.2021.280.
28. Sankaran, C., *Power quality*, CRC press, Boca Raton, 2017, https://doi.org/10.1201/9781420041026.
29. Sahoo, S., Recent trends and advances in power quality, in: *Power Quality in Modern Power Systems*, pp. 337–358, 2021.
30. Li, J. and Liu, L., Conceptual Design of Mechatronics System Based on Intelligent Control. *J. Phys. Conf. Ser.*, 1982, 1, 012113, 2021, IOP Publishing.
31. De Silva, C.W., *Mechatronics: an integrated approach*, CRC press, Boca Raton, 2004.
32. Habib, M.K., Mechatronics engineering the evolution, the needs and the challenges, in: *IECON 2006-32nd Annual Conference on IEEE Industrial Electronics*, IEEE, 2006.

33. Thramboulidis, K., A cyber–physical system-based approach for industrial automation systems. *Comput. Ind.*, 72, 92–102, 2015.
34. Christoulakis, F. and Thramboulidis, K., IoT-based integration of IEC 61131 industrial automation systems: The case of UML4IoT, in: *2016 IEEE 25th International Symposium on Industrial Electronics (ISIE)*, IEEE, 2016.
35. Syfert, M. *et al.*, Integrated approach to diagnostics of failures and cyber-attacks in industrial control systems. *Energies*, 15, 17, 6212, 2022.
36. Isermann, R., Modeling and design methodology for mechatronic systems. *IEEE/ASME Trans. Mechatron.*, 1, 1, 16–28, 1996.
37. Shyr, W.-J., Su, T.-J., Lin, C.-M., Development of remote monitoring and a control system based on PLC and webaccess for learning mechatronics. *Int. J. Adv. Rob. Syst.*, 10, 2, 97, 2013.
38. Harrison, R., Vera, D., Ahmad, B., Engineering methods and tools for cyber–physical automation systems. *Proc. IEEE*, 104, 5, 973–985, 2016.
39. Foradis, T., Thramboulidis, K., From Mechatronic Components to Industrial Automation Things: An IoT Model for Cyber-Physical Manufacturing Systems. *Journal of Software Engineering and Applications,* 10, 734–753, 2017.
40. Leitão, P. *et al.*, Integration patterns for interfacing software agents with industrial automation systems, in: *IECON 2018-44th Annual Conference of the IEEE Industrial Electronics Society*, IEEE, 2018.
41. Bangemann, T. *et al.*, State of the art in industrial automation, in: *Industrial Cloud-Based Cyber-Physical Systems: The IMC-AESOP Approach*, pp. 23–47, 2014.
42. Thramboulidis, K., Model-integrated mechatronics-toward a new paradigm in the development of manufacturing systems. *IEEE Trans. Ind. Inf.*, 1, 1, 54–61, 2005.
43. Brown, A.S., Who owns mechatronics? *Mech. Eng.*, 130, 06, 24–29, 2008.
44. Isermann, R., Mechatronic systems—Innovative products with embedded control. *Control Eng. Pract.*, 16, 1, 14–29, 2008.
45. Kozák, Š., Multidisciplinary approach and dual education in control engineering for mechatronics. *IFAC-PapersOnLine*, 49, 6, 52–56, 2016.
46. Hossain, A. and Zaman, T., HMI design: An analysis of a good display for seamless integration between user understanding and automatic controls, in: *2012 ASEE Annual Conference & Exposition*, 2012.
47. Mendes, J.M. *et al.*, Service-oriented process control using high-level Petri nets, in: *2008 6th IEEE International Conference on Industrial Informatics*, IEEE, 2008.
48. Kaltjob, P.O., *Mechatronic Systems and Process Automation: Model-Driven Approach and Practical Design Guidelines*, CRC Press, Boca Raton, 2018.
49. Huda, S. *et al.*, A malicious threat detection model for cloud assisted internet of things (CoT) based industrial control system (ICS) networks using deep belief network. *J. Parallel Distrib. Comput.*, 120, 23–31, 2018.
50. Checcozzo, R. *et al.*, RI-MACS: an innovative approach for future automation systems. *Int. J. Mechatron. Manuf. Syst.*, 2, 3, 242–264, 2009.

51. Isermann, R., Designs and Specification of Mechatronic Systems, in: *Springer Handbook of Automation*, pp. 287–313, Springer, U.K. London, 2023.
52. Gheorghe, G., Concepts and mechatronics and cyber-mixmechatronics constructions, integrated in COBOT type technology platform for intelligent industry (4.0), in: *Proceedings of the International Conference of Mechatronics and Cyber-MixMechatronics–2019 3*, Springer, 2020.
53. Moore, P. *et al.*, Virtual engineering: an integrated approach to agile manufacturing machinery design and control. *Mechatronics*, 13, 10, 1105–1121, 2003.
54. Nestinger, S.S., Chen, B., Cheng, H.H., A mobile agent-based framework for flexible automation systems. *IEEE/ASME Trans. Mechatron.*, 15, 6, 942–951, 2009.
55. Bradley, D., *Mechatronics: electronics in products and processes,* Routledge, U.K. England, 2018.
56. Bhamare, D. *et al.*, Cybersecurity for industrial control systems: A survey. *Comput. Secur.*, 89, 101677, 2020.
57. Chen, X., Application of Intelligent Control in Mechatronics System. *Eng. Adv.*, 2, 2, 182–185, 2022.

11

Study on Integrated Neural Networks and Fuzzy Logic Control for Autonomous Electric Vehicles

S. Boopathi

Department of Mechanical Engineering, Muthayammal Engineering College, Namakkal, Tamilnadu, India

Abstract

As the automotive industry undergoes a paradigm shift toward autonomous electric vehicles (AEVs), the development of advanced control systems becomes crucial to ensure safe, efficient, and reliable operation. This chapter delves into the convergence of neural networks and fuzzy logic control techniques in the context of AEVs. Neural networks, with their ability to learn complex patterns from data, and fuzzy logic, which excels in handling imprecise information, offer promising solutions for addressing the challenges associated with autonomy and energy efficiency in AEVs. This chapter explores the fundamentals of neural networks and fuzzy logic, discusses the specific control requirements of AEVs, and provides insights into the integration of these techniques to enhance autonomy. Case studies showcasing real-world applications of these advancements in AEVs further illustrate their impact on urban mobility, energy management, and fleet operations. Lastly, the chapter delves into future prospects, challenges, and emerging trends, underscoring the potential of neural networks and fuzzy logic to shape the future of autonomous electric transportation.

***Keywords*:** Sensing, path, energy management, adaptive cruise control, regenerative braking, hybrid control, fleet management, urban mobility

Email: boopasangee@gmail.com; https://orcid.org/0000-0002-2065-6539

Kolla Bhanu Prakash, Satish Kumar Peddapelli, Ivan C. K. Tam, Wai Lok Woo and Vishal Jain (eds.) Computational Intelligent Techniques in Mechatronics, (347–380) © 2024 Scrivener Publishing LLC

11.1 Introduction

In recent years, the automotive industry has been undergoing a profound transformation driven by two major trends: the rise of autonomous vehicles and the increasing adoption of electric propulsion systems. The convergence of these two transformative technologies has given birth to the autonomous electric vehicle (AEV) revolution. This chapter delves into the revolutionary landscape of AEVs, exploring how the fusion of autonomy and electrification is reshaping the way we perceive transportation, interact with vehicles, and envision the future of mobility [1].

The AEV revolution represents a pivotal shift in the automotive paradigm, where traditional vehicles are evolving into sophisticated platforms that blend cutting-edge artificial intelligence, advanced sensing technologies, and efficient electric powertrains. This transformation is set to have far-reaching impacts on various aspects of society, from urban planning and energy consumption to safety and accessibility. AEVs promise to redefine mobility by offering unparalleled levels of convenience and accessibility [2]. With autonomous capabilities, vehicles become not just modes of transportation but also platforms for work, leisure, and social interactions. The removal of the driver from the equation opens up new opportunities for passengers to engage in activities during their commutes, leading to increased productivity and improved quality of life [3].

The integration of electric powertrains with autonomous technology aligns with global efforts to reduce greenhouse gas emissions and combat air pollution. Electric propulsion systems significantly decrease tailpipe emissions, whereas the optimization of routes and driving patterns through autonomy further contributes to energy efficiency. This dual emphasis on autonomy and electrification paves the way for a more sustainable transportation ecosystem [4]. AEVs have the potential to drastically reduce traffic accidents caused by human errors. The integration of advanced sensors, such as LiDAR, cameras, and radar, enables vehicles to perceive their surroundings with unprecedented accuracy, thus enhancing collision avoidance and overall road safety. Moreover, autonomous systems can optimize driving behaviors, such as acceleration and braking, to achieve smoother traffic flow and reduce congestion [5].

The rise of AEVs calls for a reimagining of urban infrastructure and transportation planning. Designing cities that accommodate autonomous electric fleets requires rethinking road layouts, parking infrastructure, and charging networks. Additionally, the introduction of shared autonomous mobility services could lead to a reduction in the number of privately owned vehicles, thereby freeing up valuable urban space and promoting a more efficient use of resources [6]. The AEV revolution holds implications beyond technology and transportation. It has the potential to reshape industries, create new business models, and impact the job market. Companies specializing in autonomous software, electric vehicle manufacturing, and mobility services are positioned to thrive in this new ecosystem. Simultaneously, policymakers must address regulatory challenges and adapt legal frameworks to accommodate this paradigm shift [7].

Research on AEVs encompasses a wide range of interdisciplinary fields, including robotics, artificial intelligence, control systems, electric powertrains, and urban planning. Research in this area focuses on developing advanced sensor technologies, such as LiDAR, radar, cameras, and ultrasonic sensors, to enable AEVs to perceive their surroundings accurately. Object detection, scene understanding, and sensor fusion techniques are critical to ensuring safe navigation and decision-making in complex environments [8]. Artificial intelligence, particularly machine learning and deep learning, plays a central role in enabling AEVs to make real-time decisions based on sensor data. Researchers are exploring reinforcement learning, neural networks, and computer vision algorithms to enhance the autonomy and decision-making capabilities of AEVs. Efficient and safe path planning is essential for AEVs. Researchers work on algorithms that navigate AEVs through diverse scenarios, including urban environments, highways, and challenging weather conditions. Control systems aim to achieve smooth acceleration, braking, and steering to ensure passenger comfort and safety [9]. Developing intuitive interfaces and communication methods between AEVs and passengers is a critical research area. Research focuses on creating interfaces that convey the vehicle's intentions, decision-making processes, and emergency scenarios to occupants. AEVs require efficient energy management to optimize battery usage and extend driving range [10]. Researchers investigate strategies for regenerative braking, predictive energy management, and charging infrastructure optimization to maximize the benefits of electric powertrains. Ensuring the safety of AEVs is paramount. Research explores methods to validate and verify autonomous systems, including simulation testing, real-world testing, and the development of safety standards and regulations specific to AEVs. AEVs have implications for urban design and infrastructure. Researchers study how AEVs influence traffic flow, parking demand, and overall urban mobility. They also explore the integration of AEVs with public transportation systems and the design of smart cities [11]. The advent of AEVs raises ethical and societal questions. Research delves into issues such as liability in accidents involving AEVs, ethical decision-making by autonomous systems, and the impact of AEVs on job markets and transportation-related industries. AEVs challenge existing legal and regulatory frameworks. Research focuses on developing guidelines and regulations that ensure the safe deployment of AEVs on public roads while also addressing liability, data privacy, and cybersecurity concerns. Understanding consumer perceptions and acceptance of AEVs is vital for their successful adoption [12]. Researchers investigate factors influencing users' trust in autonomous systems, willingness to adopt AEVs, and preferences for mobility services. Creating realistic testing environments for AEVs is essential to validate their performance. Researchers develop simulation platforms and physical testbeds to assess AEV behavior in various scenarios without real-world risks [13].

Thus, the research landscape for AEVs is rapidly evolving, with academia, industry, and government agencies collaborating to address the multifaceted challenges and opportunities presented by this technology. Research in these areas collectively contributes to the advancement and eventual widespread adoption of AEVs, transforming the future of transportation. The fusion of autonomy and electrification has given rise to the AEV revolution, marking a pivotal moment in the history of transportation. As we journey through this chapter, we will explore the various facets of this revolution, delving into the technological advancements, challenges, and transformative impacts that AEVs bring to our roads and societies.

11.1.1 Need for Advanced Control Systems

The convergence of autonomy and electrification in the form of AEVs presents unique challenges that necessitate the development of advanced control systems. Unlike conventional vehicles, AEVs must navigate complex urban environments, make split-second decisions, and optimize energy consumption—all while ensuring passenger safety and comfort. Traditional control methods are insufficient to address these multifaceted demands, underscoring the need for innovative and sophisticated control strategies that leverage the power of computational intelligence [14].

11.1.2 Objectives of the Chapter

The primary objective of this chapter is to provide a comprehensive exploration of the role of advanced control systems, specifically focusing on the integration of neural networks and fuzzy logic, in the context of AEVs. The chapter seeks to achieve the following goals:

- To delve into the fundamentals of neural networks and fuzzy logic, elucidating their concepts, architectures, and capabilities. This foundational knowledge sets the stage for their application in AEVs.
- To examine the specific challenges faced by AEVs, including perception, path planning, energy management, and human–machine interaction. Highlight the limitations of traditional control approaches in addressing these challenges.
- To showcase how neural networks are employed in AEVs for tasks such as perception, environment mapping, decision-making, and trajectory planning. Illustrate how these networks learn from data and enhance driving performance.
- To provide insights into how fuzzy logic control is utilized to manage energy consumption, optimize driving behaviors, and enhance passenger comfort. Discuss how fuzzy logic systems handle imprecise information.
- To explore the potential of integrating neural networks and fuzzy logic to create hybrid control systems for AEVs. Discuss how these systems capitalize on the strengths of each approach to achieve enhanced autonomy and efficiency.
- To present case studies and real-world applications where neural networks and fuzzy logic have been successfully applied in AEVs. These examples highlight the tangible benefits of advanced control systems in various driving scenarios.

Through a comprehensive exploration of the integration of neural networks and fuzzy logic control in AEVs, this chapter aims to contribute to the understanding of cutting-edge control strategies that are shaping the future of autonomous electric transportation.

11.2 Fundamentals of Neural Networks and Fuzzy Logic

11.2.1 Neural Networks: Concepts

Neural networks, inspired by the structure of the human brain, have emerged as a powerful paradigm in machine learning and artificial intelligence. This section introduces the fundamental concepts and architectures of neural networks, providing an essential foundation for understanding their application in the context of AEVs [15].

11.2.1.1 Feedforward Neural Networks

Feedforward Neural Networks (FNNs), also known as Multi-Layer Perceptrons, form the basis of many neural network architectures. They consist of layers of interconnected nodes, or neurons, organized in a sequential manner. Each neuron receives inputs, performs a weighted sum, applies an activation function, and passes the result to the next layer. FNNs can approximate complex functions by adjusting the weights and biases associated with each connection. These networks excel in tasks such as pattern recognition, regression, and classification [16].

11.2.1.2 Recurrent Neural Networks

Recurrent Neural Networks (RNNs) introduce cyclic connections within the network, allowing information to flow in loops. This architecture is particularly suitable for tasks that involve sequences or time-series data, making RNNs well suited for natural language processing, speech recognition, and sequential decision-making. However, traditional RNNs suffer from the vanishing gradient problem, which hampers their ability to capture long-term dependencies. This limitation led to the development of more advanced RNN variants such as Long Short-Term Memory and Gated Recurrent Units, which address this challenge and enhance the network's memory retention [17].

11.2.1.3 Convolutional Neural Networks

Convolutional Neural Networks (CNNs) are tailored for tasks involving grid-like data, such as images and videos. CNNs are designed to automatically and adaptively learn spatial hierarchies of features from input data. They utilize convolutional layers to apply filters across the input data, enabling the network to learn local patterns and features. Pooling layers subsequently reduce the spatial dimensions while preserving the learned features. CNNs have revolutionized computer vision, enabling breakthroughs in object detection, image classification, and image generation.

These neural network architectures lay the groundwork for the application of computational intelligence techniques in AEVs. In the subsequent sections, we will delve into how these architectures can be harnessed to enhance the perception, decision-making, and control capabilities of AEVs in dynamic and complex driving environments.

11.2.2 Fuzzy Logic: Principles and Membership Functions

Fuzzy logic, a mathematical framework that models uncertainty and imprecision, provides a valuable tool for handling complex and ambiguous information in AEVs. This section introduces the principles of fuzzy logic, its foundational components, and its application in AEVs [18].

11.2.2.1 Linguistic Variables and Fuzzy Sets

Fuzzy logic introduces the concept of linguistic variables, which allows us to represent qualitative terms such as "low," "medium," and "high" with numerical values. Fuzzy sets extend traditional binary logic by allowing membership degrees to range between 0 and 1. A fuzzy set is defined by a membership function that assigns degrees of membership to elements of a universe of discourse. This enables the representation of vague and uncertain concepts, making fuzzy logic well suited for capturing human-like reasoning [19].

11.2.2.2 Fuzzy Rule–Based Systems

Fuzzy logic's power lies in its ability to model complex relationships using a set of rules. Fuzzy Rule–Based Systems consist of a rule base, a fuzzy inference engine, and a defuzzification stage. Rules are expressed in the form of "IF-THEN" statements that link fuzzy input variables to fuzzy output variables. These rules encode human knowledge and expertise in the form of linguistic terms. The fuzzy inference engine combines rules to generate fuzzy output sets, reflecting the system's response to input conditions [20].

11.2.2.3 Defuzzification Techniques

After the fuzzy inference engine produces fuzzy output sets, the final crisp output value must be obtained through a process called defuzzification. Various techniques are employed for this purpose, each with its strengths and considerations. Common defuzzification methods include centroid, mean of maxima, and weighted average. These techniques map the fuzzy output sets back to crisp output values, providing actionable information for decision-making [21].

In the context of AEVs, fuzzy logic offers a means to handle uncertain inputs, model complex driving scenarios, and facilitate human-like reasoning. By integrating fuzzy logic with neural networks and other control strategies, AEVs can make informed decisions in uncertain environments, optimize energy usage, and enhance overall driving performance. The subsequent sections of this chapter delve into the application of fuzzy logic and its integration with neural networks in the realm of AEVs.

11.3 Autonomous Electric Vehicles: Challenges and Control Requirements

The intersection of autonomy and electrification in the domain of AEVs presents a unique set of challenges that must be addressed to ensure the safe and efficient operation of these vehicles. This section delves into the key challenges that AEVs face and the specific control requirements that must be met [22]. The challenges and control requirements of AEVs is shown in Figure 11.1.

11.3.1 Control Challenges in Autonomous Electric Vehicles

AEVs navigate a dynamic and unpredictable environment, necessitating control systems that can adapt to varying road conditions, traffic scenarios, and unforeseen obstacles. Challenges include accurate perception of

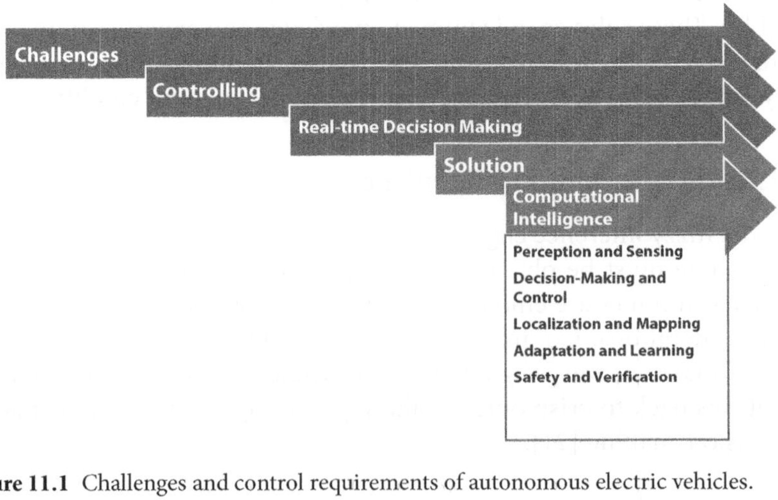

Figure 11.1 Challenges and control requirements of autonomous electric vehicles.

the surroundings, precise path planning, real-time decision-making, and ensuring passenger safety. Additionally, AEVs must grapple with uncertainties introduced by diverse weather conditions and interactions with human-driven vehicles [23].

11.3.2 Importance of Real-Time Decision-Making

The ability to make rapid and informed decisions in real-time is paramount for AEVs. Split-second choices must be made to avoid collisions, respond to unexpected events, and negotiate complex traffic situations. Real-time decision-making requires control systems that can process sensor data, analyze it for potential risks, and select appropriate actions—all within milliseconds. Failures in real-time decision-making could have severe consequences for both passenger safety and the surrounding environment [24].

11.3.3 Role of Computational Intelligence in Autonomous Vehicles

Autonomous Vehicles (AVs) represent a transformative leap in transportation, promising safer, more efficient, and convenient mobility solutions. To realize this potential, AVs must possess the ability to perceive their environment, make intelligent decisions, and navigate complex scenarios. This section delves into the pivotal role that computational intelligence plays in enabling AVs to achieve these capabilities [25].

11.3.3.1 Perception and Sensing

Perception is a fundamental challenge for AVs. The ability to accurately interpret the surrounding environment—identifying pedestrians, vehicles, road signs, and obstacles—is critical for safe navigation. Computational intelligence techniques, such as neural networks, excel at processing vast amounts of sensor data, detecting patterns, and recognizing objects. Deep learning architectures, including CNNs, enable AVs to achieve human-level accuracy in tasks like object detection and image classification [26].

11.3.3.2 Decision-Making and Control

Effective decision-making is at the heart of autonomous driving. AVs must analyze complex scenarios in real-time, predict the behavior of other road users, and select optimal actions to ensure safety and efficiency. Neural networks and reinforcement learning algorithms empower AVs to learn from past experiences and adapt to evolving situations. Fuzzy logic, on the other hand, provides a framework for modeling human-like reasoning and handling uncertain information, which is crucial for navigating ambiguous scenarios [27].

11.3.3.3 Localization and Mapping

Accurate localization and mapping are essential for AVs to understand their position and surroundings. Neural networks can aid in sensor fusion, combining data from cameras, LiDAR, and GPS to enhance localization accuracy. Furthermore, neural networks can assist in creating high-definition maps and identifying changes in the environment, ensuring AVs remain up-to-date with road conditions [28].

11.3.3.4 Adaptation and Learning

AVs operate in dynamic and evolving environments. To maintain performance in various conditions, they must continuously adapt and learn. Computational intelligence techniques facilitate adaptive learning, enabling AVs to improve their driving behavior over time. Neural networks can self-train using data collected from real-world driving experiences, enhancing their decision-making capabilities and robustness [29].

11.3.3.5 Safety and Verification

Safety is paramount in AV deployment. Computational intelligence aids in rigorous testing through simulation, enabling AVs to experience a vast range of scenarios without real-world risks. Neural networks can also be used to analyze and identify potential vulnerabilities, ensuring AVs are resilient against adversarial attacks [30].

In summary, computational intelligence, including neural networks, deep learning, and fuzzy logic, plays a multifaceted role in shaping the capabilities of AVs. These techniques empower AVs to perceive, reason, and act in complex and uncertain environments, ultimately paving the way for a safer, more efficient, and transformative era of transportation. By integrating computational intelligence into the control systems of AEVs, it becomes possible to enhance their perception, decision-making, and overall performance [31]. The subsequent sections of this chapter delve into the specific ways in which neural networks and fuzzy logic are harnessed to meet the challenges of AEVs, showcasing their potential to revolutionize autonomous transportation.

11.4 Neural Network–Based Control for Autonomous Electric Vehicles

The various contributions in the neural network–based control for AEVs are illustrated in Figure 11.2 and elaborated in the following.

11.4.1 Perception and Sensing Using Neural Networks

Perception, the ability to interpret and understand the environment, is a critical component of AVs' decision-making process. Neural networks have emerged as powerful tools for enhancing perception by enabling AVs to accurately detect and interpret various elements of their surroundings. This section explores the application of neural networks in perception tasks, focusing on object detection, recognition, and sensor fusion for comprehensive environmental perception [32].

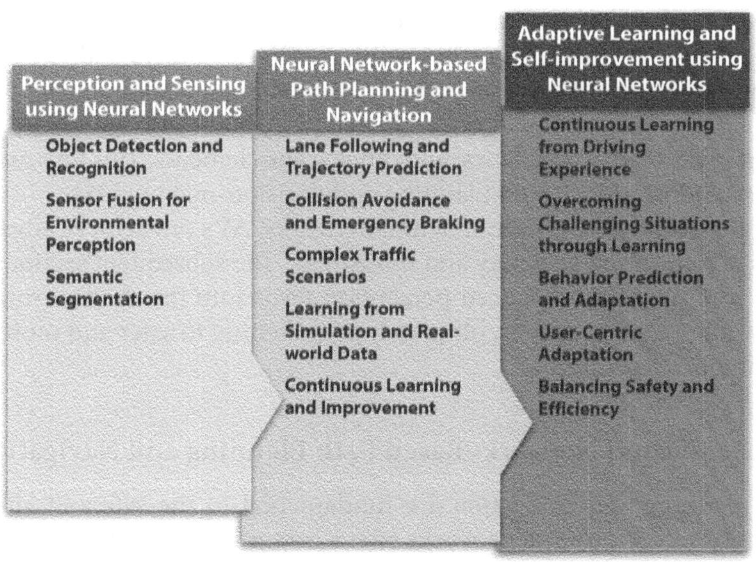

Figure 11.2 Neural network–based control for autonomous electric vehicles.

11.4.1.1 Object Detection and Recognition

Object detection and recognition involve identifying and categorizing objects within an AV's field of view. Neural networks, particularly CNNs, excel at this task by learning hierarchical features from visual data. By analyzing images or video frames captured by cameras, neural networks can accurately detect pedestrians, vehicles, cyclists, traffic signs, and other relevant objects. This capability is crucial for predicting and responding to the behavior of different road users [33].

11.4.1.2 Sensor Fusion for Environmental Perception

AVs rely on a variety of sensors, such as cameras, LiDAR, radar, and ultrasonic sensors, to perceive their environment accurately. Sensor fusion, the integration of data from multiple sensors, enhances perception reliability and robustness. Neural networks play a pivotal role in fusing data from diverse sensors, enabling AVs to generate a coherent representation of the environment. By combining visual, depth, and motion information, neural networks enable AVs to accurately estimate object positions, velocities, and trajectories [34].

11.4.1.3 Semantic Segmentation

In addition to object detection and recognition, semantic segmentation is a crucial task in perception. It involves assigning a specific class label to each pixel in an image, thereby enabling AVs to understand the spatial layout of the scene. Neural networks designed for semantic segmentation, such as fully convolutional networks, allow AVs to differentiate between road surfaces, sidewalks, lanes, and other scene elements. This fine-grained understanding aids in path planning and decision-making [35].

By harnessing the power of neural networks, AVs can achieve advanced levels of perception accuracy and robustness. The subsequent sections will delve into how this enhanced perception feeds into the decision-making and control aspects of AVs, ultimately contributing to safer and more reliable autonomous driving.

11.4.2 Neural Network–Based Path Planning and Navigation

Path planning and navigation are fundamental to the safe and efficient operation of AVs. Neural networks offer a powerful approach to enhancing these critical aspects by enabling AVs to make informed decisions about their trajectories, routes, and interactions with the environment. This section explores the application of neural networks in path planning and navigation for AVs [36].

11.4.2.1 Lane Following and Trajectory Prediction

Neural networks can assist AVs in maintaining proper lane positioning and trajectory prediction. By analyzing sensor data, including images from cameras and information from LiDAR and radar sensors, neural networks can accurately identify lane markings, road edges, and surrounding vehicles. This information is used to generate optimal trajectories that keep the AV within the desired lane and anticipate future vehicle movements [37].

11.4.2.2 Collision Avoidance and Emergency Braking

One of the primary concerns in AV navigation is avoiding collisions with obstacles and other vehicles. Neural networks can analyze real-time sensor data to detect potential collision risks and predict the behavior of other road users. In critical situations, neural networks can trigger emergency braking or evasive maneuvers to ensure the safety of passengers and surrounding traffic [36, 37].

11.4.2.3 Complex Traffic Scenarios

Navigating through complex traffic scenarios, such as intersections, merges, and roundabouts, requires intricate decision-making. Neural networks equipped with knowledge about traffic rules, regulations, and human driving behavior can assist AVs in making context-aware decisions. These networks consider various factors, such as right-of-way rules and pedestrian interactions, to determine the appropriate actions in complex traffic situations.

11.4.2.4 Learning from Simulation and Real-World Data

Neural networks in path planning and navigation can benefit from both simulation data and real-world driving experiences. Simulations allow AVs to experience a wide range of scenarios, including rare and challenging ones, without physical risks. Neural networks trained on simulation data can then be fine-tuned using real-world data to adapt to the intricacies of actual driving environments.

11.4.2.5 Continuous Learning and Improvement

Neural networks can enable AVs to engage in continuous learning and improvement. By analyzing driving data, identifying areas for enhancement, and adapting their decision-making strategies, neural network–based systems can evolve over time, becoming more adept at handling complex and dynamic driving situations.

Incorporating neural network–based path planning and navigation systems empowers AVs to navigate efficiently, avoid collisions, and handle intricate traffic scenarios with a level of precision and adaptability that is essential for safe and reliable autonomous driving.

11.4.3 Adaptive Learning and Self-Improvement Using Neural Networks

One of the remarkable capabilities of neural networks is their capacity to adapt and improve through learning from data. In the context of AVs, adaptive learning using neural networks enables AVs to enhance their driving performance, decision-making, and overall behavior over time. This section explores how neural networks facilitate adaptive learning and self-improvement in AVs [38].

11.4.3.1 Continuous Learning from Driving Experience

Neural networks can learn from real-world driving experiences to refine their performance. As AVs accumulate data from various driving scenarios, neural networks analyze these data to adapt to different road conditions, traffic patterns, and user preferences. This continuous learning process allows AVs to improve their navigation, decision-making, and interaction with other road users.

11.4.3.2 Overcoming Challenging Situations Through Learning

Challenging scenarios, such as adverse weather conditions or complex traffic interactions, can be addressed through adaptive learning. Neural networks can be exposed to simulated or real-world data representing these challenging situations. By learning from these scenarios, AVs become better equipped to handle similar challenges in the future, making them more robust and capable of safe operation under diverse conditions.

11.4.3.3 *Behavior Prediction and Adaptation*

Neural networks can predict the behavior of other road users and adapt accordingly. By analyzing historical driving data, neural networks can identify patterns in how pedestrians, cyclists, and other vehicles behave. This predictive capability enables AVs to anticipate the intentions of other road users and adjust their actions in response, contributing to smoother and safer interactions.

11.4.3.4 *User-Centric Adaptation*

Neural networks can also adapt to individual passenger preferences and driving styles. Through continuous learning, AVs can learn about passengers' comfort preferences, desired speed levels, and preferred routes. This user-centric adaptation enhances the passenger experience and fosters greater user trust in AV technology.

11.4.3.5 *Balancing Safety and Efficiency*

Adaptive learning in AVs involves striking a balance between safety and efficiency. Neural networks must learn to make decisions that prioritize passenger safety while also ensuring efficient and smooth driving. The ability to adjust this balance based on the specific driving environment and passenger preferences is a key aspect of adaptive learning.

Incorporating adaptive learning using neural networks allows AVs to evolve and improve their performance over time, ultimately leading to safer, more reliable, and more user-friendly autonomous driving experiences.

11.5 Fuzzy Logic Control for Energy-Efficient Driving

The various activities in the Fuzzy logic control for energy-efficient driving system are shown in Figure 11.3 and explained in the following.

11.5.1 Fuzzy Logic–Based Energy Management

Efficient energy management is crucial for maximizing the driving range and overall performance of AEVs. Fuzzy logic offers a robust approach to optimizing energy consumption by taking into account various driving conditions, battery state, and user preferences. This section explores the application of fuzzy logic in energy management for AEVs [39].

11.5.1.1 Battery State-of-Charge Control

Maintaining an optimal state of charge (SoC) in the battery is essential for both performance and longevity. Fuzzy logic–based energy management systems monitor the battery's SoC and adjust power distribution to ensure that the battery remains within safe operating limits. By considering factors such as regenerative braking, driving speed, and terrain, these systems make intelligent decisions to prevent overcharging or excessive discharging [40].

11.5.1.2 Optimal Power Distribution for Efficiency

Fuzzy logic can optimize the distribution of power between the electric motor and the battery for maximum energy efficiency. The system considers factors such as driving conditions, traffic flow, and desired speed to determine the most efficient power allocation. Fuzzy logic allows for real-time adjustments to power distribution based on changing driving scenarios, contributing to reduced energy consumption and extended driving range.

11.5.1.3 Dynamic Load Management

Fuzzy logic–based energy management extends beyond the powertrain to manage other vehicle systems that consume energy, such as air conditioning and heating. By considering the vehicle's overall energy budget and user preferences, the system can dynamically adjust the operation of these systems to balance comfort and energy efficiency.

11.5.1.4 User Preferences and Adaptive Control

Fuzzy logic energy management systems can adapt to user preferences and driving styles. By analyzing historical driving data and considering inputs from the driver, the system can tailor its energy management strategy to align with the driver's behavior and preferences. This adaptive approach enhances user satisfaction while optimizing energy usage.

11.5.1.5 Integration with Traffic and Route Information

Fuzzy logic energy management systems can integrate with real-time traffic and route information. By considering traffic congestion, road gradients, and upcoming intersections, the system can adjust power distribution to ensure optimal energy usage throughout the journey. This integration improves both energy efficiency and driving experience [41].

Fuzzy logic-based energy management systems enable AEVs to make intelligent decisions regarding power distribution, battery SoC, and energy-consuming systems. By optimizing energy usage based on real-time inputs and dynamic conditions, these systems contribute to increased driving range, improved efficiency, and a more sustainable driving experience.

11.5.2 Fuzzy Logic–Based Adaptive Cruise Control

Adaptive Cruise Control (ACC) is a fundamental feature in AEVs that enhances driving comfort and safety by adjusting the vehicle's speed to maintain a safe following distance from the vehicle ahead. Fuzzy logic offers a versatile and adaptive approach to implementing ACC systems that can handle varying traffic conditions and user preferences. This section explores the application of fuzzy logic in ACC for AEVs [42].

11.5.2.1 Following Distance Regulation

Fuzzy logic–based ACC systems use sensor data, such as radar or LiDAR measurements, to assess the distance between the AEV and the lead vehicle. By considering factors such as traffic density and road conditions, the system adjusts the vehicle's speed to maintain a safe following distance. Fuzzy logic allows for flexible tuning of the following distance based on driver preferences and traffic dynamics.

11.5.2.2 Traffic Scenario Adaptation

In complex traffic scenarios, such as merging lanes or congested areas, fuzzy logic–based ACC systems can adapt their behavior to ensure smooth and efficient traffic flow. The system assesses the behavior of surrounding vehicles, road geometry, and traffic patterns to adjust the AEV's speed and following distance. Fuzzy logic allows for nuanced decision-making in challenging situations [43].

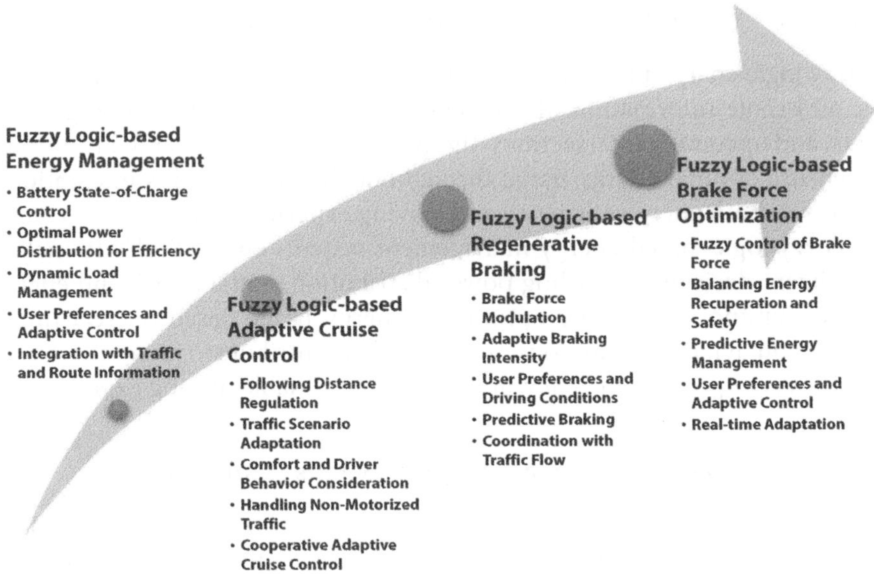

Figure 11.3 Fuzzy logic control for energy-efficient driving.

11.5.2.3 Comfort and Driver Behavior Consideration

Fuzzy logic–based ACC systems can be designed to consider driver comfort and behavior. By analyzing driver inputs, acceleration patterns, and braking tendencies, the system adapts its control strategy to align with the driver's driving style. This personalization enhances the overall driving experience and user satisfaction.

11.5.2.4 Handling Non-Motorized Traffic

Fuzzy logic–based ACC systems can handle interactions with non-motorized road users, such as pedestrians and cyclists. By analyzing sensor data and road context, the system adjusts its speed and following distance to accommodate the presence of non-motorized traffic, ensuring safe coexistence on the road.

11.5.2.5 Cooperative Adaptive Cruise Control

Cooperative ACC (CACC) involves communication between vehicles to achieve coordinated behavior. Fuzzy logic–based CACC systems can interpret data from connected vehicles to maintain consistent and efficient traffic flow. The system can adjust its speed based on information about the speed of surrounding vehicles, optimizing traffic synchronization [44].

Fuzzy logic–based ACC systems enable AEVs to navigate traffic with enhanced safety and comfort. By considering various factors and dynamically adapting to changing scenarios, these systems contribute to smoother traffic flow, reduced driver fatigue, and improved overall driving experience.

11.5.3 Fuzzy Logic–Based Regenerative Braking

Regenerative braking is a crucial aspect of electric vehicles that allows them to recover energy during deceleration and braking, thereby increasing overall energy efficiency. Fuzzy logic offers a sophisticated approach to optimizing regenerative braking systems by considering various driving conditions, road geometry, and user preferences. This section explores the application of fuzzy logic in regenerative braking for AEVs [45].

11.5.3.1 Brake Force Modulation

Fuzzy logic–based regenerative braking systems adjust the amount of regenerative braking force applied based on factors such as vehicle speed, road gradient, and battery SoC. The system optimizes the balance between regenerative braking and traditional friction braking to achieve efficient energy recovery while ensuring passenger comfort and safety.

11.5.3.2 Adaptive Braking Intensity

Fuzzy logic–based regenerative braking systems adapt the intensity of regenerative braking based on the driving scenario. For example, when descending a steep hill, the system can increase regenerative braking to maximize energy recovery. Conversely, in low-friction conditions, the system can reduce regenerative braking to prevent wheel lockup.

11.5.3.3 User Preferences and Driving Conditions

Fuzzy logic–based regenerative braking systems can be personalized based on user preferences and driving conditions. By analyzing historical driving data and driver inputs, the system tailors its regenerative braking strategy to align with the driver's behavior and preferences. This customization enhances the driving experience and user satisfaction.

11.5.3.4 *Predictive Braking*

Fuzzy logic–based regenerative braking systems can anticipate upcoming road conditions and intersections to optimize energy recovery. By considering GPS data, road maps, and sensor information, the system adjusts regenerative braking force in advance, ensuring efficient energy capture during planned deceleration events.

11.5.3.5 *Coordination with Traffic Flow*

Fuzzy logic–based regenerative braking systems can coordinate with traffic flow to achieve smoother driving and energy recovery. By analyzing data from surrounding vehicles and traffic signals, the system adjusts regenerative braking to align with traffic dynamics, contributing to improved traffic synchronization and energy efficiency.

Fuzzy logic–based regenerative braking systems empower AEVs to optimize energy recovery while maintaining driving comfort and safety. By adapting regenerative braking strategies to various driving scenarios and user preferences, these systems contribute to extended driving range, reduced energy consumption, and a more sustainable driving experience.

11.5.4 Fuzzy Logic–Based Brake Force Optimization for Maximizing Energy Recuperation

Optimizing brake force for energy recuperation is a critical aspect of electric vehicles' efficiency. Fuzzy logic provides a versatile approach to achieving optimal brake force modulation, enabling AEVs to maximize energy recuperation while ensuring safety and comfort. This section explores how fuzzy logic can be utilized to optimize brake force and enhance energy recuperation in AEVs [46].

11.5.4.1 *Fuzzy Control of Brake Force*

Fuzzy logic–based systems can dynamically adjust brake force based on a range of inputs, including vehicle speed, battery SoC, road conditions, and driver behavior. By analyzing these inputs and considering the desired energy recuperation level, the system determines the optimal amount of regenerative braking force to apply.

11.5.4.2 Balancing Energy Recuperation and Safety

Fuzzy logic enables AEVs to strike a balance between maximizing energy recuperation and ensuring passenger safety. The system considers factors such as vehicle dynamics, road surface conditions, and potential wheel lockup. By adapting brake force based on real-time conditions, the system prevents excessive energy recuperation that could compromise vehicle stability [47].

11.5.4.3 Predictive Energy Management

Fuzzy logic–based systems can predict upcoming road features, such as hills, curves, and intersections, to optimize energy recuperation. By analyzing GPS data, road maps, and sensor information, the system anticipates deceleration events and adjusts brake force in advance, enhancing energy capture during planned slowdowns.

11.5.4.4 User Preferences and Adaptive Control

Fuzzy logic systems can adapt brake force modulation to align with driver preferences and behavior. By learning from historical driving data and driver inputs, the system tailors its brake force strategy to accommodate varying driving styles. This personalized approach improves the driving experience and user satisfaction [48].

11.5.4.5 Real-Time Adaptation

Fuzzy logic–based brake force optimization operates in real-time, continuously adjusting brake force based on changing driving conditions. Whether driving in urban environments or on highways, the system dynamically adapts to traffic scenarios, terrain changes, and other factors to achieve optimal energy recuperation.

11.6 Integration of Neural Networks and Fuzzy Logic for Enhanced Autonomy

11.6.1 Combined Perception and Control using Neural-Fuzzy Systems

Neural-fuzzy systems combine the learning capabilities of neural networks with the reasoning abilities of fuzzy logic, creating a powerful framework for AEVs to perceive the environment and make intelligent decisions [49]. This integration offers several advantages.

11.6.1.1 Enhanced Perception

Neural-fuzzy systems can process sensor data using neural networks to extract meaningful features and patterns. Fuzzy logic then interprets this information, enabling the AEV to understand the environment in terms of linguistic variables. This enhanced perception facilitates more accurate object detection, recognition, and scene understanding [50]. Integration of neural networks and fuzzy logic for enhanced autonomy is shown in Figure 11.4.

11.6.1.2 Adaptive Decision-Making

By fusing neural networks and fuzzy logic, AEVs can make adaptive decisions in complex scenarios. Neural networks learn from vast datasets, capturing both typical and rare situations. Fuzzy logic handles uncertainty, allowing the system to reason and adapt to unexpected events. This combined approach empowers AEVs to handle a wide range of driving situations effectively.

11.6.1.3 Human-Like Reasoning

Fuzzy logic enables human-like reasoning by modeling imprecise and vague information. Neural networks contribute by learning from historical data and mimicking human intuition. The integration of these techniques creates a decision-making system that considers both precise data and qualitative knowledge, enhancing the AEV's ability to navigate dynamic environments.

11.6.1.4 Autonomous Learning and Improvement

Neural-fuzzy systems can continuously learn and improve over time. Neural networks refine their models using data, whereas fuzzy logic adapts rules based on new information. This autonomous learning ensures that the AEV remains up-to-date with changing road conditions, traffic patterns, and user preferences.

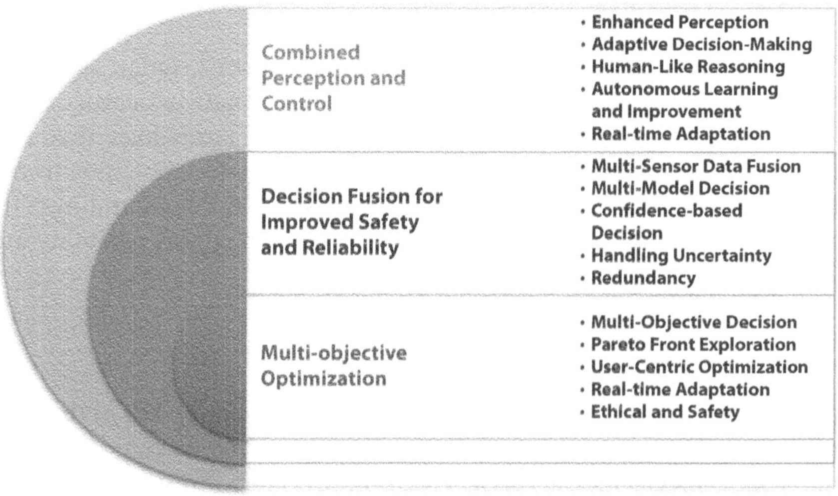

Figure 11.4 Integration of neural networks and fuzzy logic for enhanced autonomy.

11.6.1.5 Real-Time Adaptation

The integration of neural networks and fuzzy logic enables real-time adaptation. As conditions change, the system adjusts its perception, decision-making, and control strategies in a seamless manner. This real-time adaptability is crucial for safe and efficient autonomous driving [51].

The combination of neural networks and fuzzy logic in a unified neural-fuzzy system capitalizes on their respective strengths, resulting in AEVs with enhanced perception, adaptive decision-making, and human-like reasoning capabilities. The subsequent sections of this chapter will delve into more ways this integration can propel AEVs toward higher levels of autonomy and intelligence.

11.6.2 Decision Fusion for Improved Safety and Reliability

In the pursuit of advancing the safety and reliability of AEVs, the integration of neural networks and fuzzy logic extends beyond individual capabilities. Decision fusion, a powerful concept, involves combining outputs from multiple sources to make more informed and robust decisions. This section explores how decision fusion, facilitated by the integration of neural networks and fuzzy logic, can enhance the safety and reliability of AEVs [52].

11.6.2.1 Multi-Sensor Data Fusion

AEVs rely on various sensors such as cameras, LiDAR, radar, and ultrasonic sensors to perceive their environment. Neural networks process data from these sensors to extract valuable information. Fuzzy logic then synthesizes this information, considering uncertainty and ambiguity, to create a comprehensive and accurate representation of the surroundings. The fusion of data from multiple sensors enhances perception reliability and reduces the risk of misinterpretation.

11.6.2.2 Multi-Model Decision-Making

Neural networks and fuzzy logic can generate different models of the environment or driving scenarios. These models capture varying degrees of complexity, uncertainty, and risk. Decision fusion involves assessing these models' outputs and selecting the most appropriate course of action. By leveraging both techniques, AEVs can make decisions that balance safety, efficiency, and human-like reasoning.

11.6.2.3 Confidence-Based Decision Fusion

Fuzzy logic can assess the confidence level of neural network predictions and other inputs. By assigning degrees of belief to different sources of information, the system determines the reliability of each source. This confidence-based approach ensures that decisions are made based on the most reliable information, reducing the impact of erroneous inputs.

11.6.2.4 Handling Uncertain Situations

In scenarios with ambiguous or conflicting information, decision fusion becomes invaluable. Fuzzy logic can weigh the inputs and outputs from neural networks and other sources, considering the uncertainty inherent in each. This enables AEVs to make well-informed decisions, even in situations where no single source of information provides a clear solution [53].

11.6.2.5 Redundancy and Fault Tolerance

Decision fusion offers redundancy and fault tolerance. If one sensor or neural network encounters a malfunction or provides inaccurate information, then the integrated system can rely on other sources to compensate.

Fuzzy logic, with its ability to reason under uncertainty, aids in detecting anomalies and selecting appropriate alternative sources.

The integration of neural networks and fuzzy logic for decision fusion empowers AEVs to make more reliable and robust decisions. By leveraging the strengths of each technique, AEVs enhance their perception accuracy, decision-making adaptability, and overall safety, propelling them closer to achieving the levels of autonomy and reliability needed for widespread adoption [54].

11.6.3 Multi-Objective Optimization Using Hybrid Approaches

Achieving the optimal balance between various objectives is a complex challenge in AEVs, where safety, efficiency, and user experience are all critical considerations. Hybrid approaches that combine the strengths of neural networks and fuzzy logic offer a powerful means of tackling multi-objective optimization problems. This section explores how hybrid techniques can enhance decision-making in AEVs for improved overall performance [55–57].

11.6.3.1 Multi-Objective Decision Formulation

Neural networks can learn intricate patterns from vast datasets, whereas fuzzy logic can handle qualitative information and uncertainty. By formulating multi-objective decisions that encompass factors like energy efficiency, travel time, and passenger comfort, hybrid systems consider both quantitative and qualitative aspects for well-rounded optimization.

11.6.3.2 Pareto Front Exploration

Hybrid approaches facilitate the exploration of the Pareto front—the set of solutions that represent different trade-offs between objectives. Neural networks, trained on diverse data, can estimate potential outcomes, whereas fuzzy logic evaluates the desirability of these outcomes based on human-like reasoning. The integration of these techniques helps AEVs identify optimal solutions that cater to various preferences and requirements.

11.6.3.3 User-Centric Optimization

Hybrid techniques allow AEVs to consider user preferences and adapt their decisions accordingly. Neural networks learn from historical driving data to understand user behavior, whereas fuzzy logic captures imprecise user preferences. This user-centric approach ensures that optimization aligns with individual expectations, enhancing user satisfaction.

11.6.3.4 Real-Time Adaptation

The integration of neural networks and fuzzy logic enables real-time adaptation to changing conditions. As AEVs encounter different traffic scenarios, road conditions, and user requirements, hybrid systems adjust their optimization strategies seamlessly. This real-time adaptability ensures optimal decision-making in dynamic environments.

11.6.3.5 Ethical and Safety Considerations

Multi-objective optimization often involves ethical and safety considerations. Hybrid systems can integrate neural networks to analyze potential risks and outcomes, whereas fuzzy logic evaluates the moral implications of different choices. This fusion of techniques enables AEVs to make decisions that prioritize safety, ethics, and overall system objectives.

By combining the capabilities of neural networks and fuzzy logic, hybrid approaches enable AEVs to navigate the complex landscape of multi-objective optimization. These techniques empower vehicles to make informed decisions that balance diverse objectives, ultimately leading to safer, more efficient, and user-centric autonomous driving experiences.

11.7 Case Studies and Applications

This section delves into real-world case studies and applications that demonstrate the practical implementation and benefits of the integration of neural networks and fuzzy logic in AEVs. These examples showcase how these techniques can enhance various aspects of AEV operations and urban mobility.

11.7.1 Autonomous Electric Fleet Management

Autonomous electric fleets present an opportunity to revolutionize transportation systems. By combining neural networks and fuzzy logic, fleet management can be optimized to improve traffic flow, route planning, and overall efficiency [23, 24, 58].

11.7.1.1 Traffic Flow Optimization

Neural networks analyze traffic data to predict congestion and potential bottlenecks. Fuzzy logic interprets this data, considering uncertain factors such as weather and road conditions. By integrating these techniques, fleet management systems can dynamically adjust routes and dispatch vehicles to minimize congestion and maintain smooth traffic flow.

11.7.1.2 Dynamic Route Planning

Fuzzy logic can consider real-time traffic updates and user preferences to adapt routes dynamically. Neural networks learn from historical data to predict likely traffic patterns. This hybrid approach enables fleet managers to plan efficient routes that account for both immediate conditions and long-term trends.

11.7.2 Urban Mobility Solutions and Ride-Sharing

Autonomous electric vehicles are poised to reshape urban mobility and ride-sharing services. The integration of neural networks and fuzzy logic contributes to enhancing user experience and optimizing fleet utilization [43, 53, 54].

11.7.2.1 User Experience Enhancement

Neural networks capture user preferences and behaviors from past trips, whereas fuzzy logic interprets qualitative feedback. By combining these insights, ride-sharing systems can personalize the in-car experience, adjusting climate control, entertainment, and lighting to match individual preferences.

11.7.2.2 Fleet Utilization Optimization

Neural networks analyze historical demand data, whereas fuzzy logic handles uncertain factors such as events and sudden changes in demand. This integration empowers ride-sharing services to optimize fleet deployment, ensuring that vehicles are distributed efficiently to meet passenger demand while minimizing idle time.

These case studies underscore the practical impact of integrating neural networks and fuzzy logic in AEVs. By optimizing fleet management, enhancing user experiences, and contributing to urban mobility solutions, these techniques showcase the potential for AEVs to revolutionize transportation and improve the quality of life for individuals and communities.

11.8 Future Prospects and Challenges

Looking ahead, the integration of neural networks and fuzzy logic holds great promise for the future of AEVs. However, this integration also presents certain challenges and considerations that need to be addressed to ensure safe, reliable, and successful deployment [8].

11.8.1 Ethical and Legal Considerations

As AEVs become more autonomous, ethical dilemmas arise. Neural networks and fuzzy logic may need to navigate complex moral decisions, such as the "trolley problem." Additionally, legal frameworks must evolve to accommodate autonomous systems and their decision-making mechanisms, ensuring accountability and liability in case of accidents.

11.8.2 Scalability and Real-World Deployment Challenges

While neural networks and fuzzy logic offer powerful capabilities, ensuring their scalability to handle the vast array of driving scenarios is crucial. Developing models that can generalize across diverse conditions is challenging. Furthermore, deploying these systems in real-world environments requires addressing issues related to sensor variability, hardware limitations, and edge cases [25, 26, 58].

11.8.3 Emerging Trends in Control and Autonomy

The field of control and autonomy is rapidly evolving. New trends, such as reinforcement learning and neuro-fuzzy systems, are emerging to further enhance AEV capabilities. Integrating these trends with existing neural networks and fuzzy logic approaches can lead to more sophisticated and adaptable autonomous systems. In navigating these future prospects and challenges, continued research, collaboration among various disciplines, and adherence to safety standards are imperative. The fusion of neural networks and fuzzy logic offers tremendous potential to revolutionize transportation, but a holistic and cautious approach is necessary to harness this potential effectively [2, 6].

11.9 Conclusions

The integration of neural networks and fuzzy logic marks a significant leap forward in the development of AEVs. This synergistic combination unlocks a realm of possibilities, enhancing perception, decision-making, and control capabilities and propelling AEVs toward higher levels of autonomy and intelligence. Neural networks excel in learning complex patterns from data, whereas fuzzy logic adeptly handles uncertainty and qualitative reasoning. Together, they create a harmonious partnership that empowers AEVs to navigate diverse and dynamic road scenarios, making decisions that balance safety, efficiency, and user preferences. As AEVs evolve, the potential applications of this integration are vast. From energy management to ACC, from regenerative braking to multi-objective optimization, the collaboration between neural networks and fuzzy logic paves the way for a safer, greener, and more user-centric transportation landscape. Yet, this journey is not without its challenges. Ethical dilemmas, legal frameworks, scalability, and real-world deployment pose intricate hurdles that require thoughtful consideration. The field of control and autonomy is in constant flux, with emerging trends like reinforcement learning reshaping its landscape.

In conclusion, the fusion of neural networks and fuzzy logic is a transformative force that ushers in a new era of autonomy for electric vehicles. By capitalizing on their complementary strengths and addressing challenges with prudence, AEVs have the potential to redefine transportation, making it safer, more efficient, and more accessible for all.

List of Abbreviations

AEVs	Autonomous Electric Vehicles
ACC	Adaptive Cruise Control
GPS	Global Positioning System
LiDAR	Light Detection and Ranging
SoC	State of Charge
CACC	Cooperative Adaptive Cruise Control

References

1. Qiu, W., Hao, Q., Annamareddy, S.H.K., Xu, B., Guo, Z., Jiang, Q., Electric vehicle revolution and implications: ion battery and energy. *Eng. Sci.*, 20, 100–109, 2022.

2. Hawkins, J. and Nurul Habib, K., Integrated models of land use and transportation for the autonomous vehicle revolution. *Transp. Rev.*, 39, 1, 66–83, 2019.
3. Fredriksson, G., Roth, A., Tagliapietra, S., Veugelers, R., Is the European automotive industry ready for the global electric vehicle revolution? Bruegel Policy Contribution, 2018.
4. Porter, L. *et al.*, The autonomous vehicle Revolution: Implications for planning/The driverless city?/autonomous vehicles–a planner's response/autonomous vehicles: Opportunities, challenges and the need for government action/three signs autonomous vehicles will not lead to less car ownership and less car use in car dependent cities–a case study of Sydney, Australia/planning for autonomous vehicles? Questions of purpose, place and pace/ensuring good governance: The role of planners in the development of autonomous vehicles/putting technology in its place. *Plan. Theory Pract.*, 19, 5, 753–778, 2018.
5. Pizzuto, L., Thomas, C., Wang, A., Wu, T., and others, How China will help fuel the revolution in autonomous vehicles, *McKinsey & Company*, 11, 1–13, 2019.
6. Banerjee, S., Autonomous vehicles: a review of the ethical, social and economic implications of the AI revolution. *Int. J. Intell. Unmanned Syst.*, 9, 4, 302–312, 2020.
7. Morris, M. *et al.*, Working Paper 1: Electricity storage and electric vehicles. *Policy & Regulatory Landscape Review Series*, 1, 1–46, 2019.
8. Roberts, C., Easy Street for Low-Carbon Mobility? The Political Economy of Mass Electric Car Adoption, in: *Electrifying Mobility: Realising a Sustainable Future for the Car*, pp. 13–31, Emerald Publishing Limited, Leeds LS1 4DL, United Kingdom, 2022.
9. Babu, B., Suresh, J., Kamalakannan, N., Meenatchi, S., Karthik, S., Boopathi, S., Economic impacts and reliability evaluation of battery by adopting Electric Vehicle. In: *International Conference on Power, Energy, Control and Transmission Systems (ICPECTS)*, IEEE, vol. 44, pp. 1–6. 2022.
10. Shaikh, P.W. and Mouftah, H.T., Intelligent charging infrastructure design for connected and autonomous electric vehicles in smart cities, in: *2021 IFIP/IEEE International Symposium on Integrated Network Management (IM)*, IEEE, pp. 992–997, 2021.
11. El Arayshi, M., Amato, G., Pischiutta, F., A new monitoring strategy and automatic speed reduction for PI slip control in autonomous electric vehicles with in-wheel motors: Experimental results, in: *2022 International Symposium on Power Electronics, Electrical Drives, Automation and Motion (SPEEDAM)*, IEEE, pp. 820–825, 2022.
12. Liu, X., Zhao, Y., Sadek, A., Qiao, C., Autonomous Electric Vehicles as Mobile Green Energy Sources, in: *2023 IEEE 24th International Conference on High Performance Switching and Routing (HPSR)*, IEEE, pp. 210–216, 2023.

13. Kovacs, A. and Vajk, I., Integrated Path Planning and Lateral-Longitudinal Control for Autonomous Electric Vehicles, in: *2021 AEIT International Conference on Electrical and Electronic Technologies for Automotive (AEIT AUTOMOTIVE)*, IEEE, pp. 1–6, 2021.
14. Abou Setta, I.G., Shehata, O.M., Awad, M.A., Multivariate prediction of correct lane for autonomous electric vehicle using deep learning models, in: *2020 8th International Conference on Control, Mechatronics and Automation (ICCMA)*, IEEE, pp. 127–130, 2020.
15. Lee, J. and Wang, L., A method for designing and analyzing automotive software architecture: A case study for an autonomous electric vehicle, in: *2021 International Conference on Computer Engineering and Artificial Intelligence (ICCEAI)*, IEEE, pp. 20–26, 2021.
16. Flynn, J. and Giannetti, C., Using convolutional neural networks to map houses suitable for electric vehicle home charging. *AI*, 2, 1, 135–149, 2021.
17. Chandrika, V., Sivakumar, A., Krishnan, T. S., Pradeep, J., Manikandan, S., Boopathi, Theoretical Study on Power Distribution Systems for Electric Vehicles, in: *Intelligent Engineering Applications and Applied Sciences for Sustainability*, IGI Global, pp. 1–19, 2023.
18. Lazaroiu, G.C. and Roscia, M., Fuzzy logic strategy for priority control of electric vehicle charging. *IEEE Trans. Intell. Transp. Syst.*, 23, 10, 19236–19245, 2022.
19. Liu, Y., Huang, B., Yang, Y., Lei, Z., Zhang, Y., Chen, Z., Hierarchical speed planning and energy management for autonomous plug-in hybrid electric vehicle in vehicle-following environment. *Energy*, 260, 125212, 2022.
20. Younes, D., Boudiaf, M., Karim, N., Energy management based hybrid fuel cell/battery for electric vehicle using type 2 fuzzy logic controller. *Int. J. Adv. Stud. Comp. Sci. Eng.*, 12, 1, 18–23, 2023.
21. Halima, N.B., Hadj, N.B., Chaieb, M., Neji, R., Energy management of parallel hybrid electric vehicle based on fuzzy logic control strategies. *J. Circuits Syst. Comput.*, 32, 01, 2350007, 2023.
22. Singh, S. and Saini, B.S., Autonomous cars: Recent developments, challenges, and possible solutions. *IOP Conf. Ser.: Mater. Sci. Eng.*, 012028, IOP Publishing, 1022, 1, 2021.
23. Damaj, I.W., Serhal, D.K., Hamandi, L.A., Zantout, R.N., Mouftah, H.T., Connected and autonomous electric vehicles: Quality of experience survey and taxonomy. *Veh. Commun.*, 28, 100312, 2021.
24. Rassõlkin, A., Vaimann, T., Kallaste, A., Kuts, V., Digital twin for propulsion drive of autonomous electric vehicle, in: *2019 IEEE 60th International Scientific Conference on Power and Electrical Engineering of Riga Technical University (RTUCON)*, IEEE, pp. 1–4, 2019.
25. Venkitaraman, A.K. and Kosuru, V.S.R., A review on autonomous electric vehicle communication networks-progress, methods and challenges. *World J. Adv. Res. Rev.*, 16, 3, 013–024, 2022.

26. Nascimento, A.M. *et al.*, A systematic literature review about the impact of artificial intelligence on autonomous vehicle safety. *IEEE Trans. Intell. Transp. Syst.*, 21, 12, 4928–4946, 2019.
27. Quan, H., Khosravi, A., Yang, D., Srinivasan, D., A survey of computational intelligence techniques for wind power uncertainty quantification in smart grids. *IEEE Trans. Neural Networks Learn. Syst.*, 31, 11, 4582–4599, 2019.
28. Elsisi, M. and Ebrahim, M.A., Optimal design of low computational burden model predictive control based on SSDA towards autonomous vehicle under vision dynamics. *Int. J. Intell. Syst.*, 36, 11, 6968–6987, 2021.
29. Rajawat, A.S., Rawat, R., Barhanpurkar, K., Shaw, R.N., Ghosh, A., Vulnerability analysis at industrial internet of things platform on dark web network using computational intelligence, in: *Computationally intelligent systems and their applications*, pp. 39–51, 2021.
30. Olayode, O., Tartibu, L., Okwu, M., Application of Artificial Intelligence in Traffic Control System of Non-autonomous Vehicles at Signalized Road Intersection. *Proc. CIRP*, 91, 194–200, 2020.
31. Del Ser, J., Osaba, E., Sanchez-Medina, J.J., Fister, I., Bioinspired computational intelligence and transportation systems: A long road ahead. *IEEE Trans. Intell. Transp. Syst.*, 21, 2, 466–495, 2019.
32. Fatemimoghadam, A., Yan, Y., Iyer, L.V., Kar, N.C., Permanent Magnet Synchronous Motor Drive Using Deep-Neural-Network-Based Vector Control for Electric Vehicle Applications, in: *2022 International Conference on Electrical Machines (ICEM)*, IEEE, pp. 2358–2364, 2022.
33. Xie, J., Xu, X., Wang, F., Tang, Z., Chen, L., Adaptive coordination control strategy for path-following of DDAEV with neural network based moving weight coefficients. *Trans. Inst. Meas. Control*, 45, 7, 1282–1297, 2023.
34. Eagon, M.J., Kindem, D.K., Selvam, H.P., Northrop, W.F., Neural Network-Based Electric Vehicle Range Prediction for Smart Charging Optimization. *J. Dyn. Syst. Meas. Contr.*, 144, 1, 011110, 2022.
35. El Hajjami, L., Mellouli, E.M., Berrada, M., Neural network based sliding mode lateral control for autonomous vehicle, in: *2020 1st International Conference on Innovative Research in Applied Science, Engineering and Technology (IRASET)*, IEEE, pp. 1–6, 2020.
36. Swain, S.K., Rath, J.J., Veluvolu, K.C., Neural network based robust lateral control for an autonomous vehicle. *Electronics*, 10, 4, 510, 2021.
37. Vodovozov, V., Aksjonov, A., Petlenkov, E., Raud, Z., Neural network-based model reference control of braking electric vehicles. *Energies*, 14, 9, 2373, 2021.
38. Zavala, E., Franch, X., Marco, J., Berger, C., Adaptive monitoring for autonomous vehicles using the HAFLoop architecture. *Enterp. Inf. Syst.*, 15, 2, 270–298, 2021.
39. Parra, A., Tavernini, D., Gruber, P., Sorniotti, A., Zubizarreta, A., Pérez, J., On nonlinear model predictive control for energy-efficient torque-vectoring. *IEEE Trans. Veh. Technol.*, 70, 1, 173–188, 2020.

40. Wang, J., Gao, S., Wang, K., Wang, Y., Wang, Q., Wheel torque distribution optimization of four-wheel independent-drive electric vehicle for energy efficient driving. *Control Eng. Pract.*, 110, 104779, 2021.
41. Štefek, A., Pham, V.T., Krivanek, V., Pham, K.L., Optimization of fuzzy logic controller used for a differential drive wheeled mobile robot. *Appl. Sci.*, 11, 13, 6023, 2021.
42. Kumar, M., Kumar, B., Rani, A., Fuzzy Logic Based Optimal Flux Estimator for Energy Efficient Operation of Electric Vehicle, in: *Proceedings of International Conference on Sustainable Computing in Science, Technology and Management (SUSCOM)*, Amity University Rajasthan, Jaipur-India, 2019.
43. Parra, A., Zubizarreta, A., Pérez, J., An energy efficient intelligent torque vectoring approach based on fuzzy logic controller and neural network tire forces estimator. *Neural Comput. Appl.*, 33, 15, 9171–9184, 2021.
44. Guo, J., Li, W., Wang, J., Luo, Y., Li, K., Safe and energy-efficient car-following control strategy for intelligent electric vehicles considering regenerative braking. *IEEE Trans. Intell. Transp. Syst.*, 23, 7, 7070–7081, 2021.
45. Prince, A., Hati, S., Chakrabarti, P., Abawajy, J.H., Keong, N.W., Development of energy efficient drive for ventilation system using recurrent neural network. *Neural Comput. Appl.*, 33, 14, 8659–8668, 2021.
46. Yang, J., Zhang, T., Hong, J., Zhang, H., Zhao, Q., Meng, Z., Research on driving control strategy and Fuzzy logic optimization of a novel mechatronics-electro-hydraulic power coupling electric vehicle. *Energy*, 233, 121221, 2021.
47. Garg, R., Sharma, A., Yadav, K., Kaur, M., Rana, A., Fuzzy Logic Based Power Train Controller For Different Hybrid Electric Vehicle Configurations, in: *2023 3rd International Conference on Intelligent Technologies (CONIT)*, IEEE, pp. 1–5, 2023.
48. Bathala, P., Srivasthav, B.S., Reddy, Y.S.S., Deepa, K., Estimation of regenerative braking force in electric vehicles for maximum energy recovery, in: *2020 IEEE 17th India Council International Conference (INDICON)*, IEEE, pp. 1–7, 2020.
49. Singh, K.V., Bansal, H.O., Singh, D., Feed-forward modeling and real-time implementation of an intelligent fuzzy logic-based energy management strategy in a series–parallel hybrid electric vehicle to improve fuel economy. *Electr. Eng.*, 102, 967–987, 2020.
50. Arunprasad, V., Gupta, B., Karthikeyan, T., Ponnusamy, M., Hybrid neuro-fuzzy-genetic algorithms for optimal control of autonomous systems. *ICTACT J. Soft Comput.*, 13, 3015–3020, 2023.
51. Maged, N.A., Hasanien, H.M., Ebrahim, E.A., Tostado-Véliz, M., Turky, R.A., Jurado, F., Optimal Real-time implementation of fuzzy logic control strategy for performance enhancement of autonomous microgrids. *Int. J. Electr. Power Energy Syst.*, 151, 109140, 2023.

52. Ibrahim, A., Abosekeen, A., Azouz, A., Noureldin, A., Enhanced Autonomous Vehicle Positioning Using a Loosely Coupled INS/GNSS-Based Invariant-EKF Integration. *Sensors*, 23, 13, 6097, 2023.
53. Shafieenejad, I., Rouzi, E.D., Sardari, J., Araghi, M.S., Esmaeili, A., Zahedi, S., Fuzzy logic, neural-fuzzy network and honey bees algorithm to develop the swarm motion of aerial robots. *Evol. Syst.*, 13, 2, 319–330, 2022.
54. Taghavifar, H. and Rakheja, S., Path-tracking of autonomous vehicles using a novel adaptive robust exponential-like-sliding-mode fuzzy type-2 neural network controller. *Mech. Syst. Signal Process.*, 130, 41–55, 2019.
55. Boopathi, S., Experimental investigation and multi-objective optimization of cryogenic Friction-stir-welding of AA2014 and AZ31B alloys using MOORA technique. *Elsevier: Mater. Today Commun.*, 33, 104937, 2022.
56. Boopathi, S., Sasikumar, C., Myilsamy, S., Application of TOPSIS Optimization Technique in the Micro-Machining Process. In: *Trends, Paradigms, and Advances in Mechatronics Engineering*, IGI Global, pp. 162–187, 2023.
57. Boopathi, S. and Sivakumar, K., Experimental investigation and parameter optimization of near-dry wire-cut electrical discharge machining using multi-objective evolutionary algorithm. *Springer: Int. J. Adv. Manuf. Technol.*, 67, 9–12, 2639–2655, 2013.
58. Vdovic, H., Babic, J., Podobnik, V., Automotive software in connected and autonomous electric vehicles: A review. *IEEE Access*, 7, 166365–166379, 2019.

12

Advancing Mechatronics Through Artificial Intelligence

Pawan Whig[1]*, Jhansi Bharathi Madavarapu[2], Venugopal Reddy Modhugu[3], Balaram Yadav Kasula[4] and Ashima Bhatnagar Bhatia[1]

[1]*Vivekananda Institute of Professional Studies - Technical Campus, Delhi, India*
[2]*Information Technology, University of the Cumberlands, Williamsburg, KY, USA*
[3]*Independent Researcher, Fairfax, USA*
[4]*University of The Cumberlands, Williamsburg, KY, USA*

Abstract

This book chapter explores the transformative impact of integrating artificial intelligence (AI) techniques within the field of mechatronics, showcasing the synergistic potential that emerges when advanced AI technologies are harnessed to enhance the functionality, adaptability, and intelligence of mechatronic systems. Through an examination of key case studies and cutting-edge research, this chapter highlights how AI-driven approaches are revolutionizing various aspects of mechatronics, from autonomous robotics and smart manufacturing to real-time decision-making and predictive maintenance. By delving into the convergence of AI and mechatronics, this chapter provides insights into the challenges, opportunities, and future directions of this dynamic interdisciplinary frontier.

Keywords: Artificial intelligence, mechatronics, cognitive integration, autonomous systems, robotics, smart manufacturing, decision-making

12.1 Introduction

In the rapidly evolving landscape of technology, the convergence of artificial intelligence (AI) and mechatronics has emerged as a driving force behind the development of intelligent and adaptive systems. Mechatronics,

**Corresponding author*: pawanwhig@gmail.com

Kolla Bhanu Prakash, Satish Kumar Peddapelli, Ivan C. K. Tam, Wai Lok Woo and Vishal Jain (eds.) Computational Intelligent Techniques in Mechatronics, (381–400) © 2024 Scrivener Publishing LLC

an interdisciplinary field that amalgamates mechanics, electronics, and computer science, has traditionally focused on creating machines and systems with enhanced functionality through the integration of these domains. On the other hand, AI, particularly with the advancements in machine learning and deep learning, has enabled machines to simulate human-like cognitive functions such as perception, reasoning, learning, and decision-making. The marriage of these two realms, AI and mechatronics, gives rise to a new era where machines not only possess physical capabilities but also demonstrate cognitive intelligence, enabling them to interact with their environment and adapt to changing conditions with remarkable proficiency.

12.1.1 Background and Motivation

Historically, mechatronics has been responsible for remarkable innovations, from manufacturing robots that enhance industrial productivity to advanced medical devices that improve healthcare outcomes [1]. However, the capabilities of traditional mechatronic systems are often constrained by pre-programmed responses and limited adaptability. As environments become more complex and dynamic, the need for systems that can perceive, learn, and make decisions autonomously becomes increasingly evident [2–5]. This is where the integration of AI techniques becomes a game changer.

Artificial intelligence, particularly machine learning and neural networks, has demonstrated remarkable prowess in tasks like image and speech recognition, natural language understanding, and even strategic gameplay. These accomplishments have raised the question: Can these AI techniques be harnessed to augment mechatronic systems and imbue them with intelligence that goes beyond mere programming? The answer is a resounding yes.

12.1.2 Scope and Objectives of the Chapter

This chapter delves into the intricate fusion of AI and mechatronics, aiming to provide a comprehensive understanding of the transformative potential this integration holds. The primary objective is to showcase how AI, by endowing mechatronic systems with cognitive capabilities, is reshaping industries and applications that range from manufacturing and robotics to healthcare and transportation. We will explore the synergies between AI and mechatronics through illustrative case studies, emphasizing the remarkable achievements and technological advancements that result from this fusion. The literature review with main focus is shown in Table 12.1.

Table 12.1 Literature review.

Reference	Title	Journal/ Conference	Main focus
Kuru, K., Ansell, D., Hughes, D., Watkinson, B. J., Gaudenzi, F., Jones, & DeGoede, C. (2023)	Treatment of nocturnal enuresis using miniaturised advanced mechatronics with artificial intelligence	IEEE Journal of Translational Engineering in Health and Medicine	Application of advanced mechatronics with AI in healthcare for treating nocturnal enuresis
Kuru, K., & Yetgin, H. (2019)	Transformation to advanced mechatronics systems within new industrial revolution: A novel framework in automation of everything (AoE)	IEEE Access	Framework for integrating advanced mechatronics in the industrial automation paradigm
Demir, H., & Sarı, F. (2020)	The Effect of Artificial Intelligence and Industry 4.0 on Robotic Systems	Engineering on Energy Materials, Iksad Publications	Exploration of the impact of AI and Industry 4.0 on robotic systems
Tanie, K. (1999)	Mechatronics-Past, present and future	1999 IEEE/ ASME International Conference on Advanced Intelligent Mechatronics	Historical overview and future prospects of mechatronics

(*Continued*)

Table 12.1 Literature review. (*Continued*)

Reference	Title	Journal/ Conference	Main focus
Nguyen, A. T., Dinh, T. Q., Chong, J., Iwasaki, M., Precup, R. E., & Ruderman, M. (2023)	Guest editorial introduction to the special issue on "Emerging control and automation technologies for advanced mechatronic systems"	Control Engineering Practice	Introduction to special issue focusing on control and automation of advanced mechatronic systems
Rzevski, G. (2003)	On conceptual design of intelligent mechatronic systems	Mechatronics	Conceptual design of intelligent mechatronic systems
Mohebbi, A., Achiche, S., & Baron, L. (2019)	Integrated and concurrent detailed design of a mechatronic quadrotor system using a fuzzy-based particle swarm optimization	Engineering Applications of Artificial Intelligence	Detailed design optimization of a mechatronic quadrotor system using fuzzy-based particle swarm optimization
Hingu, P. R., & Panchal, D. N.	Industry 4.0–Advance To the Mechatronic and Management with Approach of Artificial Intelligence	-	Application of Industry 4.0 and AI in mechatronics and management

(*Continued*)

Table 12.1 Literature review. (*Continued*)

Reference	Title	Journal/ Conference	Main focus
Dirican, C. (2015)	The impacts of robotics, artificial intelligence on business and economics	Procedia- Social and Behavioral Sciences	Examination of the impacts of robotics and AI on business and economics
Vladareanu, L. (2020)	Advanced intelligent control through versatile intelligent portable platforms	Sensors	Application of advanced intelligent control using portable platforms
Shi, Y., Shen, C., Fang, H., & Li, H. (2017)	Advanced control in marine mechatronic systems: A survey	IEEE/ASME transactions on mechatronics	Survey of advanced control methods in marine mechatronic systems

Furthermore, this chapter will delve into the challenges and opportunities that arise when marrying AI with mechatronics. Technical challenges include ensuring the reliability and safety of AI-driven systems, addressing the inherent uncertainties in dynamic environments, and creating methods for explainable AI that can provide insights into decision-making processes. Additionally, ethical considerations come to the forefront, as intelligent machines raise questions about accountability, privacy, and human-AI collaboration. As we peer into the future, the chapter will conclude by discussing emerging trends and potential trajectories for this interdisciplinary field. From collaborative human–machine systems that leverage the strengths of both AI and mechatronics to the emergence of ethical and responsible AI, the possibilities are both exciting and daunting.

In essence, this chapter strives to uncover the layers of potential that unravel when AI and mechatronics intertwine. By exploring the foundations, exemplifying real-world successes, and dissecting challenges, we aim to equip readers with a holistic perspective on the realm of AI-driven mechatronics. As this integration propels us into an era of unparalleled innovation, understanding its implications becomes not just a matter of technological curiosity, but a necessity for envisioning the future of intelligent machines and systems. In the following sections, we will build upon this introduction by delving into the foundational concepts of both mechatronics and AI, elucidating how their confluence leads to cognitive integration, and exploring the transformative potential through real-world case studies and future trajectories.

12.2 Foundations of Mechatronics and Artificial Intelligence

To embark on a journey into the integration of AI and mechatronics, it is essential to first understand the bedrock upon which these two disciplines are built. Mechatronics, at its core, represents a crossroads where mechanics, electronics, and computer science converge to create intelligent and interactive systems [6]. On the other hand, AI, particularly in recent years, has revolutionized the way machines process information and make decisions, leading to advancements that mirror human cognitive abilities. As we delve into the foundations of these two fields, we begin to see the threads that, when woven together, create a fabric of innovation, adaptability, and intelligence [7].

12.2.1 Mechatronics: Where Physicality Meets Computation

Mechatronics is a discipline born out of the recognition that complex systems, especially those involving mechanical components, can achieve new levels of functionality and performance when coupled with electronic and computational elements [8–10]. It is an interdisciplinary endeavor that capitalizes on the synergy between mechanics, electronics, control theory, and computer science. The foundational principles of mechatronics encompass the design, analysis, and control of systems that combine mechanical components with sensors, actuators, and embedded computing systems.

Mechatronic systems often span a wide range of applications. From industrial robotics that execute precision tasks on manufacturing assembly lines to medical devices that assist in surgeries, mechatronics has led to transformative breakthroughs [11–15]. The integration of sensors allows machines to perceive their environment, whereas actuators facilitate physical interaction with it. Control algorithms, enabled by embedded computing, govern the behavior of the system in real time, making it adaptable to varying conditions.

12.2.2 Artificial Intelligence: Pinnacle of Machine Cognition

Artificial intelligence, while a concept that dates back decades, has truly come into its own in recent years, thanks, in large part, to the advancements in machine learning and deep learning. AI refers to the capability of machines to imitate human-like cognitive functions such as learning from data, reasoning through complex problems, recognizing patterns, and making decisions. It is about imparting intelligence to machines so that they can perform tasks that previously demanded human intervention.

At the heart of AI's recent successes are neural networks—computational models inspired by the structure of the human brain. Machine learning techniques, powered by large datasets and sophisticated algorithms, enable neural networks to learn and generalize from examples. This has led to breakthroughs in image and speech recognition, natural language processing, and even tasks like playing complex games.

12.2.3 Confluence of Forces: Cognitive Integration

The integration of mechatronics and AI brings about a new paradigm of cognitive integration. It is not just about endowing machines with physical capabilities; it is about instilling them with cognitive abilities that enable them to perceive, learn, reason, and adapt. This convergence opens doors to systems that can sense their environment, make decisions based on data analysis, and optimize their actions to achieve desired outcomes.

Consider an autonomous drone equipped with AI-driven computer vision. Such a drone can navigate through complex environments, avoiding obstacles, recognizing targets, and adapting its flight path based on real-time data. The marriage of mechatronics and AI empowers this drone to perform tasks that require both physical dexterity and cognitive acumen, which would have been unimaginable in the absence of this integration.

12.2.4 Toward Intelligent Autonomy

The potential of AI and mechatronics integration extends beyond individual systems. Entire ecosystems of interconnected intelligent devices can collaborate to achieve objectives that were once reserved for human operators. Smart factories, for instance, embody this synergy by integrating AI-driven predictive maintenance, robotic automation, and data analytics to optimize manufacturing processes.

As we witness the emergence of AI-driven autonomous vehicles, we see the pinnacle of cognitive integration. These vehicles navigate, make decisions, and adapt to road conditions, seamlessly combining mechatronic control with AI-powered perception and decision-making.

The foundations of mechatronics and AI, although rooted in distinct disciplines, converge in a manner that redefines what machines are capable of achieving. Mechatronics gives machines the physical capabilities to interact with the world, whereas AI imparts them with cognitive prowess. This union gives rise to systems that learn from data, reason through complex scenarios, and autonomously adapt to changing environments. As we progress through this chapter, we will delve deeper into the synergistic potential of AI and mechatronics, exploring how this confluence shapes industries, advances technologies, and paves the way for a future where intelligence and adaptability coexist in machines like never before.

12.3 Synergies Between Artificial Intelligence and Mechatronics

In the realm of technological advancement, the fusion of AI and mechatronics represents a harmonious marriage that transcends the limitations of individual disciplines. The synergy between these two fields not only enhances the capabilities of machines but also imbues them with a level of intelligence and adaptability that opens new frontiers of possibilities. This chapter delves into the intricate web of synergies that emerge when AI and mechatronics interweave, showcasing how this convergence catalyzes breakthroughs in diverse domains, from autonomous robotics to smart manufacturing and beyond.

12.3.1 Enhanced Adaptability Through AI

At the heart of the synergies between AI and mechatronics lies the concept of adaptability. Traditional mechatronic systems, while proficient in predefined tasks, often struggle to cope with dynamic and evolving environments. AI, with its ability to process vast amounts of data and learn patterns, offers a solution to this limitation. By infusing mechatronic systems with AI-driven learning algorithms, machines become capable of adapting to changing conditions, identifying anomalies, and optimizing their actions in real time.

Consider the case of an AI-powered industrial robot. Equipped with sensors that provide real-time feedback, this robot can dynamically adjust its movements to ensure precision in manufacturing processes. If a new component is introduced or if there are variations in the production line, then the robot can autonomously recalibrate its actions based on the data it gathers, ensuring consistent performance without the need for manual reprogramming.

12.3.2 Real-Time Decision-Making and Control

One of the most remarkable outcomes of the AI-mechatronics integration is the empowerment of machines to make informed decisions in real time. Traditional mechatronic systems rely on pre-programmed instructions, limiting their responses to anticipated scenarios. However, when AI enters the picture, machines can process data from sensors and external sources, interpret the information, and make decisions that optimize outcomes.

In autonomous vehicles, this synergy is strikingly evident. A self-driving car navigates through complex traffic scenarios by constantly analyzing data from cameras, lidar, radar, and GPS. These data are processed by AI algorithms that recognize objects, predict trajectories, and make split-second decisions to ensure the vehicle's safe passage. The real-time nature of these decisions exemplifies the transformative impact of combining AI's cognitive abilities with mechatronics' control mechanisms.

12.3.3 Cognitive Robotics and Autonomous Systems

Perhaps one of the most intriguing outcomes of the AI-mechatronics synergy is the emergence of cognitive robotics and autonomous systems. These systems combine physical embodiment with cognitive intelligence, leading to entities that not only interact with their environment but also possess the ability to understand and reason about it. The convergence of AI and mechatronics has led to robots that learn from experience, adapt their behaviors, and even communicate with humans in natural language.

Consider the field of healthcare. Surgical robots, guided by AI-powered image analysis, can assist surgeons in complex procedures with unprecedented precision. These robots can interpret medical images in real time, provide suggestions, and execute delicate maneuvers with human-like dexterity. This blend of physicality and intelligence not only enhances surgical outcomes but also reflects the potential of AI-driven mechatronics to reshape the future of healthcare.

12.3.4 Breaking Boundaries in Smart Manufacturing

The manufacturing industry stands at the forefront of reaping the benefits of AI-mechatronics integration. Smart manufacturing, characterized by the interconnectedness of machines and data-driven decision-making, leverages AI to optimize production processes. Predictive maintenance, a hallmark of this synergy, involves AI algorithms analyzing sensor data to anticipate equipment failures before they occur, thereby minimizing downtime and maximizing efficiency.

Furthermore, collaborative robots, or cobots, exemplify how AI-mechatronics convergence is transforming manufacturing. These robots work alongside human operators, adapting their actions based on real-time interactions. This collaboration is enabled by AI algorithms that enable the robot to sense human intent and adjust its movements accordingly, enhancing safety and productivity in manufacturing environments.

The synergies between AI and mechatronics represent a paradigm shift in the capabilities of machines. The combination of physical embodiment with cognitive intelligence creates systems that can adapt, learn, reason, and interact with their environment in ways previously thought impossible. From enhancing adaptability and real-time decision-making to enabling cognitive robotics and revolutionizing industries like manufacturing, this convergence transcends the sum of its parts.

As we navigate through this chapter, the depths of these synergies will become even more apparent. Through illuminating case studies, we will explore how this fusion has led to transformative advancements across various sectors. Additionally, we will unravel the challenges and opportunities that accompany this dynamic integration, shedding light on both the potential and the responsibilities that come with creating machines that possess not only the power of mechanics but also the brilliance of cognition.

12.4 Case Studies: AI-Driven Advances in Mechatronics

The marriage of AI and mechatronics has given rise to a series of remarkable advancements across a spectrum of industries. These case studies provide compelling insights into the transformative power of this integration, demonstrating how AI's cognitive capabilities synergize with mechatronics' physical prowess to reshape the boundaries of what machines can achieve.

12.4.1 Smart Manufacturing and Industrial Automation

In the realm of manufacturing, the fusion of AI and mechatronics has paved the way for smart factories characterized by increased efficiency, reduced downtime, and optimized production processes. Consider the case of a leading automotive manufacturer. By deploying AI-driven predictive maintenance, the company can monitor the condition of its assembly line robots in real time. Sensors gather data on factors such as motor temperature, vibration, and wear, which is then processed by AI algorithms. These algorithms can predict the likelihood of equipment failures and trigger maintenance actions before issues escalate. As a result, the manufacturer minimizes unplanned downtime, reduces maintenance costs, and ensures seamless production.

Furthermore, AI-powered vision systems are revolutionizing quality control. Cameras equipped with machine learning algorithms can inspect products on the assembly line, identifying defects with high accuracy. This real-time analysis ensures that only flawless products move forward in the manufacturing process, maintaining product quality and reducing waste.

12.4.2 Self-Learning Robotic Systems

The synergy between AI and mechatronics has breathed new life into the field of robotics, enabling machines to learn and adapt to their environment. Take the example of a warehouse automation robot. Traditionally, these robots were programmed with specific paths and tasks. However, by integrating AI, these robots can navigate complex environments, adapt to changes in warehouse layouts, and even collaborate with human workers.

AI-driven reinforcement learning algorithms allow robots to learn from their interactions with the environment. Through trial and error, they identify the most efficient paths for tasks like picking and packing items. This self-learning capability empowers robots to continuously optimize their actions, leading to increased efficiency and adaptability in dynamic warehouse environments.

12.4.3 Predictive Maintenance and Prognostics

The convergence of AI and mechatronics is dramatically transforming the realm of maintenance, particularly in critical infrastructure and transportation. In the aviation industry, for instance, engine failure can have catastrophic consequences. However, with AI-driven prognostics, airlines can predict potential engine malfunctions before they occur.

By analyzing a multitude of data points collected during flight, including engine temperature, pressure, and performance metrics, AI algorithms can identify subtle deviations from normal behavior. This predictive capability enables airlines to schedule maintenance proactively, preventing costly and dangerous in-flight engine failures.

12.4.4 Autonomous Vehicles: A Driving Force

Perhaps one of the most visible outcomes of the AI-mechatronics integration is the development of autonomous vehicles. These vehicles leverage AI's ability to process complex sensor data and make real-time decisions to navigate safely through varied environments.

A prominent case study is the evolution of self-driving cars. These vehicles are equipped with an array of sensors, including cameras, lidar, radar, and ultrasonic sensors. AI algorithms process the data from these sensors to identify road signs, traffic signals, pedestrians, and other vehicles. By analyzing this data, the AI system can make split-second decisions about accelerating, braking, and changing lanes, all while adhering to traffic rules and ensuring passenger safety.

The case studies presented here offer glimpses into the transformative impact of AI-mechatronics synergy. From enhancing manufacturing processes and revolutionizing robotics to ensuring the safety of critical infrastructure and propelling autonomous vehicles, the integration of AI's cognitive abilities with mechatronics' physicality is rewriting the possibilities of technological advancement. As we journey through this chapter, we will continue to uncover more examples of how this synergy is shaping industries and pushing the boundaries of innovation.

12.5 Challenges and Opportunities

The integration of AI and mechatronics, while holding immense promise, also presents a landscape of challenges and opportunities that shape the trajectory of this interdisciplinary frontier. As we delve into this dynamic interplay, we confront technical, ethical, and collaborative considerations that guide the evolution of AI-driven mechatronics.

12.5.1 Technical Challenges: Reliability and Safety

One of the central technical challenges in AI-driven mechatronics lies in ensuring the reliability and safety of intelligent systems. As machines become more autonomous and decision-making becomes increasingly complex, the potential for unexpected behaviors or errors becomes a critical concern. Designing fail-safe mechanisms, building redundancy, and developing error-detection algorithms are essential to mitigating risks and guaranteeing the safe operation of AI-mechatronic systems, especially in domains like autonomous vehicles and medical robotics.

Additionally, AI algorithms often operate in uncertain environments, where accurate predictions and responses are challenging. Adapting AI to handle uncertainty and variability without compromising safety is a continual technical hurdle that requires sophisticated modeling and validation techniques.

12.5.2 Ethical Considerations: Accountability and Transparency

The infusion of AI into mechatronic systems brings forth ethical considerations that extend beyond technical boundaries. Intelligent machines capable of independent actions raise questions of accountability. Who is responsible when an AI-driven autonomous vehicle is involved in an accident? Should it be the manufacturer, the software developer, or the user? Establishing frameworks for accountability that align with legal and ethical standards is a challenge that necessitates interdisciplinary collaboration between AI experts, ethicists, and policymakers.

Moreover, transparency in AI-driven decisions is crucial. The "black box" nature of deep learning algorithms can make it difficult to understand why a certain decision was made. This opacity raises concerns about bias, discrimination, and the potential for erroneous decisions. The development of explainable AI—a field that aims to provide human-interpretable explanations for AI decisions—becomes pivotal in ensuring trust and ethical use of AI in mechatronics.

12.5.3 Collaborative Intelligence: Human–Machine Interaction

The collaboration between humans and AI-driven mechatronics introduces an opportunity-rich landscape. Systems that leverage the strengths of both humans and machines can achieve remarkable outcomes. However, designing interfaces and interactions that seamlessly facilitate this collaboration is a complex challenge. Human–machine interfaces must be intuitive, adaptable, and capable of interpreting human intent while communicating the machine's actions and intentions effectively.

For instance, collaborative robots (cobots) are designed to work side by side with human operators in manufacturing environments. Ensuring the safety of human workers while optimizing the productivity of the robot requires intricate sensor systems and advanced control algorithms. Balancing the capabilities of both entities and orchestrating their harmonious interaction presents a multidisciplinary challenge that spans robotics, AI, and human factors engineering.

12.5.4 Opportunities for Innovation and Progress

Amidst these challenges, the integration of AI and mechatronics holds unparalleled opportunities for innovation. The ability of machines to learn and adapt in real time opens doors to autonomous systems capable of optimizing resource utilization, adapting to unpredictable conditions, and making decisions beyond the realm of pre-programmed instructions. In healthcare, AI-driven medical devices can enhance diagnosis and treatment, improving patient outcomes. In agriculture, precision mechatronics powered by AI can revolutionize crop monitoring and management, addressing global food security challenges.

Furthermore, AI-mechatronics collaboration contributes to the ongoing advancement of Industry 4.0, wherein smart factories are characterized by interconnected machines that communicate and optimize production processes. The digital twin concept, which involves creating virtual replicas of physical machines, allows for simulations, predictive analytics, and optimization, reducing downtime and boosting efficiency.

The challenges and opportunities presented by the convergence of AI and mechatronics encapsulate the complexity and potential of this interdisciplinary journey. Overcoming technical hurdles while adhering to ethical principles will be key to realizing the full potential of AI-driven mechatronics. The collaborative intelligence of human–machine systems promises innovation and advancement across domains, whereas responsible development and design practices will shape the ethical foundation of this exciting and transformative field. As we continue our exploration, we uncover not just the capabilities of AI and mechatronics but also the responsibilities that accompany this powerful fusion.

12.6 Future Directions and Trends

The intersection of AI and mechatronics has already redefined technological possibilities, but the journey is far from over. As we peer into the future, several key directions and trends emerge that are set to shape the landscape of AI-driven mechatronics, ushering in a new era of innovation and collaboration.

12.6.1 AI-Enabled Mechatronic Innovations

The future promises a proliferation of AI-enabled mechatronic innovations that will disrupt industries and create new opportunities. Intelligent robots will become more adaptable and capable of performing intricate tasks that require both physical dexterity and cognitive acumen. Manufacturing processes will be further optimized through real-time data analysis, resulting in increased efficiency and reduced waste. Smart cities will integrate AI-driven mechatronics to enhance urban living, from autonomous transportation systems to intelligent infrastructure management.

In healthcare, surgical procedures will be revolutionized by AI-guided robotic systems that augment the capabilities of human surgeons, leading to improved precision and patient outcomes. Similarly, space exploration will see advancements with autonomous rovers capable of navigating complex terrains on distant planets, adapting to unforeseen challenges, and conducting scientific missions with minimal human intervention.

12.6.2 Collaborative Intelligence in Human–Machine Systems

Collaborative intelligence, where humans and machines seamlessly work together, will take center stage as shown in Figure 12.1. This goes beyond mere coexistence; it is about synergizing the unique strengths of each entity to achieve outcomes that neither could accomplish alone. Collaborative robots will continue to evolve, adapting their actions based on human feedback and intent. Teleoperation systems will allow humans to remotely control intelligent machines in hazardous environments, from disaster-stricken areas to deep-sea exploration.

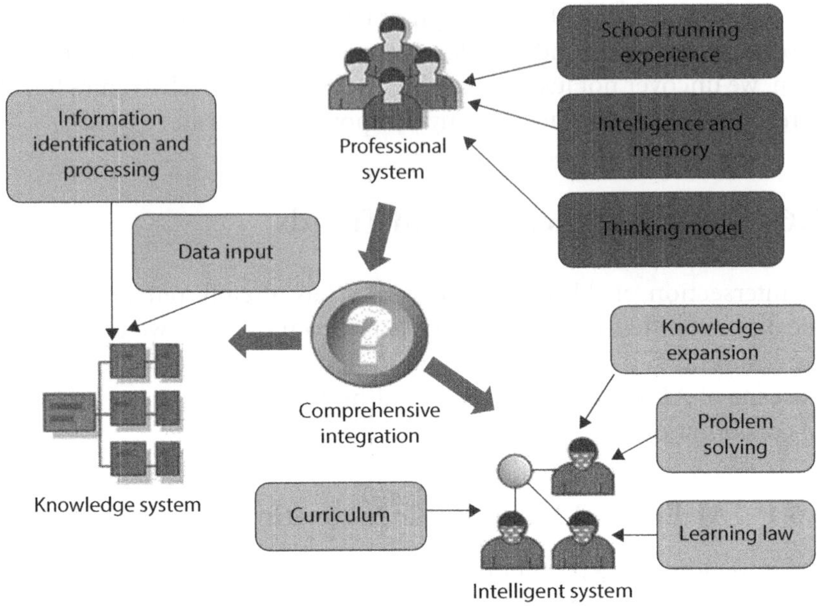

Figure 12.1 Collaborative intelligence in human–machine systems.

In healthcare, AI-driven robotic assistants will become integral parts of surgical teams, enhancing the skills of surgeons and improving patient care. Similarly, industries like construction and agriculture will harness collaborative intelligence to optimize complex tasks, reducing human labor while maintaining precision and safety.

12.6.3 Ethical and Responsible AI in Mechatronics

As AI-driven mechatronics become increasingly prevalent, ethical considerations will play a pivotal role in shaping their development and deployment. There will be a growing emphasis on designing AI systems that are not only efficient and capable but also transparent and accountable. Explainable AI techniques will be refined to provide insights into AI-driven decisions, ensuring that humans can understand and trust the choices made by intelligent machines.

Furthermore, regulations and guidelines surrounding AI ethics will evolve. Governments and international organizations will work together to establish frameworks that govern the responsible use of AI, particularly in safety-critical domains such as autonomous vehicles and healthcare.

Privacy concerns will also be addressed, as AI systems collect and process vast amounts of data.

12.6.4 Empowering Edge Computing

The proliferation of Internet of Things devices and the need for real-time processing will drive the integration of AI at the edge—closer to where data are generated. AI algorithms will be optimized to run on resource-constrained devices, enabling rapid decision-making without relying on cloud-based processing. This is especially crucial in applications where low latency is paramount, such as autonomous vehicles and industrial automation.

Edge AI will enable autonomous drones to make split-second decisions in complex environments, and wearable medical devices will provide real-time health monitoring and intervention. This trend also reduces the need for constant data transmission to central servers, enhancing data privacy and security.

The future of AI-driven mechatronics is a realm of boundless possibilities. As we journey forward, we will witness innovations that integrate AI's cognitive prowess with mechatronics' physical capabilities, giving rise to autonomous systems that adapt, learn, and interact in ways we can only imagine today. Collaborative intelligence will redefine human–machine relationships, whereas ethical considerations and responsible development will ensure that the benefits of AI-mechatronics convergence are harnessed for the betterment of society. The convergence of AI and mechatronics is not just a technological advancement; it is a transformation that holds the potential to reshape industries, revolutionize how we interact with machines, and create a future where intelligence and adaptability are inseparable attributes of machines.

12.7 Conclusion

The fusion of AI and mechatronics has ushered in a new era of technological possibilities, where machines transcend their traditional roles to become intelligent, adaptive, and collaborative entities. Throughout this chapter, we have explored the intricate interplay between AI and mechatronics, uncovering the synergies, challenges, and future directions that shape this dynamic convergence. From the foundations of mechatronics, which bridge mechanics, electronics, and computer science, to the cognitive capabilities of AI that mirror human cognition, we have seen how these disciplines, once distinct, intertwine to create systems that possess both physicality and intelligence. The case studies presented vividly demonstrate the transformative impact of this integration, spanning domains like manufacturing, robotics, healthcare, and transportation. Yet, with progress comes responsibility. We have encountered technical challenges, from ensuring reliability and safety to designing transparent and ethical AI systems. The collaborative potential of human–machine systems not only opens doors to innovation but also prompts questions about accountability and seamless interaction. As AI-driven mechatronics continues to evolve, addressing these challenges becomes pivotal to its responsible development.

12.8 Future Scope

The future unfolds with a promise of AI-enabled mechatronic innovations that will reshape industries, redefine human–machine relationships, and lead to unprecedented collaborative achievements. Collaborative intelligence, driven by the strengths of both humans and machines, holds potential across various domains, from healthcare to space exploration. As ethical and responsible AI takes center stage, frameworks and regulations will shape the path forward, ensuring that AI-mechatronics integration aligns with societal values.

In this era of digital transformation, edge computing emerges as a powerful trend, enabling real-time decision-making and enhancing data privacy. As we conclude this exploration, we are reminded that the journey of AI-driven mechatronics is not just about technology; it is about harnessing the capabilities of machines to enhance human life, improve industries, and shape a future where intelligent autonomy is not only possible but also responsible.

References

1. Kuru, K., Ansell, D., Hughes, D., Watkinson, B.J., Gaudenzi, F., Jones, M., DeGoede, C., Treatment of Nocturnal Enuresis using miniaturised advanced mechatronics with Artificial Intelligence. *IEEE J. Transl. Eng. Health Med.*, 1–16, 2023.
2. Kuru, K. and Yetgin, H., Transformation to advanced mechatronics systems within new industrial revolution: A novel framework in automation of everything (AoE). *IEEE Access*, 7, 41395–41415, 2019.

3. Demir, H. and Sarı, F., The Effect of Artificial Intelligence and Industry 4.0 on Robotic Systems, in: *Engineering on Energy Materials*, pp. 51–72, Iksad Publications, Ankara, 2020.
4. Tanie, K., Mechatronics-Past, present and future, in: *1999 IEEE/ASME International Conference on Advanced Intelligent Mechatronics (Cat. No. 99TH8399)*, pp. 2–2, IEEE, 1999, September.
5. Nguyen, A.T., Dinh, T.Q., Chong, J., Iwasaki, M., Precup, R.E., Ruderman, M., Guest editorial introduction to the special issue on "Emerging control and automation technologies for advanced mechatronic systems". *Control Eng. Pract.*, *136*, 105532, 2023.
6. Rzevski, G., On conceptual design of intelligent mechatronic systems. *Mechatronics*, *13*, 10, 1029–1044, 2003.
7. Mohebbi, A., Achiche, S., Baron, L., Integrated and concurrent detailed design of a mechatronic quadrotor system using a fuzzy-based particle swarm optimization. *Eng. Appl. Artif. Intell.*, *82*, 192–206, 2019.
8. Hingu, P.R. and Panchal, D.N., Industry 4.0–Advance to the Mechatronic and Management with Approach of Artificial Intelligence. *The International Journal of Engineering and Science (IJES)*, 10, 4, 32–361, 2021
9. Dirican, C., The impacts of robotics, artificial intelligence on business and economics. *Proc. Soc. Behav. Sci.*, *195*, 564–573, 2015.
10. Vladareanu, L., Advanced intelligent control through versatile intelligent portable platforms. *Sensors*, *20*, 13, 3644, 2020.
11. Shi, Y., Shen, C., Fang, H., Li, H., Advanced control in marine mechatronic systems: A survey. *IEEE/ASME Trans. Mechatron.*, *22*, 3, 1121–1131, 2017.
12. Whig, P., Kouser, S., Velu, A., Nadikattu, R.R., Fog-IoT-Assisted-Based Smart Agriculture Application, in: *Demystifying Federated Learning for Blockchain and Industrial Internet of Things*, pp. 74–93, IGI Global, USA, 2022.
13. Whig, P., Nadikattu, R.R., Velu, A., COVID-19 pandemic analysis using application of AI, in: *Healthcare Monitoring and Data Analysis Using IoT: Technologies and Applications*, p. 1, 2022.
14. Whig, P., Velu, A., Bhatia, A.B., Protect Nature and Reduce the Carbon Footprint With an Application of Blockchain for IIoT, in: *Demystifying Federated Learning for Blockchain and Industrial Internet of Things*, pp. 123–142, IGI Global, USA, 2022.
15. Whig, P., Velu, A., Naddikatu, R.R., The Economic Impact of AI-Enabled Blockchain in 6G-Based Industry, in: *AI and Blockchain Technology in 6G Wireless Network*, pp. 205–224, Springer, Singapore, 2022.

13

Computational Intelligent Techniques in Mechatronics: Emerging Trends and Case Studies

Anita Mohanty[1]*, Ambarish G. Mohapatra[1], Subrat Kumar Mohanty[2], Bright Keswani[3] and Sasmita Nayak[4]

[1]Electronics Engineering, Silicon Institute of Technology, Silicon University, Bhubaneswar, Odisha, India
[2]Electronics and Communication Engineering, Einstein Academy of Technology and Management, Bhubaneswar, Odisha, India
[3]Department of Computer Science and Engineering, Poornima University, Jaipur, India
[4]Mechanical Engineering, Government College of Engineering, Kalahandi, Odisha, India

Abstract

This chapter investigates the emerging trends and presents compelling case studies in the realm of computational intelligent (CI) techniques applied in mechatronics. Mechatronics, an interdisciplinary arena integrating mechanical engineering, electronics, and computer science, faces complex challenges in system design and control. To address these challenges, CI techniques, like artificial neural networks, evolutionary multi-objective optimization, reinforcement learning, swarm intelligence, explainable artificial intelligence (AI), human–robot interaction (HRI), and biologically inspired robotics, are becoming increasingly predominant. The chapter highlights the real-world applications of these techniques in various mechatronic tasks, including robotic control, adaptive systems, and autonomous vehicles, demonstrating their latent to improve system performance and adaptability. Furthermore, the chapter highlights the significance of explainable AI in safety-critical mechatronics, ensuring transparency and trust in AI-driven decisions. Furthermore, it delves into the intriguing area of HRI and emotional intelligence, showcasing how AI enhances interaction and emotional understanding

**Corresponding author*: anitamohanty776@gmail.com

Kolla Bhanu Prakash, Satish Kumar Peddapelli, Ivan C. K. Tam, Wai Lok Woo and Vishal Jain (eds.) Computational Intelligent Techniques in Mechatronics, (401–444) © 2024 Scrivener Publishing LLC

between humans and machines. Through these illustrative case studies, this chapter offers valuable insights into the transformative impact of CI techniques, shaping the future of intelligent and pioneering mechatronic systems.

***Keywords*:** Computational intelligent techniques, mechatronics, emerging trends, case studies, ANNs, reinforcement learning, robotics

13.1 Introduction to Mechatronics and Computational Intelligence

Mechatronics is a multidisciplinary field at the crossroads of mechanical engineering, electronics, and computer science, aiming to outline intelligent systems that seamlessly integrate mechanical components, sensors, actuators, and computational elements [1]. This combination results in novel devices capable of efficient and precise tasks. On the other hand, computational intelligence (CI) involves harnessing artificial intelligence techniques like neural networks, fuzzy logic, genetic algorithms (GAs), and swarm intelligence (SI) to solve complex problems. By merging the principles of mechatronics and CI, we can develop cutting-edge systems that possess both physical prowess and cognitive abilities, initializing the creation of sophisticated autonomous systems able to adapt and excel in various dynamic environments.

13.1.1 Outline of Mechatronics and Its Interdisciplinary Nature

Mechatronics integrates mechanical engineering, electronics, and computer science to forge integrated systems that transcend traditional limits [2]. This amalgamation melds mechanical components, sensors, actuators, and algorithms, giving rise to intelligent devices with physical and cognitive expertise. Bridging hardware and software, mechatronics extends robotics, automation, and biomedicine, and fosters innovation through interdisciplinary collaboration, pushing technological boundaries.

13.1.2 Introduction to Computational Intelligence Techniques

In the realm of artificial intelligence and data science, CI methods stand out as a captivating domain. CI mimics human cognitive processes, encompassing diverse strategies for intelligent data analysis and decision-making. Core to CI, neural networks replicate interconnected neurons for tasks such as image recognition and language processing [3, 4]. Fuzzy logic addresses data uncertainty, GAs evolve solutions, and SI imitates collective behavior. Evolutionary strategies (ESs) optimize solutions using genetic principles, forming a versatile toolkit for real-world challenges spanning finance, healthcare, and engineering. Leveraging CI unveils intricate patterns, informs choices, and tackles evolving complexity.

13.1.3 Importance of CI in Solving Mechatronics Challenges

Mechatronics, the fusion of mechanics, electronics, and computing, reshapes intelligent system design. In this intricate field, CI assumes a central role, in addressing complex challenges. CI employs neural networks, fuzzy logic, GAs, and SI, amplifying mechatronics' potential. Abundant data are processed for real-time decisions. Neural networks learn patterns, fuzzy logic handles uncertainty, and GAs optimize parameters. Swarm intelligence facilitates teamwork coordination. CI ensures adaptability, efficiency, and robustness, empowering machines to learn, respond, and optimize. This symbiosis advances systems and propels autonomous smart machines in diverse industries.

13.2 Artificial Neural Networks (ANNs) in Mechatronics

Artificial neural networks (ANNs) are fundamental to CI, mimicking the neural structure of the human brain [5]. In mechatronics, ANNs drive transformative change, revolutionizing intelligent system design, operation, and optimization. Within the mechatronics realm, where mechanics converge with sensors and actuators, ANNs proficiently bridge physical and computational realms. They process extensive sensory data, enabling real-time learning, pattern recognition, and adaptive responses—essential for dynamic scenarios like autonomous vehicles navigating complex traffic or robotic arms adapting in manufacturing.

ANNs span diverse applications in mechatronics, benefiting robotics, industrial automation, and healthcare. Despite their prowess in unveiling hidden data relationships, ANNs pose challenges. Complex network design and training demand expertise and resources. Achieving reliability and interpretability, particularly in safety-critical contexts, remains a continuous research endeavor.

13.2.1 Fundamentals of Artificial Neural Networks

Artificial neural networks (ANNs) underpin modern CI, emulating the intricate neural pathways of the human brain to process data, learn, and predict outcomes. ANNs play a vital role in machine learning, data science, robotics, and mechatronics, driving transformative advancements in technology.

Comprising interconnected nodes or "neurons," ANNs process inputs, perform computations, and yield outputs across layers [6]. The perceptron, a basic unit, combines inputs with weighted sums using an activation function, imbuing ANNs with the ability to capture complex data relationships through non-linearity.

ANNs typically comprise three main types of layers as shown in Figure 13.1.

- **Input Layer:** This layer receives raw data, which could be images, text, sensor readings, or any form of structured information.
- **Hidden Layers:** Hidden layers process information through interconnected nodes, where each connection has an associated weight. These layers enable ANNs to capture intricate patterns and representations within the data.
- **Output Layer:** The final layer produces the network's prediction or classification. The number of nodes in this layer depends on the task at hand, such as binary classification, multi-class classification, or regression.

Training an ANN involves fine-tuning connection weights to minimize the gap between predicted and actual outputs [7]. This optimization employs diverse algorithms, with the backpropagation algorithm being a pivotal technique. Backpropagation adjusts weights by computing gradients and iteratively updating them through gradient descent. ANNs exhibit varying architecture and complexity. Shallow networks with few hidden layers suffice for simpler tasks, whereas deep neural networks (DNNs) with numerous layers excel in complex domains like image recognition, language processing, and gaming.

While revolutionizing AI and machine learning, ANNs present challenges like optimal structure design, countering overfitting, and enhancing interpretability. ANNs mirror human brain connections, enabling machines to learn, generalize, and tackle complex tasks; reshaping industries; and showcasing CI's potential.

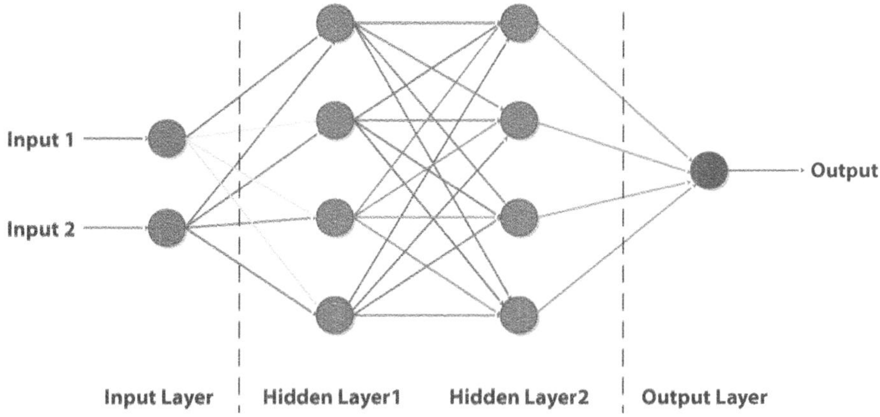

Figure 13.1 Architecture of ANN.

13.2.2 Applications of ANNs in Mechatronic System Modeling

ANNs empower mechatronics, transforming complex system modeling. They enhance accuracy, adaptability, and efficiency, reshaping engineering. The different applications of ANN in mechatronic systems are shown in Figure 13.2.

- **System Identification:** ANNs model intricate system relationships, aiding control strategy optimization.
- **Sensor Fusion:** ANNs merge sensor data for robust perception, vital for robotics and autonomy.
- **Control System Design:** ANNs adapt complex control functions for precision, vital in robotics and automation.
- **Fault Detection:** ANNs spot anomalies, crucial for reliability in aerospace and automation.
- **Energy Efficiency:** ANNs optimize energy usage, vital for renewables and electric vehicles (EVs).

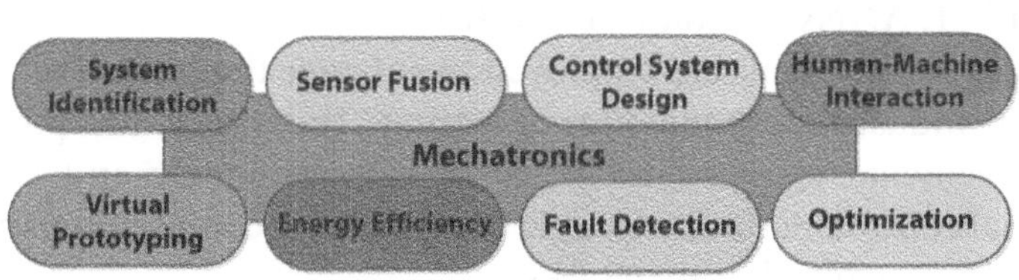

Figure 13.2 Applications of ANNs in mechatronic system modeling.

- **Virtual Prototyping:** ANNs simulate systems, accelerating design and fine-tuning.
- **Human–Machine Interaction:** ANNs enhance user interaction, enabling natural communication.
- **Optimization:** ANNs fine-tune parameters, optimizing performance and behavior.

ANN integration in mechatronics balances expertise, data, and resources. Advancing technology widens mechatronic possibilities, offering smarter, adaptive solutions aligned with industry demands.

13.2.3 ANN-Based Control in Mechatronic Systems

ANNs have transformed mechatronic systems, introducing innovative control strategies that merge CI, mechanical engineering, and electronics. This approach, termed ANN-based control, revolutionizes complex system design, operation, and optimization, unlocking heightened adaptability, efficiency, and precision. Traditional controls often struggle with nonlinear, intricate mechatronic systems [8, 9]. ANNs offer a data-driven alternative, learning strategies from observations for adaptable, optimized performance across applications.

ANN-based control excels in dynamic environments, continuously learning from sensors to adapt responses. It benefits robotics, automation, autonomous vehicles, and smart manufacturing. In challenging scenarios like uncertain robotics or external disturbances, ANNs learn intricate strategies and ensure stability. Moreover, ANN-based control enables human-like decision-making and behavior through data and rule-based training, enhancing user-friendliness. Yet, integrating ANN-based control entails overcoming data, training, and interpretability hurdles. Robust datasets and interpretable methods are crucial, ensuring trust and safety in critical applications.

13.2.4 Case Studies: ANNs for Robotic Control and Fault

Case Study 1: ANNs for Robotic Control

In this case study, ANNs are employed to enhance the control of a robotic arm in an industrial manufacturing setting. The goal is to optimize the arm's movements to pick and place objects accurately and efficiently. Traditional control methods struggle to capture the complex dynamics and nonlinearities involved in such tasks.

Using ANNs, a dataset of various arm configurations, object placements, and corresponding sensor data is collected [10]. The ANN is trained to map sensor inputs (such as camera images and joint positions) to control commands that drive the robotic arm's movements. As the ANN learns from the dataset, it exposes intricate patterns and relationships, allowing the arm to adapt its movements established on real-time sensory information [11].

The result is a highly adjustable robotic arm that can navigate different object shapes, sizes, and orientations with better precision. The ANN-based control plan learns from its interactions and improves its performance over time, making it suitable for tasks that need flexibility, like material handling in manufacturing or sorting operations in logistics.

Case Study 2: ANNs for Fault Diagnosis in Autonomous Vehicles

In this case study, ANNs are used to raise the fault diagnosis capabilities of an autonomous vehicle. Autonomous vehicles must uninterruptedly monitor their systems to find any faults or anomalies that could impact safety or performance [12]. Traditional rule–based techniques can be limited in detecting complex and subtle issues.

Using ANNs, a comprehensive dataset of vehicle sensor readings, vehicle dynamics, and historical fault data is gathered. The ANN is trained to classify various fault scenarios based on sensor inputs. As the ANN learns from the data, it becomes proficient at identifying even subtle deviations from normal behaviour [13].

When a vehicle encounters faults like sensor malfunctions or mechanical issues, the ANN-based system swiftly detects and responds. It can adjust vehicle behavior, alert maintenance teams, or initiate safe stops. This adaptive approach excels in identifying multifaceted faults, surpassing rule-based methods. As autonomous vehicles rise, this fault diagnosis system becomes pivotal for passenger safety and vehicle reliability. ANNs prove their prowess in control and fault diagnosis across both case studies, showcasing their ability to merge mechanical systems with intelligent decision-making, creating smarter and more efficient solutions in robotics and autonomous systems.

13.3 Reinforcement Learning in Mechatronics

Reinforcement learning (RL) revolutionizes mechatronics by imparting decision-making skills through trial and error [14]. Extending across mechanical engineering, electronics, and computer science, RL dynamically optimizes intricate systems. By learning from experience and using rewards as feedback, RL adapts mechatronics to uncertain environments, excelling where precise models falter. For instance, RL empowers robots to navigate new terrain, learning efficient movement and obstacle avoidance from real-time sensor data [15].

The integration of RL in mechatronics encompasses several key aspects:

- **State Representation:** Defining the state of the system, which encapsulates relevant information from sensors and components, is crucial. This state serves as the basis for decision-making within the RL framework.
- **Action Space:** The set of possible actions that the system can take based on the given state. In mechatronics, this could involve adjusting motor speeds, changing trajectories, or altering control parameters.
- **Reward Design:** Constructing a reward function that quantifies the desirability of different outcomes is essential. This function guides the RL algorithm in learning the optimal actions by seeking to maximize cumulative rewards.
- **Exploration vs. Exploitation:** Balancing the exploration of new actions to discover their effects with exploiting the actions that have shown promise in yielding higher rewards.

Reinforcement learning (RL) enables robots to learn intricate movements and behaviors through trial and error, tackling tasks like object grasping, walking on uneven surfaces, and agile maneuvers. It optimizes mechatronic systems, enhancing energy efficiency, resource allocation, and manufacturing processes. RL is vital for autonomous vehicles and drones navigating diverse environments safely and efficiently. It also aids in designing adaptive assistive devices for rehabilitation and improved mobility.

13.3.1 Introduction to Reinforcement Learning (RL)

Reinforcement learning (RL) is a foundational pillar of artificial intelligence, providing machines with a distinctive framework to learn and decide through environmental interaction. Rooted in trial and error learning, RL spans diverse fields—robotics, gaming, finance, and healthcare. Fundamentally, RL is influenced by behavioral psychology's reward-punishment concepts. Here, an "agent" interacts with an "environment," acting and gaining feedback via rewards or penalties. The agent aims to learn actions for maximum cumulative rewards.

The fundamental components of RL include the following:

- **State:** The current situation or context in which the agent finds itself. It encapsulates all relevant information necessary for decision-making.
- **Action:** The choices available to the agent to influence the environment. Actions can be discrete or continuous, depending on the problem.
- **Policy:** The strategy or decision-making algorithm employed by the agent to map states to actions. The goal is to learn an optimal policy that leads to the highest rewards.
- **Reward:** A scalar feedback signal that indicates the immediate benefit or cost associated with taking a particular action in a specific state.
- **Value Function:** A function that estimates the expected cumulative reward the agent can achieve from a given state while following a specific policy.
- **Exploration vs. Exploitation:** The balance between trying new actions to gather information (exploration) and selecting actions that are known to yield high rewards (exploitation).
- **Learning Algorithm:** The mechanism through which the agent updates its policy or value estimates based on experience and feedback.

Reinforcement learning (RL) algorithms are categorized as model-free and model-based. Model-free methods, including Q-learning and SARSA, learn the best policy without an environment model. Model-based approaches construct an environment model for action planning. RL excels in games, robotic tasks, algorithmic trading, portfolio optimization, and personalized healthcare resource allocation.

13.3.2 RL Algorithms and Their Applications in Mechatronics

Reinforcement learning (RL) algorithms offer a groundbreaking method for crafting intelligent systems that learn and adjust through environmental interaction. In mechatronics, RL integration sparks inventive solutions for intricate problems, fostering innovation in mechatronic system design, control, and optimization [16]. Numerous notable RL algorithms and their applications in mechatronics encompass:

- **Q-Learning and Deep Q-Networks (DQNs):** Q-learning is a foundational RL method for acquiring optimal actions through a state-action value function. In mechatronics, DQN, extending Q-learning with DNNs, proves beneficial. It controls robotic arms, optimizes manufacturing, and devises adaptive control strategies for mechanical systems.
- **Proximal Policy Optimization (PPO) and Trust Region Policy Optimization (TRPO):** These algorithms directly optimize policies, enabling intricate behavior learning. In mechatronics, PPO and TRPO contribute to robotic locomotion control, autonomous vehicle path planning, and mechatronic system energy optimization.
- **Deep Deterministic Policy Gradients (DDPGs):** DDPG merges DNNs and policy gradients for continuous action domains. In mechatronics, it governs drones, manipulators, and underwater vehicles, enhancing navigation and task completion.
- **Model-Based RL:** This constructs environment dynamics models for decision-making. In mechatronics, it optimizes robotic trajectories, predicts industrial machinery maintenance, and simulates complex systems pre-deployment.
- **Multi-Agent RL:** Addressing systems of interacting agents, it aids cooperative robotics, swarm robotics, and decentralized mechatronic control.
- **Hierarchical RL:** Learning actions across different levels, it adapts controllers to tasks and cultivates complex robot behaviors with variable autonomy in mechatronics.

RL algorithms find diverse and impactful applications in mechatronics. They enable robots to learn intricate actions such as locomotion, grasping, and manipulation without explicit programming [17]. RL aids in adaptive control strategies for self-driving cars, drones, and underwater vehicles, enhancing navigation efficiency. It optimizes industrial manufacturing, control systems, and maintenance, minimizing downtime. RL contributes to personalized prosthetics, rehabilitation devices, and assistive tools for tailored patient adaptation.

13.3.3 RL for Autonomous Systems and Decision-Making

Reinforcement learning (RL) has solidified its role as a cornerstone, shaping the realm of autonomous systems and intelligent decision-making.

In an age marked by technology's pursuit of autonomy and adaptability, RL propels the advancement of machines capable of informed decisions, adept navigation in intricate surroundings, and amplifying human capacities.

a. Autonomous Systems

Autonomous systems, encompassing self-driving cars, drones, and robots, demand adept decision-making in complex, uncertain environments. Unlike rule-based methods, RL thrives, enabling autonomous systems to learn optimal behaviors through interactions with their surroundings, showcasing exceptional adaptability and efficiency [18]. RL guides self-driving cars in skilfully navigating bustling streets, evading obstacles, and reacting to unexpected scenarios. Drones, powered by RL, explore uncharted terrains, optimize data collection routes, and adjust to varying wind conditions. In search and rescue missions, RL-equipped robots master hazardous terrain traversal, locate survivors, and determine the best rescue paths.

b. Decision Making

The sphere of decision-making, spanning finance, healthcare, and intricate industrial systems, entails sifting through vast data to reach optimal choices. RL has revolutionized decision-making, infusing adaptability, learning, and optimization [19]. In finance, RL algorithms optimize investments by learning from data and trends, maximizing returns while minimizing risks. It tailors treatment plans using patient data, optimizing drug dosages, and adapting to health changes. It also optimizes smart grid energy use, learning patterns, and balancing distribution for efficient demand-supply. Also, it fine-tunes inventory, orders, and routes in intricate supply networks.

13.3.4 Case Studies: Reinforcement Learning in Autonomous Vehicles

a. Case Study 1: RL for Autonomous Vehicle Navigation

In this case study, RL is applied to enhance the navigation capabilities of autonomous vehicles. The goal is to create a self-driving car that can efficiently navigate complex urban environments while adhering to traffic rules and ensuring passenger safety. Using RL, the autonomous vehicle is equipped with sensors such as cameras, lidar, and radar to perceive its surroundings [20]. The RL agent learns to make driving decisions by interacting with a simulated situation, receiving rewards for staying in the correct lane, following traffic signals, escaping collisions, and reaching the destination within a reasonable time. As the RL agent interacts with the environment, it learns optimal driving behaviors that exploit cumulative rewards. It learns to deal with various scenarios, such as merging onto highways, traversing through intersections, and responding to pedestrians and cyclists. Over the period, the RL algorithm refines its driving rule, enabling the autonomous vehicle to handle complex real-world driving situations.

The outcome is an autonomous vehicle that can navigate urban streets, highways, and intersections with human-like decision-making. The RL-based navigation system adjusts to different traffic patterns, road layouts, and unexpected happenings, making it a promising technology for safe and reliable self-driving cars.

b. Case Study 2: RL for Energy-Efficient Autonomous Electric Vehicles
In this case study, RL is used to improve the energy efficiency of autonomous EVs, aiming to extend the EV's range while maintaining peak performance and adhering to user preferences. Through RL, the autonomous EV acquires information and learns to control acceleration, braking, and energy regeneration systems, maximizing efficiency [21]. The RL agent factors in road gradients, traffic, and user speed and comfort preferences, striking an equilibrium between regenerative braking and energy consumption. The RL agent relates to a simulated environment, emulating real-world driving conditions. It learns adaptive driving styles that lower energy usage without compromising safety or comfort. With experience, the EV improves its energy-efficient driving policy, optimizing energy usage in various scenarios. The consequence is an intelligent EV that manages energy, extending the range and decreasing recharging frequency. This improves economic viability and promotes sustainable transportation by curbing energy consumption and emissions. Both case studies underscore RL's potential for autonomous vehicles. By enabling experiential learning and adaptive decision-making, RL advances self-driving cars that operate efficiently and safely in diverse environments.

13.4 Evolutionary Algorithms for Mechatronic System Design

The fusion of mechanical engineering, electronics, and computer science in mechatronics necessitates inventive design approaches that embrace the intricate interplay between physical components and computational elements. In this domain, Evolutionary algorithms (EAs) have emerged as a potent asset, introducing a nature-inspired method to optimize and innovate mechatronic systems [22, 23]. EAs derive inspiration from natural evolution's progression—where successive generations undergo selection, reproduction, and mutation to adapt to their surroundings. In mechatronics, EAs entail generating populations of potential solutions that evolve through iterations, seeking optimal designs. This proves particularly advantageous when conventional optimization methods struggle within intricate, nonlinear, and multi-dimensional design realms.

Key Elements of Evolutionary Algorithms

- **Initialization:** A population of potential solutions is generated randomly or based on expert knowledge. Each solution represents a candidate design for the mechatronic system.
- **Fitness Evaluation:** Solutions are evaluated based on a fitness function that quantifies their performance or quality. The fitness function could involve factors such as efficiency, accuracy, cost, or safety.
- **Selection:** Solutions with higher fitness scores are more likely to be selected for reproduction, mimicking the concept of "survival of the fittest" in nature.
- **Crossover and Mutation:** Reproduction involves combining genetic information from selected solutions through mechanisms like crossover and introducing small changes (mutations) to create new solutions.
- **Iterative Evolution:** The process of selection, reproduction, and mutation is repeated for multiple generations, allowing the population to evolve and converge toward better solutions.

Evolutionary algorithms (EAs) optimize mechatronic system parameters like motor speeds, gear ratios, and control parameters, achieving performance goals while considering trade-offs. EAs design robot morphologies, optimizing locomotion and manipulation capabilities. They determine sensor and actuator locations to maximize information gathering. EAs optimize mechanical components, meeting performance metrics and constraints. They fine-tune control algorithms for stability, responsiveness, and robustness.

Benefits and Challenges

EAs excel in exploring varied design realms, revealing unique solutions, and adjusting to evolving needs. Yet, they can demand substantial computational resources and parameter refinement. Hybrid strategies, blending EAs with other optimization methods or machine learning techniques, are frequently employed to tackle these complexities.

13.4.1 Genetic Algorithms and Their Optimization Applications

Genetic algorithms (GAs) represent a potent optimization method inspired by natural evolution's principles. Rooted in the survival of the fittest and genetic inheritance, GAs span engineering, economics, biology, and art [24, 25]. Their knack for navigating intricate search spaces and uncovering optimal solutions positions them as invaluable tools for demanding optimization challenges. Crucial GA components are as follows:

- **Initialization:** A population of possible solutions, termed "individuals" or "chromosomes," is randomly formed or guided by domain knowledge. Each individual presents a potential optimization answer.
- **Fitness Evaluation:** Individuals undergo assessment via a fitness function, quantifying their performance or quality toward problem goals. The fitness function steers the algorithm toward superior solutions.
- **Selection:** Better-fitting individuals are more likely to be chosen for reproduction, echoing natural selection.
- **Crossover:** In reproduction, genetic details from selected individuals combine, creating new offspring with traits from both parents.
- **Mutation:** For diversity and to prevent early convergence, some individuals experience mutation—minor genetic changes.
- **Iteration:** Selection, crossover, and mutation repeat over generations, evolving the population toward improved solutions.

Genetic algorithms optimize intricate engineering designs—structures, aerodynamics, mechanics—balancing performance and constraints. They refine model and algorithm parameters (e.g., controls and machine learning) for peak effectiveness and adaptability. GAs optimize resources (people, equipment, and finances) for optimal outcomes in manufacturing, projects, and logistics. In data analysis and machine learning, GAs pinpoint pertinent features from vast datasets, boosting accuracy and efficiency. GAs fine-tune investment portfolios, selecting assets to maximize returns and minimize risks, factoring correlations and historical data.

Benefits and Challenges

The versatility of GAs lies in their ability to handle complex, nonlinear, and multidimensional optimization problems, often encountered in real-world scenarios. GAs can explore vast solution spaces and discover non-obvious, innovative solutions. However, they might require careful parameter tuning, and convergence to a globally optimal solution is not guaranteed.

13.4.2 Evolutionary Strategies in Mechatronics

Mechatronics, an amalgamation of mechanical engineering, electronics, and computer science, necessitates inventive strategies to grapple with integrated system intricacies. Evolutionary strategies (ES) emerge as a potent optimization method within this domain, drawing inspiration from nature's workings [26]. ES provides a distinct framework to optimize mechatronic systems, producing solutions adept in intricate, nonlinear, and high-dimensional designs [27].

a. Core Evolutionary Strategies Principles:

- **Population-Based Optimization:** Like GAs, ES employs a population of potential solutions (individuals) for iterative enhancements toward optimal solutions.
- **Mutation-Centric Approach:** ES relies primarily on mutation, not just crossover, to create new solutions. Mutation introduces random adjustments, refining search space exploration.
- **Self-Adaptation:** ES often incorporates self-adaptation, dynamically tuning mutation process parameters during optimization for enhanced exploration finesse.

- **Black-Box Optimization:** ES excel when mathematical models or gradients are absent or resource-intensive. They treat optimization as a "black box," relying on fitness evaluations.

b. Evolutionary Strategies in Mechatronics Applications

- **Mechanical Design:** ES optimizes mechanical components—the shape of aerodynamic surfaces, material layouts, and structures—meeting performance with constraints.
- **Robotics:** ES designs robot forms, determine optimal joint configurations, and fine-tune control parameters for locomotion and manipulation.
- **Control System Tuning:** ES refines control algorithms, ensuring stability, responsiveness, and robustness in mechatronic systems across conditions.
- **Sensor and Actuator Placement:** ES optimize sensor and actuator arrangement, enhancing mechatronic system observability and controllability.
- **Parameter Optimization:** ES optimizes complex model parameters—finite element simulations, neural networks, and dynamic systems—in mechatronics.

c. Benefits and Challenges
Evolutionary strategies offer several benefits in mechatronics:

- **Robustness:** ES is robust against noisy fitness evaluations and can handle non-differentiable and discontinuous objective functions.
- **Global Optimization:** The mutation-based approach of ES enhances their ability to escape local optima and explore the solution space globally.
- **Domain Independence:** ES is applicable to various domains without requiring extensive domain-specific knowledge.

13.4.3 Multi-Objective Optimization with Evolutionary Algorithms

In real-world situations, optimization often involves conflicting objectives that must be simultaneously addressed [28]. Multi-objective optimization (MOO) aims to find solutions that balance these objectives. Evolutionary algorithms (EAs) offer a potent approach to MOO, discovering diverse and well-balanced solutions along the Pareto front [29].

a. Key Aspects of Multi-Objective Optimization with EAs

- **Pareto Dominance:** Solutions are compared using Pareto dominance, assessing if one solution is better in at least one objective without being worse in any.
- **Pareto Front:** The set of non-dominated solutions forms the Pareto front, showcasing optimal trade-offs.
- **Diversity Preservation:** EAs in MOO maintain solution diversity across the Pareto front for varied viable options.
- **Convergence and Spread:** Convergence approximates the true Pareto front, while spread prevents clustering.

b. Applications of Multi-Objective Optimization With EAs

- **Engineering Design:** Balancing factors like cost and performance in aerospace design for fuel efficiency and aerodynamics.
- **Manufacturing:** Optimizing processes with cost, quality, and resource utilization for efficiency and quality.
- **Portfolio Management:** Crafting investment portfolios for maximum returns, minimal risks, and diversification.
- **Urban Planning:** Optimizing transportation, green spaces, and energy efficiency for sustainable cities.
- **Healthcare:** Optimizing treatment plans considering patient outcomes, cost, and resource allocation.

c. Benefits and Challenges

Multi-objective optimization with EAs offers several advantages:

- **Trade-Off Analysis:** It provides decision-makers with a range of optimal trade-offs, allowing them to make informed decisions based on their preferences.
- **Efficiency:** EAs can explore the Pareto front effectively without relying on analytical gradients, making them suitable for complex and non-linear problems.
- **Diverse Solutions:** EAs encourage diversity among solutions, providing a comprehensive view of the solution space.

13.4.4 Case Studies: Evolutionary Algorithms for Mechatronic Design

a. Case Study 1: Robotic Arm Design Optimization

In this case study, an EA is employed to optimize the design of a robotic arm for a specific task [30]. The aim is to find the optimal parameters for the arm's kinematic construction, such as the lengths and angles of its joints, to maximize its range and dexterity while minimizing energy consumption.

The EA starts by creating an initial population of robotic arm designs with random parameters. Each design is assessed using a fitness function that considers reach, dexterity, and energy efficiency. Designs that do better are selected for reproduction, with crossover and mutation operations introducing variations in their parameters. The process repeats over multiple generations, allowing the EA to reach toward designs that excel in the specified objectives.

The result is a set of improved robotic arm designs that strike a balance between reach, dexterity, and energy efficiency. These designs can be utilized for various tasks such as assembly, material handling, or medical procedures, demonstrating how EAs can contribute to creating mechatronic systems tailored to exact requirements.

b. Case Study 2: Autonomous Drone Morphology Design

In this case study, an EA is utilized to optimize the morphology of an autonomous drone for improved flight performance. The aim is to realize a drone configuration that boosts flight stability, agility, and endurance while considering constraints like payload capacity and energy consumption [31].

The EA starts by creating a population of drone configurations, where factors like wing span, rotor positions, and fuselage shape are different. Each configuration is assessed using a fitness function considering flight stability, agility in various maneuvers, and energy efficiency. Over generations of selection, crossover, and mutation, the EA grows the drone configurations to excel in the specified objectives.

The outcome is a set of drone morphologies optimized for particular flight characteristics. These optimized designs enable drones to navigate challenging environments, perform complex tasks, and extend their flight range. The case study highlights how EAs can revolutionize the design of mechatronic systems by exploring diverse design spaces and identifying innovative solutions that meet complex performance criteria.

In both case studies, EAs showcase their potential to optimize mechatronic system designs, tailoring them to specific objectives while accounting for intricate interplays between mechanical, electronic, and computational components. EAs provide a pathway to unlocking new levels of efficiency, performance, and innovation in various mechatronic applications.

13.5 Emerging Trends in Mechatronics with Computational Intelligence

Mechatronics, merging mechanical engineering, electronics [32], and computer science, propels innovation through these trends as shown in Figure 13.3:

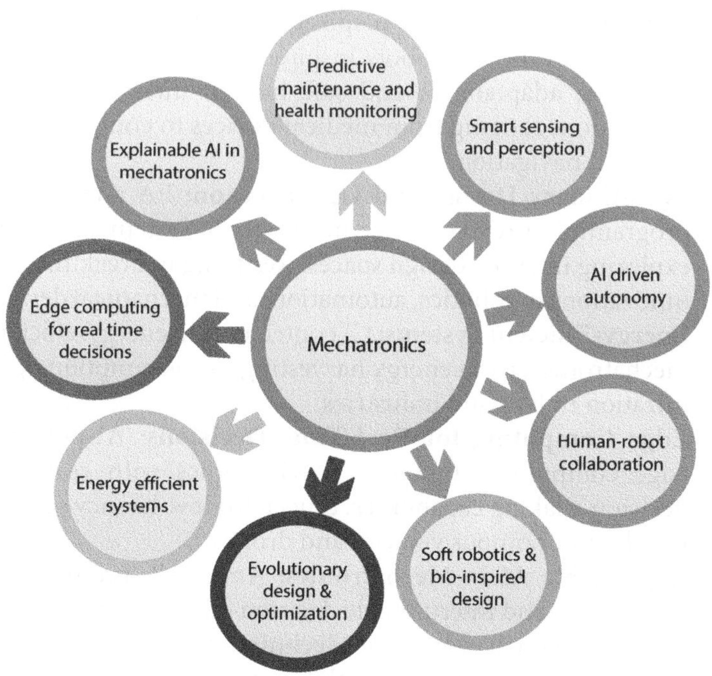

Figure 13.3 Emerging trends in mechatronics with computational intelligence.

- **AI-Driven Autonomy:** Robots, drones, and vehicles gain autonomy by merging AI and ML, adapting to environments, and making intelligent choices.
- **Smart Sensing and Perception:** CI enhances perception capabilities through computer vision and sensor fusion. This enables systems to identify objects, interpret environments, and decide in real time. Vital for robotics, surgery, quality control, and environmental monitoring.
- **Predictive Maintenance and Health Monitoring:** Mechatronics coupled with CI gives rise to predictive maintenance systems. These leverage sensor data, machine learning, and IoT to foresee equipment failures and optimize upkeep.
- **Human–Robot Collaboration:** Natural language processing and human–machine interaction enable mechatronic systems to collaboratively interact with humans. Facilitating safer, efficient human–robot partnerships across healthcare, logistics, and manufacturing.
- **Soft Robotics and Bio-Inspired Designs:** CI drives the development of soft robots inspired by nature. These offer flexibility, adaptability, and safe human interaction, with applications ranging from medical devices to complex environment navigation.
- **Evolutionary Design and Optimization:** EA and genetic programming revolutionize mechatronic system design by exploring intricate design spaces. Resulting in breakthrough innovations in robotics, automation, and mechanical design.
- **Energy-Efficient Systems:** CI contributes to energy-efficient mechatronics, from energy harvesting to consumption optimization in EVs and industries.
- **Edge Computing for Real-Time Decisions:** AI-powered edge computing empowers mechatronics with real-time decision-making capabilities, crucial for low-latency scenarios like autonomous vehicles and drones.
- **Explainable AI in Mechatronics:** Amid AI's prominence, transparent and interpretable decision-making gains importance. Explainable AI (XAI) techniques ensure intelligible and dependable outcomes within mechatronics.

13.5.1 Integration of AI and CI in Mechatronics

The amalgamation of AI and CI within mechatronic systems signifies a pivotal shift in engineering and technology as shown in Figure 13.4. This combination blends AI's machine learning, deep learning, and cognitive computing with CI's EAs, neural networks, and fuzzy logic [33]. This fusion brings about revolutionary outcomes spanning various industries and mechatronics applications.

- **Intelligent Decision-Making:** The fusion of AI and CI empowers mechatronics for informed decisions by processing data, learning patterns, and adapting. This is vital in areas like autonomous vehicles, industrial automation, and robotics.
- **Predictive Analytics:** AI and CI enable future predictions from historical data. In manufacturing, predictive maintenance forecasts equipment failures, reducing downtime and costs.

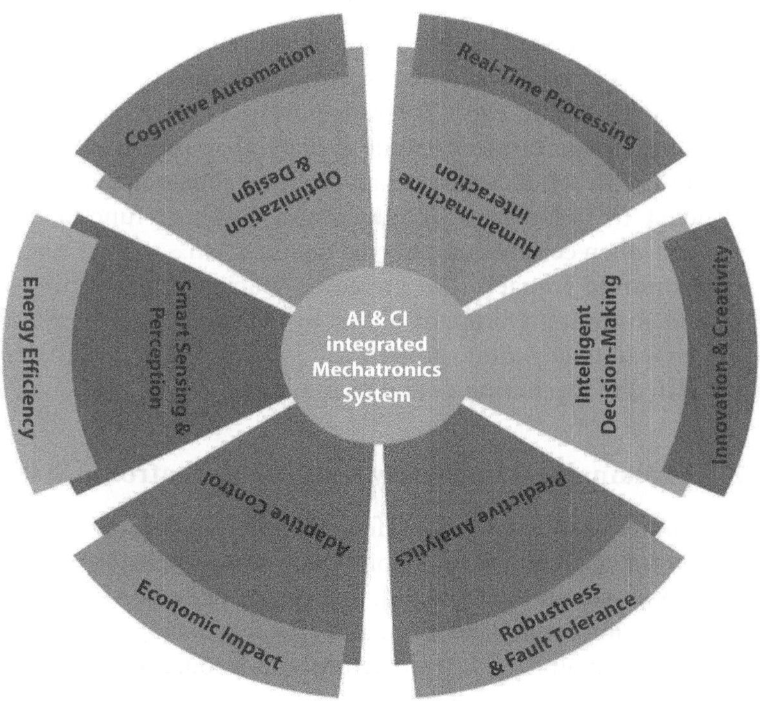

Figure 13.4 Integration AI an CI in mechatronic systems.

- **Adaptive Control:** AI and CI enable adaptable control strategies, crucial for self-tuning controllers optimizing real-time performance.
- **Smart Sensing and Perception**: AI's computer vision and pattern recognition enhance perception for accurate environment interpretation and human interaction.
- **Optimization and Design:** AI and CI optimize intricate mechatronic designs. Evolutionary algorithms refine parameters for mechanical components and control systems.
- **Human–Machine Interaction:** NLP and sentiment analysis foster human–robot collaboration and customer service.
- **Energy Efficiency:** AI and CI optimize power, manage resources, and predict energy demands, aligning with sustainability and cost goals.
- **Cognitive Automation:** AI and CI introduce cognitive abilities, fostering intelligent learning, reasoning, and interaction.
- **Real-Time Processing:** AI and CI enable edge-based real-time processing in autonomous vehicles, drones, and more.
- **Innovation and Creativity:** AI and CI spark innovation, exploring novel design, optimization, and solutions. This opens new frontiers for breakthroughs in mechatronics.
- **Robustness and Fault Tolerance:** AI and CI can contribute to the design of mechatronic systems that are more robust in the face of uncertainties and failures. These systems can adapt to unforeseen situations and continue functioning even when certain components malfunction.
- **Economic Impact:** The increased efficiency, productivity, and innovation brought about by AI and CI integration can lead to economic growth and job creation in industries related to mechatronics and technology.

13.5.2 Explainable AI in Safety-Critical Mechatronic Systems

As AI continues to integrate into safety-critical mechatronic systems, the necessity for transparency, accountability, and comprehension of AI-driven choices becomes paramount [34, 35]. The emergence of XAI becomes pivotal to establishing trust, validation, and effective management in critical settings where human lives, infrastructure, and resources are at risk.

a. The Significance of Explainable AI

In safety-critical mechatronic systems, such as autonomous vehicles, medical devices, and industrial robots, the demand for a transparent grasp of AI-induced decisions is crucial. XAI provides insight into the decision-making mechanisms of AI algorithms, rendering them comprehensible and lucid for human operators, engineers, and regulatory bodies.

b. Key Aspects of Explainable AI in Safety-Critical Mechatronics

- **Transparency:** XAI demystifies AI algorithms, offering visibility into data and processes for accurate and justifiable decisions.
- **Human-Centric Design:** XAI designs interfaces for human understanding and trust in AI decisions, catering to cognitive needs.
- **Interpretable Models:** Safety-critical AI employs interpretable models like decision trees, ensuring transparent decision-making.
- **Feature Importance:** XAI reveals input feature significance, aiding engineers in addressing biases and enhancing model performance.

XAI in safety-critical mechatronics enhances safety and accountability in self-driving cars' decisions. It clarifies AI medical device outputs for informed medical treatments. It provides insights for equipment maintenance, and process optimization in industrial setups. In aerospace, XAI ensures transparent, safety-compliant decisions for aircraft control.

c. Benefits and Challenges

- XAI cultivates trust in AI, enhancing user confidence and wider acceptance.
- It helps verify AI's compliance with safety standards in critical applications, aiding engineers and regulators.
- Amid increasing mechatronic complexity, XAI simplifies understanding of intricate decision processes.

However, challenges involve finding the right equilibrium between interpretability and performance, devising suitable visualization methods, and guaranteeing accurate and meaningful explanations.

13.5.3 Human–Robot Interaction and Emotional Intelligence

Technology advances rapidly, highlighting human–robot interaction (HRI) [36]. As intelligent machines become integrated, understanding and responding to human emotions gain significance. Incorporating emotional intelligence (EI) into robots transforms interactions, enhancing naturalness and empathy. Emotional intelligence (EI) involves perceiving, comprehending, and reacting to emotions in oneself and others [37]. In HRI, EI enables robots to discern, interpret, and respond to human emotions, enriching interactions.

a. Key Aspects of Emotional Intelligence in HRI

- **Emotion Detection:** Robots, equipped with cameras and sensors, discern facial expressions, body language, and vocal cues to deduce human emotions. Machine learning associates these cues with emotional states.
- **Emotion Interpretation:** Through natural language processing and pattern recognition, robots comprehend the context and intensity of human emotions. This enables the identification of happiness, sadness, frustration, or neutrality.
- **Empathetic Responses:** Emotionally intelligent robots respond empathetically and considerately, tailoring actions, tone, and gestures to match human emotional states.
- **Adaptation and Learning:** EI-endowed robots learn continuously from human interactions. They adjust responses over time, aligning with personal preferences and emotional cues.

Robots with EI support patient care by recognizing emotions, providing companionship, and aiding emotional well-being. In education, EI-equipped robots engage students empathetically, adapting teaching methods to emotions, and elevating learning quality. For customer interactions, EI-driven robots comprehend emotions, enhancing satisfaction. They also offer therapeutic assistance and aid individuals with disabilities while considering emotional needs.

b. Benefits and Challenges

- EI enriches HRI, fostering relatable interactions.
- Emotionally responsive robots build trust and deeper connections.

- Challenges arise in universal emotional recognition due to cultural disparities.
- Ensuring privacy, appropriate emotional handling, and prevention of manipulation are vital for robots with EI.

13.5.4 Biologically Inspired Robotics and Soft Robotics

Robotics finds inspiration in both engineering and nature's ingenious designs. Biologically inspired robotics (BIR) and soft robotics are branches that mimic nature's adaptability and efficiency [38]. They challenge rigid robot norms, drawing from biology's flexibility.

a. Biologically Inspired Robotics
This approach replicates living organisms' functional principles, enabling robots to navigate, adapt, and achieve complex tasks:

1. **Biomimetic Locomotion:** Inspired by insects, snakes, and cheetahs, and replicates their unique modes of movement, enabling robots to traverse challenging terrains with agility.
2. **Swarm Robotics:** Inspired by insects like ants and bees, and involves coordinating simple robots to achieve complex tasks, like exploration, surveillance, and environmental monitoring.
3. **Biohybrids:** Combine biological tissues with engineered components, blurring the boundary between living organisms and machines, like muscle-powered robots and cyborg insects.

b. Soft Robotics
Soft robotics departs from rigidity, using flexible materials for human-safe interactions, adaptation, and delicate handling:

- **Pneumatic Actuators:** Soft robots use air to move gently and flexibly, suited for versatile interactions.
- **Octopus-Inspired Manipulation:** Mimicking an octopus' dexterity, soft robotic arms handle objects and adapt to shapes.
- **Wearable Robotics:** Conforming to the body, they enhance mobility and support for a better quality of life.

Bio-inspired robotics (BIR) and soft robotics find versatile applications across domains. In the medical field, soft robots enhance surgical precision, minimizing risk. Biohybrids show promise for drug delivery and tissue regeneration [39]. In search and rescue scenarios, inspired robots navigate disaster zones while soft robots handle debris with care. These technologies are also well-suited for exploring challenging terrains, such as underwater or space environments. Additionally, soft exoskeletons in assistive technology provide valuable support to individuals with mobility impairments.

c. Benefits and Challenges

- BIR and soft robots excel in unstructured and dynamic environments.
- Soft robots are safer for human interaction.
- Achieving desired complexity in biologically inspired complex designs poses challenges due to the intricate designs and materials involved.
- Imitating nature demands advanced sensing for effective interactions.

13.5.5 Swarm Intelligence for Mechatronic System Control

Swarm Intelligence (SI) draws inspiration from the collective behavior of natural social organisms like ants, bees, and birds, reshaping mechatronic system control [40]. SI employs self-organization, collaboration, and emergent behavior for intricate tasks, optimization, and adaptability. Local interactions among simple agents lead to global behavior patterns, emphasizing decentralized decision-making, robustness, and scalability driven by simple rules, not central control [41].

a. Key Aspects of Swarm Intelligence in Mechatronics

1. **Decentralized Control:** Agents make decisions based on local data, eliminating central control. Suitable for vast areas or dynamic environments.
2. **Emergent Behavior:** Complex behaviors arise from agent interactions, offering solutions beyond central control.
3. **Adaptability:** SI enables collective behavior adjustments to changing conditions based on real-time information.

4. **Scalability:** Mechatronic systems using SI can expand without undue complexity, vital for large-scale coordination tasks.

Swarm intelligence (SI) is used in mechatronic system control for different applications. Swarm robots, with sensors, excel in sensing tasks like disaster response. Multi-robot coordination aids logistics, agriculture, and search missions. Swarm algorithms optimize delivery routes, energy-efficient paths, and manufacturing lines. SI enhances traffic control for smoother transportation systems, showcasing its broad impact.

b. Benefits and Challenges

- Swarm intelligence handles failures due to collective redundancy.
- Adaptive nature suits changing or unpredictable conditions.
- Developing algorithms that maintain simplicity while capturing desired behavior is challenging.
- Effective communication and congestion avoidance is crucial as swarm size increases

13.6 Real-World Case Studies

Real-world cases demonstrate how advanced technologies and innovative solutions address complex issues, driving transformative changes across domains. These examples showcase how technologies impact industries, processes, and human experiences. Here, we will explore practical applications through compelling cases.

13.6.1 Case Study 1: Adaptive Control of a Mechatronic System Using ANNs

Exploring how ANNs enhance adaptive control in mechatronic systems, this case study showcases ANNs' learning ability and flexibility. ANNs, inspired by the human brain, can adapt to changing conditions, elevating system performance.

a. Scenario
Consider an industrial robotic arm for precision assembly. Dynamic factors, such as wear and environmental changes, impact its performance. ANNs address this challenge.

b. Approach

1. **Data Collection:** Gather robotic arm data under different conditions, including load speeds and environmental factors. The collected data include input signals (position, velocity, and torque) and corresponding desired output responses.
2. **Training Phase:** Collected data are used to train a neural network controller (NNC) to grasp input signals and desired output responses.
3. **Adaptive Learning:** NNC continuously receives from the arm's sensory input, adapting its weights to improve control accuracy over time.
4. **Real-Time Control:** NNC generates signals, ensuring precise movements despite changing dynamics.

This approach is shown in Figure 13.5.

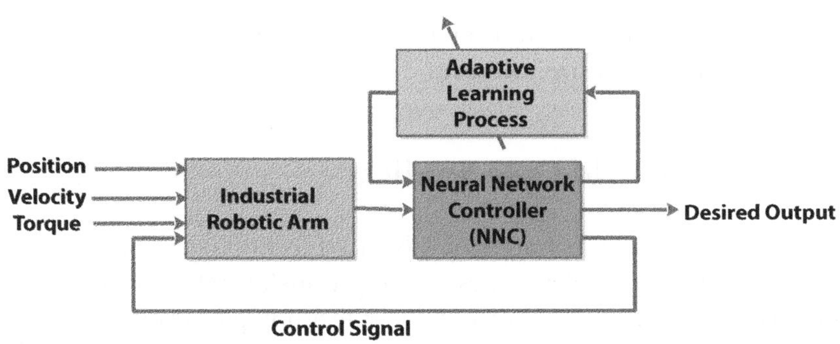

Figure 13.5 Adaptive control process using ANNs.

Table 13.1 Sample data for training ANN.

Position (degree)	Velocity (degree/s)	Torque (Nm)	Desired output
30	10	2.5	0.5
45	8	3.0	0.6
60	12	2.0	0.8
75	11	2.1	1.0
83	13	3.2	0.9

The input data to train the ANN are given in Table 13.1.

a. Benefits

1. ANNs enable the control system to adapt to variations in the mechatronic system's dynamics, ensuring consistent performance despite changing conditions.
2. ANNs can learn complex nonlinear relationships that might be challenging to model analytically, making them suitable for systems with intricate dynamics.
3. The adaptive control provided by ANNs can lead to improved accuracy, reduced errors, and enhanced efficiency in mechatronic tasks.

b. Challenges

1. Collecting representative data is essential for training an effective ANN.
2. The ANN should be carefully trained to avoid overfitting, where it learns the training data too well and fails to generalize to new situations.
3. Ensuring that the ANN can adapt to a technical challenge.

c. Conclusion

ANNs empower mechatronic systems to handle dynamic settings. Learning and adapting capabilities enhance performance, robustness, and responsiveness. This case study underscores ANNs' role in efficient, reliable industrial and automation operations.

13.6.2 Case Study 2: Reinforcement Learning for Autonomous Drone Navigation

a. Introduction

In this study, we delve into the use of RL for autonomous drone navigation. RL, a subset of machine learning, involves training agents to make decisions for maximum rewards. Implementing RL in drone navigation equips drones to adapt their flight paths in real time, ensuring safe and efficient travel in intricate environments.

b. Scenario

Imagine an autonomous drone navigating a densely populated urban area to deliver packages as shown in Figure 13.6. The drone must maneuver

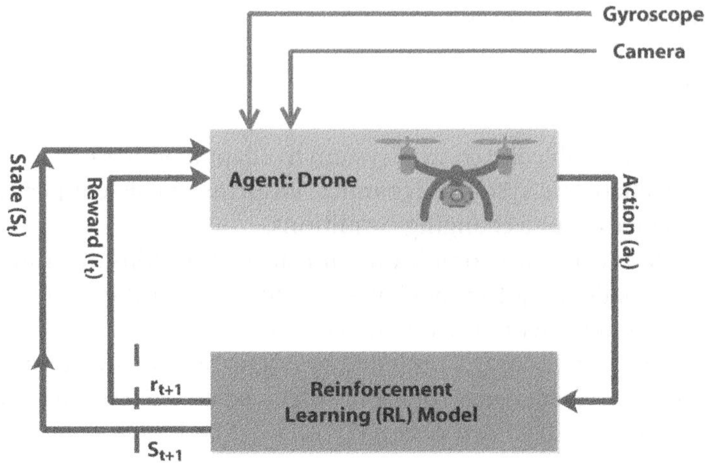

Figure 13.6 Reinforcement learning process for autonomous drone navigation.

around buildings, avoid obstacles, and optimize its path for energy efficiency and timely deliveries.

Each of its states and its description with action and reward are shown in Table 13.2.

Table 13.2 Sates and their description of an autonomous drone.

State	Action	Next state	Reward	Description
Start	Move Forward	Near Obstacle	−2 (Obstacle)	Initial movement toward an obstacle
Start	Turn Right	Start	−1 (Turn)	Initial turn, avoiding immediate obstacle collision
Near Obstacle	Move Forward	State 1	−1 (Obstacle)	Continued movement toward an obstacle

(*Continued*)

Table 13.2 Sates and their description of an autonomous drone. (*Continued*)

State	Action	Next state	Reward	Description
Near Obstacle	Turn Left	State 2	+3 (Turn)	Successful evasion of an obstacle through a turn
Near Obstacle	Turn Right	State 3	+2 (Turn)	Partial evasion, moving toward the goal
State 1	Move Forward	State 1	−1 (Obstacle)	No progress, obstacle still nearby
State 1	Turn Left	Near Obstacle	−2 (Obstacle)	Turned into the obstacle, increased proximity
State 1	Turn Right	State 2	+1 (Turn)	Successful evasion of obstacle
State 2	Move Forward	Goal	+10 (Goal)	Successful navigation toward the goal
State 2	Turn Left	State 3	+1 (Turn)	Continued evasion, moving away from the obstacle
State 2	Turn Right	State 3	+2 (Turn)	Successful evasion with better trajectory
State 3	Move Forward	Goal	+8 (Goal)	Successful navigation, improved path around obstacle
State 3	Turn Left	State 2	+1 (Turn)	Turned back toward the obstacle area

(*Continued*)

Table 13.2 Sates and their description of an autonomous drone. (*Continued*)

State	Action	Next state	Reward	Description
State 3	Turn Right	State 2	+1 (Turn)	Turned back but still moved away from the obstacle
Goal	-	-	+5 (Goal)	Training episode completed, positive terminal reward

a. Approach

1. **State Representation:** The drone's surroundings are divided into discrete states, considering its position, orientation, nearby obstacles, and environmental factors.
2. **Action Space:** The drone has actions like moving, turning, ascending, or descending, altering its state.
3. **Reward Structure:** Rewards guide the RL agent. Positive rewards favor efficient navigation, obstacle avoidance, and punctual deliveries. Negative rewards discourage collisions, energy wastage, or deviating from the optimal path.
4. **Training Phase:** The drone interacts with the environment through episodes, learning from actions, outcomes, and rewards. RL algorithms like DQNs or PPO map actions to states for higher rewards.
5. **Exploration vs. Exploitation:** The drone balances new actions (exploration) with known actions (exploitation) for optimal training.
6. **Policy Optimization:** The RL algorithm refines the drone's policy, mapping states to actions for improved navigation.

b. Benefits

1. RL enables drones to adapt navigation for changing conditions, weather, and traffic.
2. Drones navigate urban landscapes, avoiding obstacles and finding optimal paths missed by pre-programmed routes.

3. RL-trained drones learn energy-efficient paths, extending flight time for better operations.

c. Challenges

1. RL needs substantial interactions, time-consuming in real-world scenarios.
2. Safe training to prevent accidents and collisions is vital.
3. Learning in one environment and applying in another requires policy generalization.

Conclusion

Reinforcement learning empowers drones to navigate intricate, dynamic environments autonomously. By learning from experience, RL-enabled drones efficiently deliver, conduct surveillance, or assist in search and rescue, advancing autonomous aerial capabilities.

13.6.3 Case Study 3: Multi-Objective Optimization in Mechatronic Design

In this study, we explore using MOO in mechatronic design. This approach balances conflicting objectives in complex systems, creating efficient designs with multiple performance criteria. For instance, optimizing an autonomous delivery robot requires balancing speed, energy, payload, and obstacle avoidance. MOO finds solutions that strike a balance.

a. Scenario

Designing an autonomous delivery robot involves conflicting objectives as shown in Figure 13.7. MOO aims to balance speed, energy, payload, and obstacle avoidance.

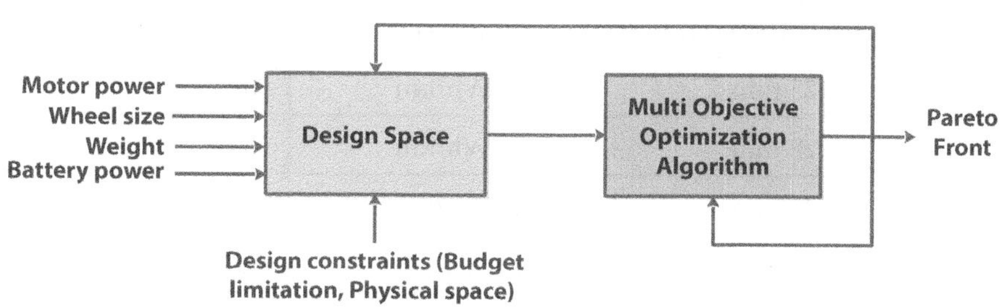

Figure 13.7 Application of multi-objective optimization technique for the design of autonomous delivery robot.

b. Approach

1. **Define Objectives:** Identify design objectives like energy efficiency and payload. Assign functions to quantify performance.
2. **Design Variables:** Choose variables influencing performance, like motor power, wheel size, sensor arrangement, and control algorithms.
3. **Constraints:** Define budget limitations or physical space restrictions.
4. **Optimization Algorithm:** Use algorithms like non-dominated sorting genetic algorithm or multi-objective particle swarm optimization. Generate a set of Pareto front solutions—trade-offs between objectives.
5. **Pareto Front Analysis:** Understand trade-offs between objectives for informed decisions.

In this example, Table 13.3 displays different solutions for an autonomous delivery robot optimization problem. The objectives include minimizing delivery time and energy consumption while maximizing customer satisfaction. Each solution is evaluated based on these objectives, and the goal is to find solutions that strike a balance between them. Solution 1, for instance, has relatively lower delivery time and customer satisfaction compared to Solution 2, but it consumes slightly more energy.

The Pareto front solutions represent a set of trade-off solutions where no solution is better in all objectives compared to another solution in the set. This allows decision-makers to make informed choices based on their

Table 13.3 Observations of an autonomous delivery robot.

Solution	**Delivery time (lower is better)**	**Energy efficiency (lower is better)**	**Customer satisfaction (higher is better)**
1	25 min	120 Wh/km	0.85
2	28 min	115 Wh/km	0.88
3	23 min	125 Wh/km	0.82
4	27 min	110 Wh/km	0.90
5	24 min	118 Wh/km	0.86

priorities and the trade-offs that they are willing to accept in the context of autonomous delivery robot design and operation.

c. Benefits

1. Multiple trade-offs let engineers pick a design fitting project priorities.
2. Optimized parameters enhance efficiency and performance.
3. Consider all criteria, not just one.

d. Challenges

1. Simultaneous optimization can be demanding and needs specialized algorithms.
2. Weighing objectives requires clear goals and input.
3. Mechatronics involves many variables, complex to explore.

e. Conclusion

Multi-objective optimization in mechatronic design takes a holistic view. Balancing multiple goals creates efficient systems meeting diverse criteria. This study shows how optimization shapes adaptable and high-performing mechatronics across domains.

13.6.4 Case Study 4: Explainable AI for Fault Diagnosis in a Robotic Arm

This study explores using XAI for robotic arm fault diagnosis. XAI focuses on creating AI models that provide understandable explanations for decisions. In safety-critical systems like robotics, XAI enhances transparency and trust.

a. Scenario

A complex industrial robotic arm handles precise tasks. AI diagnoses potential faults in arm components, crucial for safety and performance as shown in Figure 13.8.

b. Approach

1. **Data Collection:** Gather sensor and component data during normal and fault scenarios.
2. **Feature Engineering:** Extract relevant features indicating component health, like motor currents and temperatures.

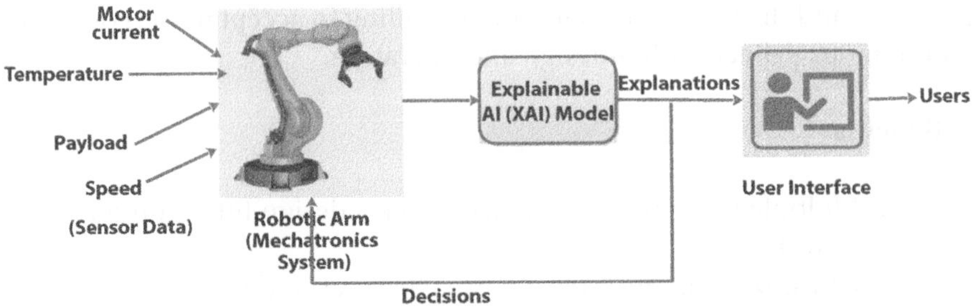

Figure 13.8 Fault diagnosis in robotic arm by using explainable AI (XAI).

3. **Model Development:** Train AI models to classify normal/faulty states based on features.
4. **Explainability Techniques:** Use XAI techniques like local interpretable model-agnostic explanations or Shapley values to explain AI decisions, highlighting influential features.
5. **User Interface:** Display diagnostic outcomes and explanations through a user interface.

The result of the above fault diagnosis system in the robot arm is given in Table 13.4.

c. Benefits

1. XAI reveals decision processes, aiding understanding.
2. Transparent explanations build trust in critical applications.
3. Explanations help identify fault causes, reducing downtime.

d. Challenges

1. Balancing accuracy and explainability can be complex.
2. Deep learning explanations need advanced techniques.
3. Clear presentation is vital.

e. Conclusion

Using XAI for robotic arm fault diagnosis enhances transparency and trust. Understandable AI decisions foster collaboration between engineers and operators, improving fault resolution, system performance, and safety in complex mechatronic setups.

Table 13.4 Results of AI diagnosis and the corresponding explanation.

Sensor data	AI diagnosis	Explanation
Current: 5A	Normal	Motor current within the expected range.
Temp: 40°C	Faulty	A high-temperature reading indicates overheating.
Pressure: 30 psi	Normal	Pressure reading is within the safe range.
Humidity: 85%	Normal	Humidity level is within the acceptable range.
Voltage: 220 V	Faulty	Low voltage detected, potential power issue.
Noise: 75 dB	Faulty	High noise level detected, possible malfunction.
Speed: 1200 RPM	Normal	Operating at the normal rotational speed.

13.6.5 Case Study 5: Emotional Intelligence in a Social Robot

This study explores EI integration in a social robot. Emotional intelligence enables robots to engage better with humans, enhancing interactions and enabling applications like companionship and therapy.

a. Scenario
Meet “Ella,” a social robot designed to interact with elderly individuals in care facilities. Ella’s EI helps her understand and respond appropriately to residents’ emotions, providing companionship and support as shown in Figure 13.9.

Table 13.5 shows some sample emotional responses of the robot based on detected emotions.

b. Approach

1. **Emotion Recognition:** Equip Ella with sensors (cameras and microphones) to perceive emotional cues like facial expressions and voice tone.

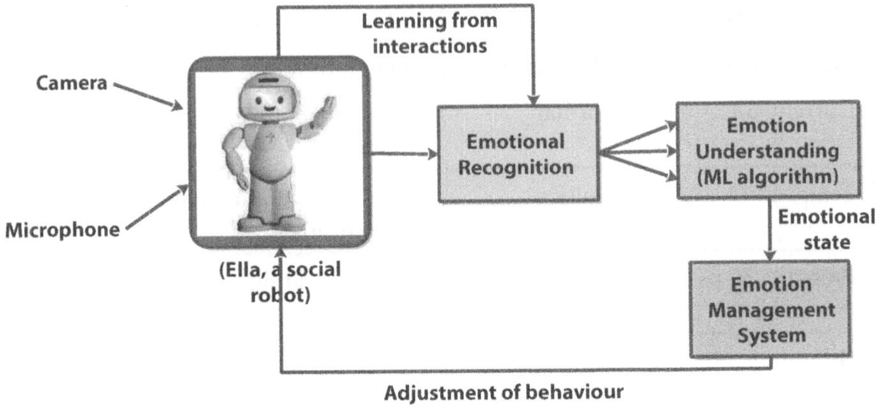

Figure 13.9 Integration of emotional intelligence into a social robot "Ella."

Table 13.5 Emotional responses of the robot based on detected emotion.

Detected emotion	Robot's response
Happiness	"It's great to see you smiling today!"
Sadness	"Is there something I can do to help?"
Anger	"I understand you might be frustrated."
Surprise	"I didn't see that coming! How can I assist you?"
Fear	"It's okay to feel scared. I'm here with you."
Disgust	"I'm sorry if something is bothering you."
Excitement	"Your excitement is contagious! What's up?"
Calmness	"I'm here to provide support whenever you need."
Confusion	"Let's work together to clear things up."
Curiosity	"Your curiosity is wonderful! How can I help?"

2. **Emotion Understanding:** Machine learning algorithms analyze sensor data to infer residents' emotional states (happiness, sadness, etc.).
3. **Emotion Management:** Ella's AI framework includes an emotional reasoning system to manage her own responses based on detected emotions.

4. **Behavioral Adaptation:** Ella adapts actions, gestures, and speech based on detected emotions (comforting words for sadness, uplifting activities).
5. **Feedback Loop:** Ella learns from interactions, replicating responses that elicit positive reactions.

c. Benefits

1. Emotional intelligence makes robot-human interactions more relatable and meaningful.
2. Emotionally intelligent robots offer companionship and support, especially where human interaction is limited.
3. These robots aid therapy, helping individuals manage stress and loneliness.

d. Challenges

1. Detecting and interpreting emotions accurately is complex due to human emotion complexity.
2. Responses must be culturally sensitive due to varied emotional expressions.
3. Ensuring privacy, consent, and proper boundaries in data collection and interactions is essential.

e. Conclusion

Emotional intelligence in social robots bridges AI and human emotions. This study showcases how robots like Ella contribute to well-being, companionship, and human-tech connections. It highlights the potential of technology to link AI with emotions, fostering innovation in various fields.

13.7 Conclusion

13.7.1 Recapitulation of Emerging Trends in CI for Mechatronics

Computational intelligence (CI) is swiftly revolutionizing mechatronics, enhancing robotic and automated systems. This integration triggers a paradigm shift, driving innovation and efficiency. The fusion encompasses pivotal trends shaping mechatronics' future. The amalgamation of AI and CI unlocks novel dimensions of intelligence, fostering dynamic learning and adaptation. Explainable AI (XAI) addresses transparency in AI decisions, nurturing trust and accountability. Emotional intelligence enriches

HRI, improving engagement. Biologically inspired robotics leverages CI for natural behavior emulation. Swarm intelligence enables decentralized collaborative problem-solving. Evolutionary algorithms (EAs) harness nature's principles for optimization. Reinforcement learning (RL) empowers autonomous systems to learn and adapt. Multi-objective optimization optimizes mechatronics with competing goals. Real-world case studies validate CI's impact, whereas continuous learning ensures adaptability, reshaping mechatronics for diverse applications.

13.7.2 Future Directions and Potential Applications

Mechatronics is dynamically evolving due to CI advancements. CI's future in mechatronics promises autonomy across sectors like self-driving cars, drones, and industrial robots. Healthcare will transform with CI-powered wearables aiding patients and assisting those with disabilities. Precision agriculture will optimize farming using AI drones. Smart manufacturing will redefine efficiency with predictive maintenance and AI-guided production. CI-enhanced robots will collaborate seamlessly with humans. Energy efficiency, evolvable hardware, and environmental monitoring are other dimensions CI will influence. Ethical and regulatory considerations must accompany responsible deployment. The fusion of CI with other fields will yield pioneering adaptive system designs.

13.7.3 Implications of Computational Intelligent Techniques in Mechatronics Advancements

CI revolutionizes mechatronics, enhancing efficiency, autonomy, and collaboration. Real-time decisions, safety, and sustainability prosper. Ethical considerations surface. CI shapes a future with advanced mechatronics, reshaping industries and human-tech interaction, fostering growth.

13.7.4 Key Impacts of CI in Mechatronics

The integration of CI within mechatronics is rapidly reshaping industries by enhancing efficiency, performance, and innovation. CI optimization has a broad impact, streamlining processes in manufacturing, logistics, and agriculture. It fosters design innovation by exploring novel solutions, resulting in optimized mechatronic systems. CI-equipped mechatronics exhibit adaptability and autonomy, navigating complex environments and learning from interactions. This integration deepens human–machine collaboration, with examples like robots aiding medical professionals and collaborating with human workers. Real-time decision-making is empowered by CI, reducing human intervention and facilitating predictive maintenance. Personalization is achieved through CI, enabling tailored behavior in devices to address individual preferences. CI enhances safety by predicting and averting hazards while also optimizing energy use for sustainability. It tackles complex problems, unveils hidden insights in data, and drives economic growth through cost-effective production. However, it also presents challenges concerning privacy, ethics, and job displacement, demanding careful consideration and regulation.

References

1. Irschik, H., Krommer, M., Nader, M., Zehetner, C., Mechatronics—The application for innovation. *Adv. Sci. Technol.*, *56*, 1–10, 2009.
2. Faria, A.C. and Barbalho, S.C.M., Mechatronics: A Study on Its Scientific Constitution and Association with Innovative Products. *Appl. Syst. Innov.*, *6*, 4, 72, 2023, https://doi.org/10.3390/asi6040072.
3. Fulcher, J., Computational Intelligence: An Introduction. *Stud. Comput. Intell. (SCI)*, 115, 3–78, 2008.
4. Mukhamediev, R., II, Popova, Y., Kuchin, Y., Zaitseva, E., Kalimoldayev, A., Symagulov, A., Levashenko, V., Abdoldina, F., Gopejenko, V., Yakunin, K., Muhamedijeva, E., Yelis, M., Review of Artificial Intelligence and Machine Learning Technologies: Classification, Restrictions, Opportunities and Challenges. *Mathematics*, *10*, 15, 2552, 2022, https://doi.org/10.3390/math10152552.
5. Knohl, T. and Unbehauen, H., Adaptive position control of electrohydraulic servo systems using ANN. *Mechatronics*, 10, 1–2, 127–143, February–March 2000.
6. Basheer, I. and Hajmeer, M.N., Artificial Neural Networks: Fundamentals, Computing, Design, and Application. *J. Microbiol. Methods*, 43, 1, 3–31, 2001, DOI: 10.1016/S0167-7012(00)00201-3.
7. Madhiarasan, M. and Louzazn, M., Analysis of Artificial Neural Network: Architecture, Types, and Forecasting Applications. Evolutionary, Neural and Fuzzy Systems for Smart and Micro Grids, vol. 2022, 2022, https://doi.org/10.1155/2022/5416722.
8. Lee, H., Jhang, S., Yu, W., Oh, J., Artificial Neural Network Control of Battery Energy Storage System to Damp-Out Inter-Area Oscillations in Power Systems. *Energies*, 12, 17, 3372, 2019, https://doi.org/10.3390/en12173372.
9. Mahar, H., Munir, H.M., Soomro, J.B., Akhtar, F., Hussain, R., Elnaggar, M.F., Kamel, S., Guerrero, J.M., Implementation of ANN Controller Based

UPQC Integrated with Microgrid. *Mathematics, 10*, 12, 1989, 2022, https://doi.org/10.3390/math10121989.

10. Liu, Z., Peng, K., Han, L. et al. Modeling and Control of Robotic Manipulators Based on Artificial Neural Networks: A Review. *Iran J. Sci. Technol. Trans. Mech. Eng.*, 47, 1307–1347, 2023. https://doi.org/10.1007/s40997-023-00596-3.
11. Chaoui, H., Sicard, P., Gueaieb, W., ANN-Based Adaptive Control of Robotic Manipulators With Friction and Joint Elasticity. *IEEE Trans. Ind. Electron.*, 56, 8, 3174 –3187, August 2009, https://doi.org/10.1109/TIE.2009.2024657.
12. Biddle, L. and Fallah, S., A Novel Fault Detection, Identification and Prediction Approach for Autonomous Vehicle Co2221ntrollers Using SVM. 4, 301–314, 2021.
13. Shen, T., Kilic, A., Thulfaut, C., Reuss, H.-C., An Intelligent Diagnostic Method for Permanent Magnet Synchronous Motors (PMSM) in the Electric Drive of Autonomous Vehicles. *European Conference on Power Electronics and Applications*, 2019, DOI: 10.23919/EPE.2019.8915161.
14. Kaelbling, L.P., Littman, M.L., Moore, A.W., Reinforcement Learning: A Survey. *JArtif. Intell. Res.*, 4, 237–285, 1996.
15. Kormushev, P., Calinon, S., Caldwell, D.G., Reinforcement Learning in Robotics: Applications and Real-World Challenges. *J. Rob.*, 2, 3, 122–148, 2013, https://doi.org/10.3390/robotics2030122.
16. Zhang, T. and Mo, H., Reinforcement learning for robot research: A comprehensive review and open issues. *Int. J. Adv. Rob. Syst.*, 18, 3, 1-22, 2021, https://doi.org/10.1177/17298814211007305.
17. Polydoros, A.S. and Nalpantidis, L., Survey of Model-Based Reinforcement Learning: Applications on Robotics. *J. Intell. Rob. Syst.*, 86, 153–173, 2017.
18. Peng, Z., Li, X., Yan, F., An Adaptive Deep Learning Model for Smart Home Autonomous System. *2020 International Conference on Intelligent Transportation, Big Data & Smart City (ICITBS)*, 2020, DOI: 10.1109/ICITBS49701.2020.00156.
19. Macaulay, M.O. and Shafiee, M., Machine learning techniques for robotic and autonomous inspection of mechanical systems and civil infrastructure. *Auton. Intell. Syst.*, 2, 8, 1–25, 2022, https://doi.org/10.1007/s43684-022-00025-3.
20. Ergun, S., A study on multi-agent reinforcement learning for autonomous distribution vehicles. *Iran J. Comput. Sci.*, 6, 297–305, 2023, https://doi.org/10.1007/s42044-023-00140-1.
21. Kim, S., Lee, U., Lee, I., Kang, N., Idle vehicle relocation strategy through deep learning for shared autonomous electric vehicle system optimization. *J. Clean. Prod.*, 333, 2022, https://doi.org/10.1016/j.jclepro.2021.130055.
22. El-Kribi, B., Houidi, A., Affi, Z., Romdhane, L., Application of multi-objective genetic algorithms to the mechatronic design of a four bar system with continuous and discrete variables. *Mech. Mach. Theory*, 61, 68–83, 2013.

23. Affi, Z., El-Kribi, B., Romdhane, L., Advanced mechatronic design using a multi-objective genetic algorithm optimization of a motor-driven four-bar system. *Mechatronics*, 17, 9, 489–500, 2007.
24. Gen, M. and Lin, L., Genetic algorithms and their applications, in: *Springer handbook of Engineering Statistics*, pp. 635–674, 2023.
25. Katoch, S., Chauhan, S.S., Kumar, V., A review on genetic algorithm: past, present, and future. *Multimedia Tools Appl.*, 80, 8091–8126, 2021.
26. Hamdani, H., Radi, B., El Hami, A., Metamodel assisted evolution strategies for global optimization of solder joints reliability in embedded mechatronic devices. *Microsyst. Technol.*, 25, 3801–3812, 2019.
27. Slowik, A. and Kwasnicka, H., Evolutionary algorithms and their applications to engineering problems. *Neural Comput. Appl.*, 32, 12363–12379, 2020.
28. Hua, Y., Liu, Q., Hao, K., Jin, Y., A Survey of Evolutionary Algorithms for Multi-Objective Optimization Problems With Irregular Pareto Fronts. *IEEE/CAA J. Autom. Sin.*, 8, 2, 303–318, 2021, DOI: 10.1109/JAS.2021.1003817.
29. Deb, K., Multi-objective Optimisation Using Evolutionary Algorithms: An Introduction, in: *Multi-objective Evolutionary Optimisation for Product Design and Manufacturing*, pp. 3–34, 2011.
30. Bugday, M. and Karali, M., Design optimization of industrial robot arm to minimize redundant weight. *Eng. Sci. Technol. Int. J.*, 22, 1, 346–352, 2019.
31. Falanga, D., Kleber, K., Mintchev, S., Floreano, D., Scaramuzza, D., The Foldable Drone: A Morphing Quadrotor That Can Squeeze and Fly. *IEEE Rob. Autom. Lett.*, 4, 2, 209–216, 2019..
32. Liagkou, V., Stylios, C., Pappa, L., Petunin, A., Challenges and Opportunities in Industry 4.0 for Mechatronics, Artificial Intelligence and Cybernetics. *Electronics*, 10, 16, 1–23, 2021, https://doi.org/10.3390/electronics10162001.
33. Guo, K., Application of Artificial Intelligence in Mechatronics. *Appl. Sci.*, 13, 158, 1–2, 2023, https://doi.org/10.3390/app13010158.
34. Schöning, J. and Pfisterer, H.-J., Safe and Trustful AI for Closed-Loop Control Systems. *Electronics*, 12, 1–15, 2023, https://doi.org/10.3390/electronics12163489.
35. Wäschle, M., Thaler, F., Berres, A., Pölzlbauer, F., Albers, A., A review on AI Safety in highly automated driving. *Front. Artif. Intell.*, 5, 1–22, 2022, doi: 10.3389/frai.2022.952773.
36. Spezialetti, M., Placidi, G., Rossi, S., Emotion Recognition for Human-Robot Interaction: Recent Advances and Future Perspectives. *Front. Robot. AI*, 7, 1–11, 2020, doi: 10.3389/frobt.2020.532279.
37. Stock-Homburg, R., Survey of Emotions in Human–Robot Interactions: Perspectives from Robotic Psychology on 20 Years of Research. *Int. J. Soc. Rob.*, 14, 11, 389–411, 2021, DOI:10.1007/s12369-021-00778-6.
38. Trivedi, D., Rahn, C.D., Kier, W.M., Walkerc, I.D., Soft robotics: Biological inspiration, state of the art, and future research. *Appl. Bionics Biomech.*, 5, 3, 99–117, 2008.

39. Coyle, S., Majidi, C., LeDuc, P., Hsia, K.J., Bio-inspired soft robotics: Material selection, actuation, and design. *Extreme Mech. Lett.*, 22, 51–59, 2018.
40. Smoczek, J. and Szpytko, J., Particle Swarm Optimization-Based Multivariable Generalized Predictive Control for an Overhead Crane. *IEEE/ASME Trans. Mechatron.*, 22, 1, 258–268, 2017, DOI: 10.1109/TMECH.2016.2598606.
41. Mohebbi, A. and Achiche, S., Integrated and concurrent detailed design of a mechatronic quadrotor system using a fuzzy-based particle swarm optimization. *Eng. Appl. Artif. Intell.*, 82, 192–206, 2019.

14

Advanced Sensing Systems in Automobiles: Computational Intelligence Approach

Mamta B. Savadatti and Ajay Sudhir Bale*

Department of ECE, New Horizon College of Engineering, Bengaluru, India

Abstract

It comes from logic that advances in technology will play a big part in state-of-the-art sensing because it will be the primary driver of the development of hardware required for applications utilizing sensing. Smart sensing has proved useful in the automotive sector, but there are still many of problems to be solved. To find complete solutions, we will need multidisciplinary teams to work together. Cities, utilities, and other service providers should all be included in standardization efforts for the sake of implementing fully integrated systems, for example, in the areas of sensor and communications infrastructures. It is inefficient to have many infrastructures (in terms of sensor systems), yet consolidating them all might be challenging. We have made an effort in bringing the recent trends in smart sensing in the automobiles. We have also proposed the way by which the drowsiness feeling of the driver is alerted to the driver to reduce the effects of road accidents.

Keywords: Sensing systems, eye blink, accidents, driver, automobile, intelligent networks

14.1 Introduction

Intelligent manufacturing's progress has had far-reaching and long-lasting effects on the future of manufacturing across the world. In the age of the Internet of Things (IoT), in the new age, industry factories improve efficiency, productivity, manageability, leadership, and visibility by fusing both physical and digital technologies, enabling the involved systems

**Corresponding author*: ajaysudhirbale@gmail.com

Kolla Bhanu Prakash, Satish Kumar Peddapelli, Ivan C. K. Tam, Wai Lok Woo and Vishal Jain (eds.) Computational Intelligent Techniques in Mechatronics, (445–470) © 2024 Scrivener Publishing LLC

more sophisticated and precise. Modern, inexpensive sensors are crucial for manufacturers and supply chains to efficiently collect data and put it to use. More data can be collected on more devices throughout the production process thanks to new kinds of cheap power/low price sensors. Despite the extensive studies that have been conducted on the subject of smart factories and their performance, processes, and implementation, most businesses still lack an in-depth understanding of the distinctions between conventional and smart factory systems and the vast array of sensors used in industry. The significance of the manufacturing sector in driving contemporary society forward cannot be overstated. The cutting-edge technologies available to Industry 4.0, the forerunner of smart factories, include [1, 2] big data analysis, machine learning, the sophisticated field of robotics 3-D printing, and cloud computing. The widespread use of computer numerical control (CNC) or robots for industry has allowed for more adaptability in production systems [3–5], whereas the use of computer-aided design (CAD) has made computer-integrated manufacturing a realistic option [6, 7]. The IoT practical applications have facilitated digital changes in manufacturing from a variety of perspectives, including improved efficiency, robotics, a focus on customers, competitive advantages, value chain enhancement, and quick returns [8, 9]. To comprehend how the IoT may be put to good use in manufacturing, it is necessary to get familiar with the many technologies, in particular sensors, that contribute to the effective operation of manufacturing enterprises in the age of Industry 4.0. Through the IoT, commonplace items may be networked with other devices to collect data, interpret that data, and provide an outcome that is informed by past experiences. To better manage the connection and adaptability of IoT devices, businesses may turn to cloud-based IoT systems [10, 11]. The IoT architecture must also support several wireless protocols and allow for the integration of additional sensor inputs (such as USB) [12]. This may also be recognized in terms of mobility, which can include the ability to use batteries, wearables, and other portable gadgets [10]. This is possible because to the development of sensors.

Transport in the modern world is fraught with difficulties, including as traffic congestion, insecure travel, and environmental damage. The use of communication and information technology in today's transportation systems is becoming more prominent and important. Sensors installed in vehicles and their uses in areas such as navigation, traffic control, and passenger entertainment are constantly being refined by the auto industry. In order to monitor traffic and the environment around them, government agencies are installing wayside infrastructure like webcams and detectors.

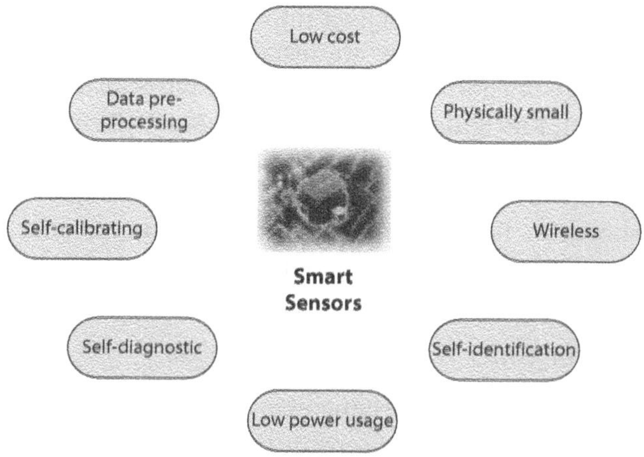

Figure 14.1 Features of a smart sensor [13].

Smart and intelligent modes of transport may be achieved by the seamless integration of cars and monitoring equipment, hence maximizing their detecting and communication abilities. Security, control of traffic, and informational applications may all benefit from having numerous sensors deployed in various aspects of an ITS, and we describe how sensor technology can be linked to the infrastructure of transportation to produce a sustainable intelligent transportation system (ITS) as in Figure 14.1 [13].

14.2 Computational Intelligence Approach

14.2.1 Sensor Technology

The use of sensors facilitates the creation of several applications for transportation for managing traffic, ensuring passenger security, and providing fun distractions. Tire pressure sensors and rear-view visibility systems are just two examples of the types of sensors and actuators that have recently become required in the production of automobiles as well as the setting up of ITSs. These systems' goals are to provide services that increase drivers' and passengers' satisfaction, enhance safety on the roads, and decreas traffic jams. The performance and health of the vehicle may be tracked by additional sensors that manufactures provide as add-ons, allowing for greater efficiency and driver support. There are now 60–100 sensors installed in the typical vehicle, but as cars get "more intelligent," that number might rise to 200 [14–16].

In [17], the writer classifies sensor into the engine, the chassis, and the body, depending on where they are installed. In another piece of study, devices in a car are categorized according to the purpose for which they are designed to assist; four groups are identified: security sensors, diagnosing sensors, utility sensors, and monitoring of the environment sensors [18]. We add two more groups to the categorization (four categories) suggested in [18], as shown in Table 14.1: sensor for driver surveillance and sensor for tracking traffic [14]. This information is also refereed from [18].

14.2.1.1 Sensor Units Within the Automobile

A vehicle's efficiency and the driving experience can be greatly improved through determining the kind of sensors that can create services that assist in addressing issues like congested roads and parking challenges, increased traffic congestion, carbon dioxide pollutants, and the number of accidents involving cars. Sensors like those shown in Figure 14.2 are becoming more commonplace in modern automobiles.

Table 14.1 Sensors in vehicles [14, 18].

Types	Highlights
Safety	Provide the framework for security measures, with an emphasis on rapidly identifying potential dangers and actual incidents.
Diagnostic	Collecting data in real time to offer data regarding the vehicle's status and efficiency is crucial for diagnosing problems.
Traffic	Keep an eye on traffic in key areas and collect data to help plan better routes.
Assistance	Keep an eye on the weather and provide passengers as well as drivers with warning and advisory services to make their travels safer and more enjoyable.
Environment	Pay close attention to accumulating information that will aid in the identification of unusual conditions and driving conduct that might impair efficiency.

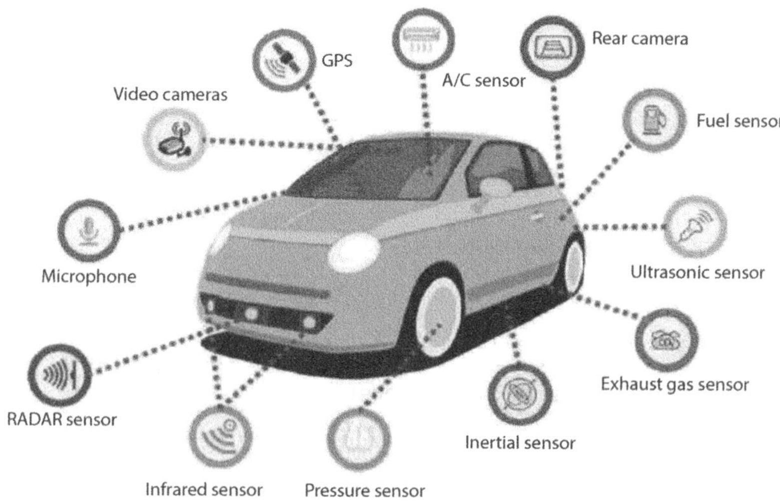

Figure 14.2 Sensor units within the automobile [14].

14.2.1.2 In-Vehicle Sensors

Traffic data gathering utilizing mechanisms positioned on roadsides has become one of the key problems for smart transportation systems, despite the automobile industry's significant investments in improving security, efficiency, and comfort in cars utilizing intra-vehicle sensors. Automated parking (e.g., linking vehicles with free parking places) and pricing for congestion are only two instances of the novel amenities made possible by the widespread use of sensors in transportation systems. The term "real" environmental information gathered by sensors is examined and processed to strengthen and fortify transportation systems.

Based on their placement, sensors are either invasive or non-intrusive [19]. Unsightly sensors are embedded into the asphalt. While they are quite precise, they come with a hefty price tag for both setup and upkeep. Invasive sensors may be broken down into three broad categories (depicted in Figure 14.3): firstly, passive magnetometers embedded in roadways and linked via wires or radio frequency transceivers to central processing units; secondly, integrated wires, or inductive loops, transmit data to receivers along the route through wired/wireless media outlets; and thirdly, pneumatic equipment tube sensors positioned across the road. This is shown in Table 14.2 [14]. The majority of traffic management systems [20] rely on this kind of sensors.

A lot of the capabilities of invasive sensors may be met by non-invasive ones and with less complications. However, weather conditions like rain,

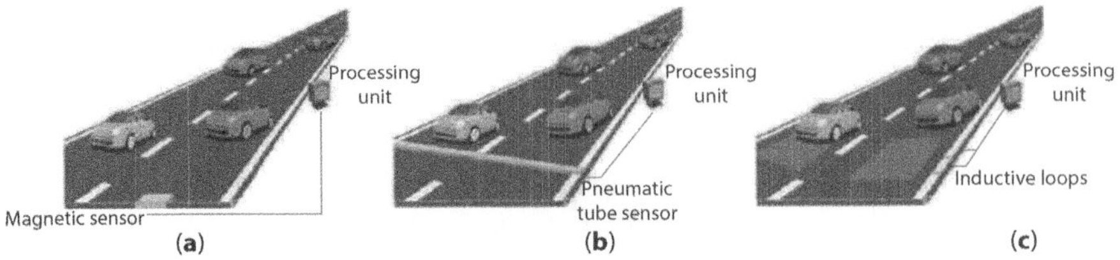

Figure 14.3 Intrusive types of sensors [14].

Table 14.2 Types of in road sensors [14].

Type	Usage and deployment
Intrusive	Useful for recording and categorizing vehicle counts and counts.
	It can identify the existence, number, and occupancy of vehicles. A gadget at the roadside records the signals sent out.
	Designed to identify the existence of cars and distinguish between those that are parked and those that are in motion.
	Automobile counting, categorization, and velocity and measurement of weight.
Non- intrusive	Automobile existence, flow rate, usage, and speed reports for each class; works over multiple roads; detects cars based on their length.
	System employ includes traffic signal management, vehicle volume and speed measurements, as well as direction movement recognition.
	Useful for tracking velocity as well as the dimensions and numbers of vehicles and available lanes.
	Keeping count of cars, their existence, and their use.
	Employed in the creation of programs that track the passing, presence, and velocity of moving vehicles.
	Measures the temperature of the surface, humidity particular, liquid layer elevation driving conditions, and grip to determine the weather.
	Typically used for the purposes of toll collection.

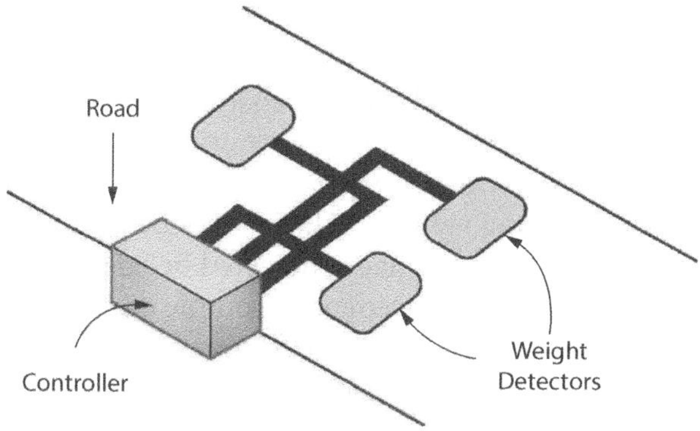

Figure 14.4 Measurement of moving mass [21].

snow, and mist have a significant impact on them. In order to make educated choices that will lead to better traffic conditions, reliable traffic data are crucial. Motorists are more inclined to notice non-intrusive detectors, prompting them to take actions such as breaking down or switching to the proper driving lane. Figure 14.4 shows the above classification.

14.2.1.3 Transport Networks with Intelligence

Traffic management is becoming more important as the number of cars on the road continues to rise rapidly; this is particularly true at crossings. The usage of signals is a common method for controlling traffic flow. These normally have unadjustable changeover interval periods (from red to a green color and to yellow). Stop-and-go driving in traffic causes considerable increases in both fuel usage and carbon emissions. It is preferable to use a system that can react to changing traffic circumstances. A technique of estimating the total number of vehicles entering a junction would provide useful data for the dynamic modification of switching interval timings in response to shifting traffic conditions. Effective techniques for detecting traffic are necessary for the implementation of such a smart system. Others have recommended doing away with traffic signals and signage altogether. In this scenario, automobiles are wirelessly linked to one another, allowing them to share information and make decisions to help each other avoid collisions. Intersection crossings are managed between cars using time slots. If a car does not make it through the junction inside its allotted time, it must wait until the next one comes around. Radio frequency identification (RFID) was suggested as a means of enabling wireless communication between devices. An automated speed-control system that takes traffic signals into account to manage the vehicle's speed, a Hall Effect sensor, was installed on each wheel, and RFID readers were mounted on the vehicle's doors. RFID tags were installed on traffic signals to facilitate interaction among the automobile and the road signs.

Induced loop sensors, which are cheap and may be buried in highways, are one means of detecting vehicles. Metals may be located with the help of loop sensors. A vehicle's presence is detected because its resonance rate rises as it gets closer to the loop. Depending on the dimensions of the automobile, the rise in periodicity may be used to roughly determine the type of car that has been detected. For traffic control purposes, these data may be utilized to make educated guesses about variables like vehicle velocity. Mass in motion is an alternative approach that may be used to determine a vehicle's weight. The sensor mats of a Weigh-in-motion (WIM) system measure the total and axle weights of passing vehicles. There are many different kinds of WIM systems; some examples include piezoelectric element systems, capacitive mats (a pair of inductive circuits and a capacitive dimensions sensor), stretching plates (which, in turn, via the WIM generate a phase change), load cells (which in turn use one load cell with two scales for every side of the car's axle), and monitoring the strain on the plate due to the changing load of the car. The installation and upkeep expenses for the aforementioned techniques are often substantial. Figure 14.4 depicts a WIM system [21].

14.2.2 UAV Sensors in Remote Automobile Sensing

Almost every modern application (including farming, logging, and mining) that requires observed data from above or sideways perspectives makes use of UAV sensors and platforms.

Although they aim to be a universal remote sensing (RS) tool, most of the techniques for processing and analyzing RS data are still case-specific [22]. Table 14.3 gives the details of the sensors.

14.2.2.1 WSN for City Traffic Control System

Modern society's reliance on automobiles has led to a dramatic rise in traffic, yet the underlying infrastructure of our roadways and transportation networks has not kept pace with this growth. This has had a negative social and economic impact on several markets throughout the globe as traffic congestion and pollution connected to vehicular traffic have worsened. Potentially life-threatening traffic bottlenecks caused by an unchanging control system should be avoided. There has been a rise in interest in using

Table 14.3 Unmanned aerial vehicle (UAV) sensors [22].

RGB Cameras	The identification and monitoring of individuals and vehicles, as well as additional visual analytics and map tasks.
Light-Weight Multispectral Camera	Analysis of images, tracking of crops and vegetation, mining, assessment of soil moisture, identifying fires and water levels, mapping of land usage and cover, etc.
Light-Weight Hyperspectral Sensors	Applications include mapping land use and cover, estimating vegetation indices and biophysical/ physiological/biochemical parameters, identifying plant diseases, assessing disaster damage, and many more.
Light-Weight Thermal Infrared Sensors	Useful for monitoring wildlife, locating volcanoes, finding forest fires, studying hydrothermal activity, measuring urban heat islands, etc.
UAV LiDAR	Analysis of plant canopies, calculation of forest carbon sequestration, mapping of historic sites, etc.

WSNs, or wireless sensor networks, for the purpose of traffic monitoring and congestion avoidance. Wireless sensor networks (WSNs) are now all the rage since they can send data quickly, need little in the way of setup or upkeep, and are low-cost and small in size.

Wireless sensor networks, RFIDs, the ZigBee VANETs as Bluetooth devices, cameras, and infrared (IR) signals have all been used by researchers in the last several decades for real-time traffic monitoring. In today's technological landscape, WSNs provide an exciting opportunity to improve the planning and creation of many traffic management applications. There are sensor nodes and gateway nodes that are in the sensor network. The role of a sensor node was to keep an eye on traffic in a certain region by collecting data from a variety of sensors that may measure things like speed, volume, density, headway, wait times, throughput, and pollution. The gateway node coordinates the dissemination of node-level traffic information to the base station.

Figure 14.5 depicts a typical WSN-based urban traffic management system (W-UTMS). With a dispersed W-UTMS application, you may (i) gather data, (ii) disseminate that data, (iii) analyze the information in order to plan the necessary activities, and (iv) put the right actions into

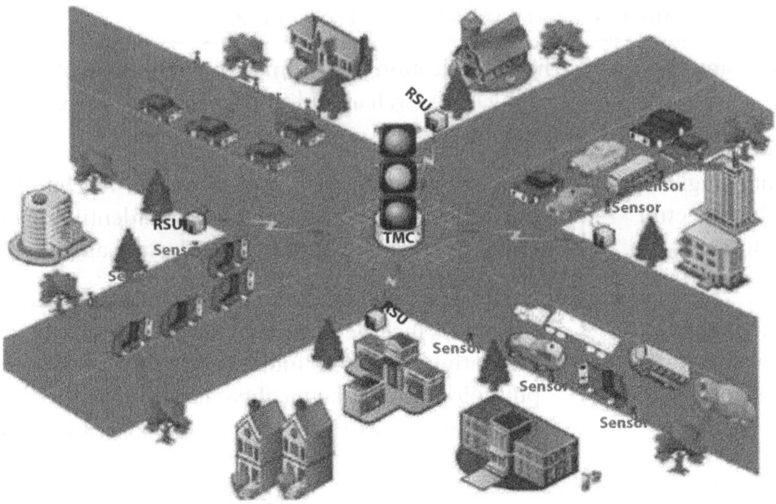

Figure 14.5 WSN-based urban TMS [23].

motion. The UTMS is outfitted of wireless sensors that allow a transportation management center (TMC), a road side unit (RSU), and on-board units (OBUs) in cars to do these tasks automatically. The RSU receives real-time traffic data from sensors that track variables including vehicle density, vehicle type, average waiting time, and pollution levels. The OBU alerts the RSU when an emergency vehicle is approaching the junction.

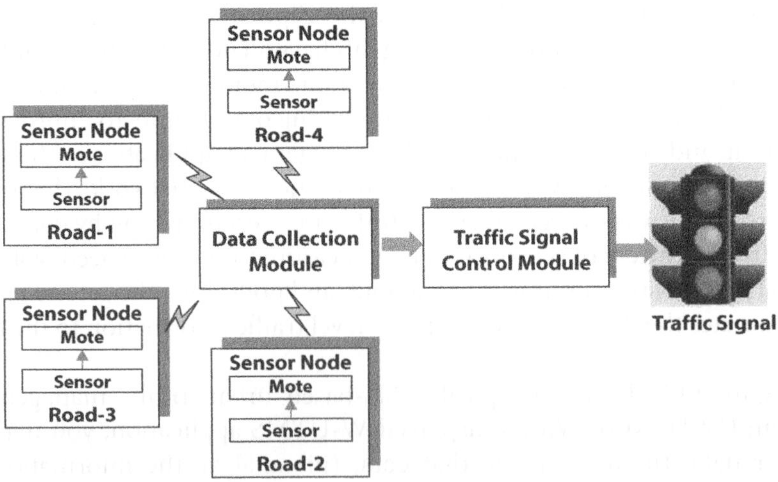

Figure 14.6 Sensors in TMS [23].

The RSU collects information from all of the sensor nodes and OBUs and then sends that information on to the TMC.

The TMC's traffic signal management module receives data from the TMC's data collecting module, which analyzes and processes the data based on the traffic characteristics. The data are analyzed, and the control module makes discerning choices to give emergency vehicles precedence at the signal lights [23]. As soon as an emergency vehicle has passed, normal service is resumed. Using this approach, congestion in cities, rescue vehicle priority, car waiting time, energy use, and safety may all be optimized as in Figure 14.6 and the technologies as in Table 14.4.

14.2.3 The Video Intelligent Car Using WSN

The development of both society and technology shows a consistent upward trajectory. People have entered the age of information, numbers, and intelligence thanks to the proliferation of electronic information technology, which has exploded in popularity over the last two decades. These days, robotic transportation for logistics and an adaptable production organization system rely heavily on intelligent vehicles and the goods based on them due to the rising importance of automation technology in modern corporate production. Intelligent design of cars is something that many nations are currently seeking [23].

An intelligent vehicle, often known as an unmanned vehicle, is a vehicle that is capable of performing a variety of tasks without human intervention, including environmental observation, planning, decision-making, and multi-scale auxiliary driving. Intelligent vehicles have the ability to avoid obstacles and follow a planned route, even in difficult driving circumstances. The smart automobile is a subfield of intelligent vehicle research that focuses on the integration of electronic components such as laptops sensors, information, communication, navigation, and autonomous control systems. It is the standard brand of cutting-edge and brand-new. Automatic barrier detection, alert, stoppage, secure distance servicing, speed and cruise control, and more are all features.

The significance of an intelligent vehicle lies in the following areas:

(1) The identification of threats and the mitigation of associated risks: engineering and military environments both provide unique hazards. The intelligent vehicle may be used to clear roadside explosives and search and eliminate mines on the battlefield. In civil applications, it may search

Table 14.4 Traffic sensing technologies [23, 23a].

Trends	Working	Special components
Inductive loop	A car or other conductive metal item may be detected by an inductive-loop sensor by measuring the decrease in loop inductance caused by currents generated in the object.	The transmitter and receiver antennas, a tag with a reader system, a computer, and a computer-read tag.
RFID	RFID is a method that uses radio waves to transfer data between the reader and an electronic label mounted on a car.	Transmission and reception antennas, a processing unit, and a control unit.
Microwave radar	Microwave radar uses radio waves to send out signals in recognition zones and pick up the echoes off of passing cars. The reflected information is analyzed to determine the vehicle's velocity and heading.	Storage devices, preamplifiers, microphones, transducers, and filters.
Acoustic	Through the use of acoustic sensors, the presence and velocity of moving vehicles may be determined simply by listening for the noises they make.	Micro-loop instruments, a control module, and a magnet probes detector.
Magnetometer	Vertical and horizontal parts of the magnetic field of the Earth may be detected using magnetometers' sensors.	Magnetic probe detector, micro loop probes, and control unit.
Magnetic	A vehicle may be detected by a magnetic sensor because a change in the magnetic field of the Earth caused by a ferrous metal item can be measured.	Micro-loop probes, a control module, and a magnetic probe detector.

for buried persons after a powerful earthquake, identify chemical leaks, etc.

(2) Highway quality assurance and failure analysis in the engineering construction industry may be completed with a simple automated cruise along the highway as part of a comprehensive safety inspection damage assessment. It has applications in the manufacturing sector, where it may be used to locate leaks and damage points in oil and gas pipelines, as well as detect mechanical damage and flaws like fractures in industrial pipelines.

(3) Armed surveillance and the environment. Recognition: military reconnaissance in modern war puts forward higher demand, which focuses on building military reconnaissance around the world and taking different effective measures to prevent the enemy's raid, and who is constantly developing multi-purpose detection equipment. Cameras, safes, laser rangefinders, night vision devices, and satellite global location receivers are just some of the gadgets onboard their cars. The range of reconnaissance may be increased, and the timeliness and accuracy of detection can be enhanced. It can also be used for spying on the adversary and collecting information.

The advancement of the wirelessly video smart car is founded on the technology of human remote-control driving, which is used in the computerized and electric car industries. It also assures the continued growth of this remote control with the maturation of Wi-Fi and Bluetooth technologies. This theory and approach are professional and trustworthy, which means it may greatly facilitate production and daily life. Therefore, there will be a massive uptick in the deployment of Bluetooth video intelligent automobile systems.

14.2.4 Effects of CO_2 Inside Cars

The creation of a driver aid powered by deep learning capable of detecting and counteracting the effects of fatigue behind the wheel. Before this tool to be of any use, however, we must first demonstrate a link between drivers' reported feelings of sleep deprivation on lengthy trips and the levels of carbon monoxide (CO_2) in their cars. The five sensors used to assess carbon monoxide (CO), carbon dioxide (CO_2), particulate matter (PM), temperature, and humidity all contribute to the multimodal signals received by the helper. The data from these sensors is sent to a server over the IoT, where a deep neural network analyzes the quality of the air within the car. An air quality anomaly detection model is constructed using a deep network that makes use of long short-term memory (LSTM), skip-generative adversarial network (GAN), and variational auto-encoder (VAE) models. The LSTM is used by the deep learning models to acquire data, whereas the GANs and VAEs are used by the semi-supervised models. The goal of this tool is to keep drivers safe by alerting them to changes in the air quality around them in real time, such as PM levels or drowsy driving [24].

Tiredness, inadequate sleep, persistent exhaustion, and elevated levels of carbon monoxide in the car are all potential contributors to driver sleepiness. When driving for extended periods of time, poor air quality within the car is a major contributor to driver fatigue as well as migraines. Driver fatigue is often brought on by stuffy cars. Sleepiness may be avoided with adequate air flow and aeration.

Sensor and algorithm-based systems for recognizing and processing driver and environmental variables have been the subject of previous research. Camera-based recognition can assess driver fatigue, but it is situationally dependent and expensive to implement widely. Many driving state detection solutions use camera or sensor-based approaches; however, they may be cumbersome to set up and might limit drivers' freedom on the road. In-car DSD systems are also often less trustworthy than rule-based AI ones. The results of DSD systems are also subject to ambiguity since the detection accuracy is dependent on the continually changing in-vehicle environment. In addition, the protocols used by different manufacturers for DSD gear render it mostly useless. Due to their high price, automated driving assistance systems (ADASs) are often only found in the costliest automobiles, putting them out of reach for many buyers. The main sources of CO_2 inside a car are as in Figure 14.7 [24].

The driver may get a headache or feel sleepy if the CO_2 level is over 2,000 parts per million. Brain damage and other symptoms of low oxygen levels may occur at concentrations over 5,000 ppm. These results are in line with those from 2012 research on driver fatigue published by the American Association for Occupational Health, which found that people in enclosed spaces with high concentrations of carbon monoxide may get headaches or become sleepy. The risks to human health from exposure to elevated levels of carbon monoxide are summarized in Table 14.5 [24].

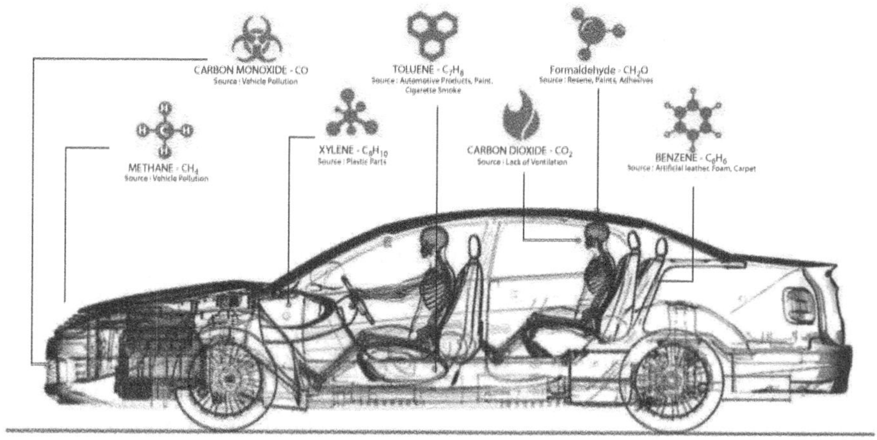

Figure 14.7 Sources of CO_2 inside cars [24].

Table 14.5 Effects of CO_2 inside vehicles [24, 24a].

CO_2 concentration (ppm)	Harm
250–350	The typical ambient pressure
350–1,000	Standard pressure for a well-ventilated space
1,000–2,000	Trouble breathing because to subpar air quality
2,000–5,000	Headaches, tiredness, lowered focus, distractedness, a faster heart rate, and a touch of nausea are possible adverse reactions
>5,000	Occupational exposure restrictions due to abnormal outside air levels, potential toxicity, and oxygen deprivation

14.2.5 Accident Prevention In-Vehicle Air Quality

Indicator of air quality (AQS): Long trips in the car cause a rise in carbon monoxide and a decrease in oxygen because of breathing. In consequence, low oxygen levels result in tiredness and impaired driving. The lack of oxygen, which CO_2 creates, weakens muscles and affects judgment. So, to enhance approaches for preventing tiredness, to improve air quality during car travels, and to reduce contamination, one must establish a self-sufficient system to track vehicle emissions.

Studies have been done recently to develop systems for filtering air pollution, tailpipe emissions, and VOCs, all contributing to poor air quality. The problem is that these filtration systems are not without their flaws. Luxury vehicles' air conditioners measure the concentration of carbon monoxide within the vehicle's interior, recirculate clean air, and sound an alarm if the CO_2 threshold is surpassed. The PM-controlling capabilities of these air conditioners have been enhanced. With the new and enhanced air conditioning system, the car's interior air is expelled if there is a lot of PM in the air, the car will pull air from the outside, filter it, and blow it back into the cabin.

These sorts of technology are already in the works. Devices like concentrations of PM2.5 as in the sensor in Figure 14.8, the use of generators, the seat replacement of filter alert systems, and the incorporated air conditioning 9AC parts like multi-function air purifiers are all gradually being phased out. This includes surface treatment equipment for microbial habitat (Mold) suppression, disinfectors in AC system parts (like heat exchangers), cluster ionizers for air purification, and the generator technology.

The work in [25] created an IoT-enabled vehicle AQS to monitor gaseous and particulate matter pollution. Also, they developed a DSD device for preventing accidents caused by sleepy driving based on data from sensors.

The work also summarizes the performance parameters of the five kinds of sensors used to compile the AQS: carbon monoxide (CO), carbon dioxide (CO_2), particulate matter (PM), temperature, and humidity. The actual value is temperature-dependent. It sets out to create an apparatus and services that can detect and report on a wide range of potential issues that might arise while a vehicle is in use, including but not limited to driver

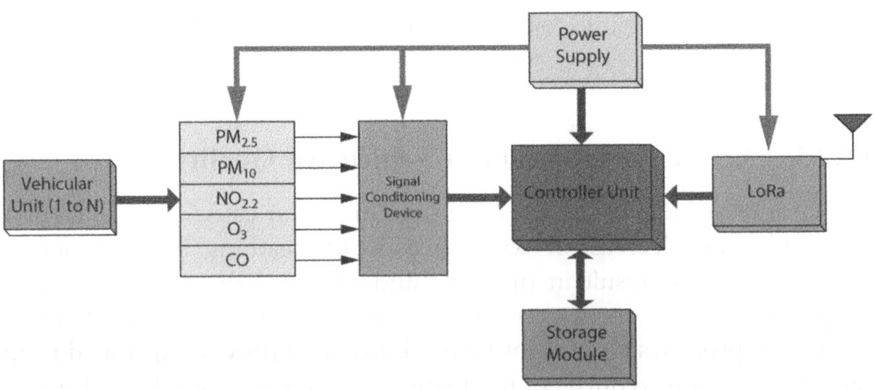

Figure 14.8 Sensor units [25].

fatigue caused by carbon dioxide levels and dangerous amounts of particulate matter within the vehicle, gas leaks, and fires [24, 25].

Sensors, processing networks, data analysis, and system monitoring are the four fundamental pillars of the IoT sensor platform (Figure 14.8). In order to improve management and reaction times, cutting-edge IoT-enabled sensor systems examine data in real time, analyze it, and send out timely alerts depending on measurements. To evaluate the data and spot abnormalities in real time, we leverage cutting-edge technologies like deep learning and networked parallel computing. For each sensor type, pre-trained deep learning systems can reliably track the health of the driver and identify any irregularities in the system's functioning. Sensor systems powered by AI have the potential to lower operational and management expenditures. In fact, these methods may lessen the need for further system checks. In addition, sensor information compensation technology may be used to lessen the initial investment, particularly for big sensors, while simultaneously enhancing the efficiency of low-cost sensors and achieving accuracy on par with that of costly precision instruments.

14.2.6 Sensors for Weather and Obstacle Detection in Vehicles

There has been a rise in the research and development of vehicle obstacle detection systems with the expansion of roads, highways, city streets, and expressways. Recent progress indicates that such a system may now automatically brake when a malfunction or collision danger is recognized, as well as identify locations and calculate distances.

Inventors have created ADAS, or advanced driver assistance systems, which have a smart processing unit that renders a precise picture of the road ahead. Adaptive cruise control, parking aid, and lane-keeping aid

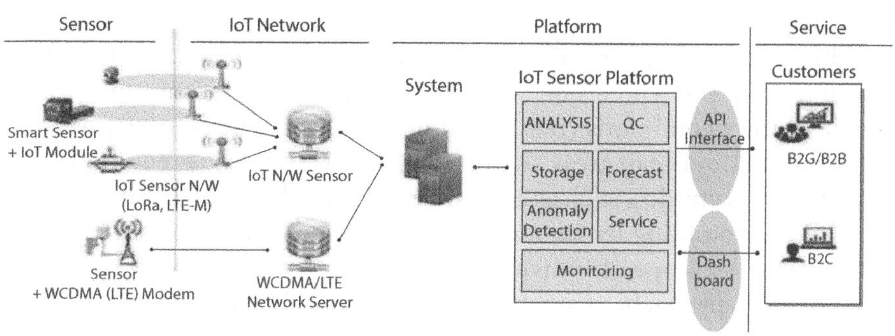

Figure 14.9 IoT sensor platform [24].

are some of the other ADAS features. Autonomous driving will become a reality with the help of ADAS. The technology will eliminate the need for human drivers, boost productivity on the roads, make driving safer, and enhance the commuter experience [26–28].

Cameras, LIDAR, RADAR, SONAR, and IR are only some of the present technical and technological solutions reviewed, and their benefits and drawbacks are discussed. Whether it is daylight or nighttime, dealing with shadows, colors, or reflections from smooth surfaces, various technologies' strengths and weaknesses are discussed in [24] as shown in Figure 14.9.

Radar is mapping from a middle distance out. In extreme conditions, it performs better than cameras and LIDAR, but it cannot identify objects because it does not have enough detail.

LIDAR offers high-resolution mapping in all directions, short and long ranges and is restricted by extreme conditions and poor target reflectivity.

Ultrasound is inexpensive and accurate for short-range measurements and is excellent for use as a parking lot assistance because of its quick reaction time and small range.

A camera's ability to read traffic signals and color buttons effectively and provide a comprehensive view of the surrounding area is contingent on the lighting conditions in which the vehicle is operated.

Infrared (IR) provides the best night vision capability of any sensor. LIDAR's ability to function in low-visibility conditions allows it to be employed at night.

Detection, surveillance, and identification of the driving environment and all surrounding environments, including vehicle and pedestrian, lane, and other objects, are improved when several technologies are fused; nevertheless, additional work is needed to produce more precise sensor fusion systems.

Because data gathered through various technologies are seldom consistent, an advanced data fusion mechanism is required for precise data analytics.

Although highly advanced, current technical improvements are not yet able to ensure 100% precision and accuracy of obstacle recognition in both harsh weather conditions (rain, snow, and fog) and certain particular scenarios in metropolitan areas. Therefore, further research is required.

Different scenarios, such as at night, in bad weather, with smoke present, or while surrounding by smooth surfaces that generate a lot of reflection, are beyond the scope of individual obstacle detection methods. This calls for a hybrid approach, using many technologies. Visual cameras, LIDAR, RADAR, SONAR, and IR are only a few examples of the technologies that contribute to the diversity of the data acquired. Accurate processing of the data is essential to provide a unified perspective; this calls for a complex data fusion method.

14.3 Methodology

14.3.1 Traditional In-Car Eye Blink Detection

Iris identification, visual engagement, and driver tiredness monitoring are just few of the many uses for eye recognition and state of consciousness (close/open) assessment. In most existing work, eye detection comes first, followed by state estimation using a different classifier. This method ignores the relationships between where someone is looking and how they are feeling. In this research, we offer a system that can detect the eyes and estimate their status at the same time. Our technique repeatedly calculates the eye's position and the likelihood that it is obscured by the eyelid using a cascade regression framework. Image characteristics from the center of the eye, as well as context image features from the eyelid and eye corners, are utilized together to estimate the eye location and openness probability at every stage of cascaded regression. It is possible to predict the most probable eye state by using the likelihood of eye opening. Due to the vast amount of annotated face photos with eye-related features needed for training, the work in [29] suggests using a hybrid of actual and synthetic images. Using this learning-by-synthesis technique, its performance is significantly enhanced [29].

Real-time eye blink detection using facial landmarks will calculate a measure called the eye aspect ratio (EAR) to create a blink detector for in-car sensing.

Traditional techniques of calculating blinks in images often utilize a mix of the following:

- Visual positioning.
- Using a threshold to identify the eye whites.
- The absence of "white" around the eyes for a certain amount of time (a sign of blinking) is detected.

Instead, the ratio of the distances between the eye's facial landmarks offers a simple yet elegant solution known as the EAR. This technique for detecting blinks in the eye is quick, accurate, and simple to use [27].

There are four ways to build eye blink detection system:

- How to tell whether a person is blinking in a video frame by analyzing the ocular aspect ratio.
- To identify face landmarks and identify blinks in video streams by writing code in Python, OpenCV, and dlib.
- Implementation to test our approach on real-world data, analyzing camera feeds and video files to identify blinks.
- Various ways that we may enhance our blink detector.

Drowsy driving is a major contributor to vehicular mishaps. If tiredness in the driver can be recognized and alert signals sent to the driver, then many lives may be saved and accidents could be reduced. Drowsiness detecting technology has come a long way in recent years.

Biological signs, vehicle behavior, and facial analysis are the three main categories.

The first class monitors vital signs including heart rate, blood pressure, and brain activity. These methods involve direct touch with the driver yet boast the highest detection accuracy. They are impractical because they are invasive [26].

The second class monitors dynamics like velocity, lateral placement, and yaw angle in moving vehicles. Although these methods may be applied discretely, they are still constrained by factors like vehicle make/model, driver skill level, and traffic circumstances. In addition, it is not cheap and needs specialized equipment.

In the third category, we have facial recognition. This form of indication is non-invasive and simple to apply, although it is not as precise as biological ones. It is useful for any car, regardless of the driver's background.

In addition, when compared to vehicle behavior analysis, it provides better accuracy and more usability. The distance between the camera and the car, as well as the available light, are two major obstacles.

Eyelid motion, pupil activity, movement of the head, and facial expression may all reflect a person's degree of awareness while driving. If a driver's tiredness, alertness, or attention can be predicted based on their eye condition, then this is a potentially useful tool. Several techniques have been developed that use eye state analysis to identify tiredness. Some of these systems for detecting tiredness employed a four-step method, beginning with face identification and progressing through eye being identified, state of vision assessment (open or closed eye), and a sleepy judgment. Traditional cameras were used in these systems, and training data were not necessary.

Face detection was skipped by certain systems. These systems were faster because they skipped this stage, but they required training data or gear like infrared (IR) illuminators, which may be harmful to the eye's health, to identify the eye directly.

In this work, we describe a simple approach for pupil's center and iris border localization based on eye map, as well as a unique technique for eye state analysis, and we provide a new four-step system for sleepiness detection utilizing face analysis. Our method can be used with any standard webcam and requires hardly any kind of training. Therefore, our technology is more reliable, user-friendly, and affordable than IR illuminators.

Unlike IR illuminators, sleepiness detection does not need any kind of training data or specialized cameras, and it is quite safe. The primary drawback of this technology is that it can only be used in daylight conditions when the eyes are not obscured by dark shades.

There are now two categories of driver fatigue detection and warning systems on the market: employing signal processing and image processing, respectively.

Input data might come from a variety of sources, such as a standard camera, an IR camera, or even bio-signals; thus it is important to know what approaches are available for each [28]. Driver drowsiness can be detected in one of two ways: either by placing a camera in front of the driver and analyzing the driver's face for signs of yawning or eye closure or by using a single detection method to determine whether or not the driver is drowsy. This study is capable of identifying left- or right-side driver fatigue. This article uses a regular camera as an input data source for real-time sleepiness diagnosis based on facial recognition using deep learning via neural network. Use a Raspberry Pi 3 with an external camera to detect eye closure, calculate the driver's drowsiness rate, and sound an alarm if any of the variables rise too high; these solutions, however, are not applicable in low-light conditions, and some of them rely solely on eye closure as a symptom of drowsiness. Infrared cameras are seldom utilized in scientific studies. One of them uses a histogram to identify the driver's face and eyes, classifying his or her eyelids as completely open, partially open, and closed. While an IR camera was utilized to capture the input picture, tiredness in the driver was only measured by the degree to which the driver's eyelids were closed.

14.3.2 Proposed Methodology

In this work, we propose the method by which the driver of any automobile is made aware of this drowsiness while driving the automobile. First,

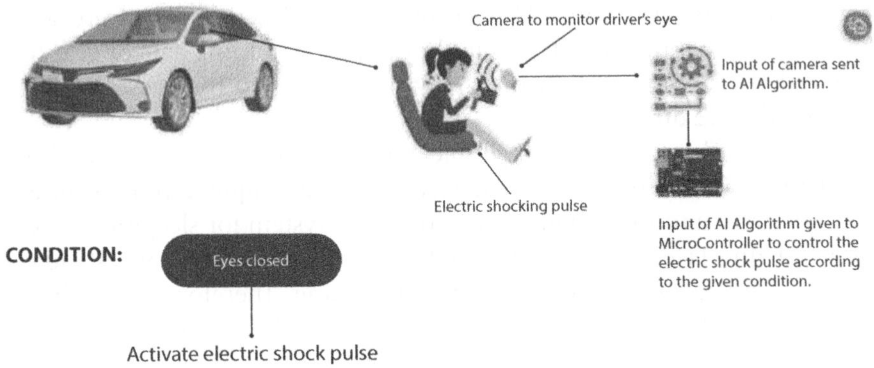

Figure 14.10 Proposed method for detecting driver drowsiness.

using image processing and the sensors installed in front of the driver's driving space, the eyes are continuously monitored for drowsiness. An algorithm is designed in such a way that when the driver closes his or her eyes for more than the threshold time, an alert is sent to the control system, which is situated in the body of the car. The sensors are also monitoring the interval at which the eyelids are closed and opened and their uniformity. Even for this, the threshold is set in the control system. The sensors are also responsible for sensing the bending of the head downward, which is a sign of drowsiness. With all these inputs, the control system makes the decision to pass the shock signals, which are impacted on the lower part of the driver's leg, to alert him or her of the drowsiness state as shown in Figure 14.10. This will help in the reduction of road accidents.

14.4 Conclusions

The development of hardware needed for applications involving sensing will undoubtedly be driven by technical progress; hence, it stands to reason that technological progress will play a significant role in advanced sensing. While smart sensing for automobile industry have been successful, many obstacles remain. Solving these problems comprehensively will need teams of experts from different fields to work together. In order to implement integrated systems with regard to of sensor and communications infrastructures, for example, standardization initiatives should take into account all concerned stakeholders, including municipalities, utilities, and other service providers. Multiple infrastructure (in regard to sensor systems) will be ineffective, but bringing all of these amenities underneath one umbrella would be difficult. In this work, we present the way by which the driver of any car is made aware of this sleepiness while driving the automobile. To begin, the eyes are constantly checked for signs of tiredness using image processing and the sensors mounted in front of the driver's driving zone. When the driver's eyes are closed for longer than the predetermined amount of time, an alarm is transmitted to the central control unit within the car's chassis. The sensors are also tracking the regularity and frequency with which the subject's eyelids are opened and closed. The control system even has a threshold for this. The sensors may also detect tiredness by looking for a downward tilt of the head. Based on these measurements, the control system decides whether or not to transmit shock impulses to the driver's lower leg in order to wake them up from their sleepiness. There will be fewer car accidents as a result of this.

References

1. Chang, W., Ellinger, A.E., Kim, K., Franke, G.R., Supply chain integration and firm financial performance: A meta-analysis of positional advantage mediation and moderating factors. *Eur. Manage. J.*, 34, 282–295, 2016.
2. Castelo-Branco, I., Cruz-Jesus, F., Oliveira, T., Assessing Industry 4.0 readiness in manufacturing: Evidence for the European Union. *Comput. Ind.*, 107, 22–32, 2019.
3. Cimini, C., Pirola, F., Pinto, R., Cavalieri, S., A human-in-the-loop manufacturing control architecture for the next generation of production systems. *J.Manuf. Syst.*, 54, 258–271, 2020.
4. Tao, F., Qi, Q., Liu, A., Kusiak, A., Data-driven smart manufacturing. *J. Manuf. Syst.*, 48, 157–169, 2018.
5. Wilkesmann, M. and Wilkesmann, U., Industry 4.0—organizing routines innovations? *VINE J. Inf. Knowl. Manage. Syst.*, 48, 238–254, 2018.
6. Shrouf, F., Ordieres, J., Miragliotta, G., Smart factories in Industry 4.0: A review of the concept and of energy management approached in production based on the Internet of Things paradigm, in: *Proceedings of the 2014 IEEE International Conference on Industrial Engineering and Engineering Management*, Bandar Sunway, Malaysia, pp. 697–701, 9–12 December 2014.
7. Kaur, J. and Kaur, K., A fuzzy approach for an IoT-based automated employee performance appraisal. *Comput. Mater. Contin.*, 53, 24–38, 2017, [Google Scholar].
8. Zanella, A., Bui, N., Castellani, A., Vangelista, L., Zorzi, M., Internet of things for smart cities. *IEEE Internet Things J.*, 1, 22–32, 2014.
9. Ghasemaghaei, M., Ebrahimi, S., Hassanein, K., Data analytics competency for improving firm decision making performance. *J. Strat. Inf. Syst.*, 27, 101–113, 2018.

10. Ortiz, A.M., Hussein, D., Park, S., Han, S.N., Crespi, N., The Cluster between Internet of Things and Social Networks: Review and Research Challenges. *IEEE Internet Things J.*, 1, 206–215, 2014.
11. Mehami, J., Nawi, M., Zhong, R.Y., Smart automated guided vehicles for manufacturing in the context of Industry 4.0. *Proc. Manuf.*, 26, 1077–1086, 2018.
12. Aydiner, A.S., Tatoglu, E., Bayraktar, E., Zaim, S., Information system capabilities and firm performance: Opening the black box through decision-making performance and business-process performance. *Int. J. Inf. Manage.*, 47, 168–182, 2019.
13. Kalsoom, T., Ramzan, N., Ahmed, S., Ur-Rehman, M., Advances in Sensor Technologies in the Era of Smart Factory and Industry 4.0. *Sensors*, 20, 6783, 2020, https://doi.org/10.3390/s20236783.
14. Guerrero-Ibáñez, J., Zeadally, S., Contreras-Castillo, J., Sensor Technologies for Intelligent Transportation Systems. *Sensors, 18*, 1212, 2018, https://doi.org/10.3390/s18041212.
15. NHTSA to Require Backup Cameras on All Vehicles, USA Today, Available online: https://www.usatoday.com/story/money/cars/2014/03/31/nhtsa-rear-view- cameras/7114531/ (accessed on 01 August 2023).
16. Automotive Sensors and Electronics Expo, Available online: http://www.automotivesensors2017.com (accessed on 01 August 2023).
17. Fleming, W.J., New automotive sensors—A review. *IEEE Sens. J.*, 8, 1900–1921, 2008.
18. Abdelhamid, S., Hassanein, H.S., Takahara, G., Vehicle as a Mobile Sensor. *Proc. Comput. Sci.*, 34, 286–295, 2014.
19. Bargagli, B., Manes, G., Facchini, R., Manes, A., Acoustic sensor network for vehicle traffic monitoring, in: *Proceedings of the First International Conference on Advances in Vehicular Systems, Technologies and Applications*, Venice, Italy, 24–29 June 2012.
20. Mathew, T.V., *Transportation Systems Engineering*, 2014, Available online: http://nptel.ac.in/downloads/105101008/ (accessed on 01 August 2023).
21. Hancke, G.P., Silva, B.D.C.e., Hancke, Jr., G.P., The Role of Advanced Sensing in Smart Cities. *Sensors*, 13, 393–425, 2013, https://doi.org/10.3390/s130100393.
22. Yao, H., Qin, R., Chen, X., Unmanned Aerial Vehicle for Remote Sensing Applications—A Review. *Remote Sens.*, 11, 1443, 2019, https://doi.org/10.3390/rs11121443.
23. Nellore, K. and Hancke, G.P.A., Survey on Urban Traffic Management System Using Wireless Sensor Networks. *Sensors*, 16, 157, 2016.

23a. Padmavathi, G, Shanmugapriya, D., Kalaivani, M., A study on vehicle detection and tracking using wireless sensor networks. *Wirel. Sens. Netw*, 2, 173–185, 2010.

24. Chung, J.-J, and Kim, H.-J., An Automobile Environment Detection System Based on Deep Neural Network and its Implementation Using IoT-Enabled

In-Vehicle Air Quality Sensors. *Sustainability*, 12, 2475, 2020, https://doi.org/10.3390/su12062475.

24a. Increased Possibility of Drowsy Driving when Running Continuously for Two Hours in a Confined Space, http://www.korea.kr/news/pressReleaseView.do?newsId=156120968 (accessed on 02 September 2023).

25. Rana, A., Rawat, A.S., Afifi, A., Singh, R., Rashid, M., Gehlot, A., Akram, S.V., Alshamrani, S.S.A., Long-Range Internet of Things-Based Advanced Vehicle Pollution Monitoring System with Node Authentication and Blockchain. *Appl. Sci.*, 12, 7547, 2022, https://doi.org/10.3390/app12157547.

26. Tabrizi, P.R. and Zoroofi, R.A., Open/Closed Eye Analysis for Drowsiness Detection. *2008 First Workshops on Image Processing Theory, Tools and Applications*, Sousse, Tunisia, pp. 1–7, 2008, doi: 10.1109/IPTA.2008.4743785.

27. Nabi, Z., Alvi, A., Mehmood, R., Strang, T., Towards Standardization of In-Car Sensors, in: *Communication Technologies for Vehicles.* Nets4Cars/Nets4Trains 2011. Lecture Notes in Computer Science, vol. 6596, A. Festag, A. Vinel, R. Mehmood, C. Rico Garcia, M. Röckl (Eds.), Springer, Berlin, Heidelberg, 2011, https://doi.org/10.1007/978-3-642-19786-4_19.

28. Tipprasert, W., Charoenpong, T., Chianrabutra, C., Sukjamsri, C., A Method of Driver's Eyes Closure and Yawning Detection for Drowsiness Analysis by Infrared Camera. *2019 First International Symposium on Instrumentation, Control, Artificial Intelligence, and Robotics (ICA-SYMP)*, Bangkok, Thailand, pp. 61–64, 2019, doi: 10.1109/ICA-SYMP.2019.8646001.

29. Gou, C., Wu, Y., Wang, K., Wang, K., Wang, F.-Y., Ji, Q., A joint cascaded framework for simultaneous eye detection and eye state estimation. *Pattern Recognit.*, 67, 23–31, 2017. ISSN 0031-3203, https://doi.org/10.1016/j.patcog.2017.01.023.

15

Design of Arduino UNO–Based Novel Multi-Featured Robot

Jaspinder Kaur[1], Rohit Anand[2*], Nidhi Sindhwani[3], Ajay Kumar Sharma[1] and Vishal Jain[4]

[1]*Department of Computer Science and Engineering, National Institute of Technology, Delhi, India*
[2]*Department of ECE, G.B. Pant DSEU Okhla-1 Campus (formerly GBPEC), New Delhi, India*
[3]*Amity Institute of Information Technology, Amity University, Noida, Uttar Pradesh, India*
[4]*Department of Computer Science and Engineering, School of Engineering and Technology, Sharda University, Greater Noida, Uttar Pradesh, India*

Abstract

This chapter is based on a novel Arduino-based multi-featured robot. The components required are Arduino UNO, ultrasonic sensor, Bluetooth module, servo motor, and a few more hardware devices. With the ever-increasing technological advancement, it has become easier to communicate from the different places; it could be nearby, few kilometers away, or very far, thousands of kilometers away. Thus, it has come to notice that there are a few places where it is hard and dangerous to explore for the human beings. Using Arduino controller, ESP32 CAM, and few sensors, we can explore such locations. Nowadays, with the increase of population, it is hard to invigilate all the students during an examination. So, this proposed model of multi-featured robot is very useful in keeping an eye, as it can send live surveillance and also any notification if human motion is detected.

***Keywords*:** Arduino UNO, bluetooth module, ultrasonic sensor, motor driver, servo motor

**Corresponding author*: roh_anand@rediffmail.com

Kolla Bhanu Prakash, Satish Kumar Peddapelli, Ivan C. K. Tam, Wai Lok Woo and Vishal Jain (eds.) Computational Intelligent Techniques in Mechatronics, (471–490) © 2024 Scrivener Publishing LLC

15.1 Introduction

As we know, the majority of Indian businesses are run by small owners, and the economy does not allow the small businesses to purchase the upmarket movers/robots/transporters to take advantage for their work. The model was thought and designed mainly because of the situations that we are currently seeing around us and how the COVID-19 taught us to stay away from each other [1–3]. The bots are developed in such a way that they can help the people in transporting items like office files, small tools, and ammunition to the army in war areas. The applications are endless as this bot can be modified in any way to give a particular user the best experience.

As time passed, single-function robots were advanced to multi-functioning robots. Nowadays, at the time of technological advancements, robots are the most sought-after challenge, creating various types from nano-sized to giant-sized, from fixed robots to moving robots, even wheeled to walking or running robots, even two- or four-legged or also eight-legged robots [4–7].

Arduino-based multi-functioning robot is designed to help the army, schools, universities, shopping malls, the small-scale businesses, and industries in transporting of goods from one place to another without the need of actual humans traveling that distance. Low cost of production and easy-to-use interface are the two key features of our model device. Arduino on its own is used as a fundamental development board for the various small-scale applications. Because of its open-source origin, it can be modified as per user's need; it provides vast diversity and freedom to operate. That is the main reason behind using this device as the brain of our proposed work is a model based on multi-functioning robots. The main objective of creating the robots all around is to be a helping hand to society [8–10]. It may be easier to visualize but is very complex in making. Thus, approved products for many jobs are in high demand due to technological advances [11–13].

Many robots are built primarily using an Arduino controller as it is easier to use or program [14–16]. Sensors are used to obtain the information. For example, ultrasonic sensors are used to find obstacles; passive infrared (PIR) sensors are used for the detection of human movements [17–19]. There are lots of areas that are yet to be explored or yet to be seen by human eyes such as time surveillance or cam-based robots [20–22]. It is also highly in demand by the military as it can help in keeping track of movements or surroundings [23, 24]. It can also be very helpful if used by common citizens as it can help increase home security or keep surveillance in classrooms [25–30]. This work has been proposed with the main intention of making it easier to invigilate exam centers through remote places with the help of live surveillance and to make note of human movement (if found) and inform the user [31–34].

The robot automatically detects the obstacles and can be used to reach humans. Its direction and speed are controlled using the L293D motor driver IC [35–38]. When a movement/barrier is detected, the data are printed on a serial monitor. In addition, live surveillance can be provided to monitor surroundings and movement [39–42].

In this chapter, a novel multi-featured robotic mechanism system with a Bluetooth module using Arduino UNO board has been presented and discussed in detail. This locking mechanism system shows a number of advantages such as precise reliability, fast fingerprint scanning speed, and very high protection to the operator and reliability.

15.2 Design Implementation

In this model, a robot is used to perform the various tasks and the most basic and common application for these kinds of robots is to transport goods and items from one place to another with some help from the humans. As an input in this specific part, we will talk about the robot fleet named multi-featured robot. The master robot is operated with the Bluetooth module about which we will elaborate later on, and the slave robots are operated via two main sensors used in every robot these days, i.e., ultrasonic sensor and infrared sensor.

15.2.1 Microcontroller

The most important part of this project, the Arduino UNO R3, is an ATmega328P-based microcontroller that includes all the components needed to hold a microcontroller; we can attach it to a PC using a USB cable, and, if we use Arduino IDE, we can easily configure this program controller [43–46]. As shown in Figure 15.1, Arduino UNO is very simple but, at the same time, an advanced-level programming board. The Arduino UNO is an easy-to-use microcontroller board that is predicated on the Microchip ATmega328P. The board is provided with several input/output pins that are used for the connections with the sensors and motors. The board has 14 digital input/output pins and six analog input/output pins [47–50]. It is coded/programmed on Arduino IDE software where

Figure 15.1 Arduino UNO R3, ATmega328P-based microcontroller.

there are an immense number of libraries available for almost all the sensors that are supported by the board. The board can also power through the USB port or the dedicated power jack which operates on voltages from 7 V to 20 V. Arduino UNO board stands for the first in the line board as it was the start of the microcontroller boards from the Arduino with the USB-ports [51–54].

15.2.2 Motor Driver

As shown in Figure 15.2, L293d IC is known as a motor driver and is a low-power O/S like any other IC. There are other ICs that may have the same/similar functions like L293d, but this may not be the case given a high power to the motor. The continuous bidirectional direct current to the motor is provided by the L293D. The current polarity may change at

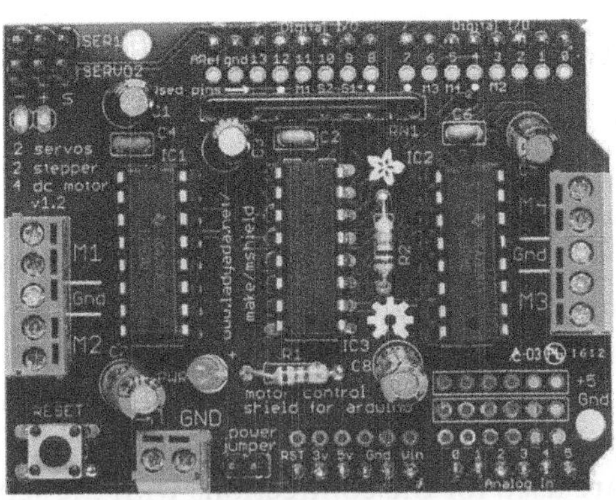

Figure 15.2 L293D motor driver shield, used to control the speed and orientation of motor.

any time without affecting any device in the circuit or itself. The benefit of L293D is that it has an internal H-bridge installed for two motors; thus, we can work on two motors using a single IC. It is mainly used for speed and direction purposes.

15.2.3 Bluetooth

The ESP32 CAM Wi-Fi Bluetooth Development Board with OV2640 Camera Module is based on the ESP32 chip as well as a camera as an addition; it may be a small-sized camera module that can be operated independently even in a minimum system. Since this work is based on IoT, this module benefits us with wireless monitoring or other wireless applications. Figure 15.3 shows the Bluetooth Development Board (ESP32 CAM Wi-Fi) [55].

15.2.4 Ultrasonic Sensor

In this work, an ultrasonic sensor is also used. This sensor uses SONAR to determine the distance to an object like the bats do. It is excellent as it offers us a no-contact range detection with stable and high accuracy in readings and easy-to-use package. It has both transmitter and receiver modules built-in [56, 57]. Figure 15.4 depicts the ultrasonic sensor [58].

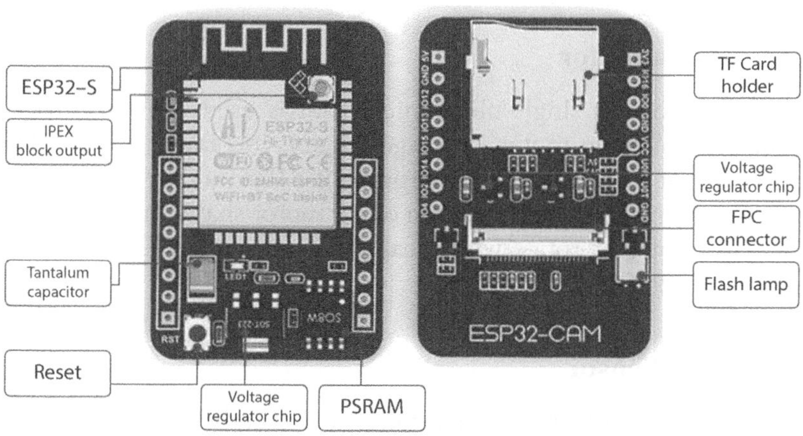

Figure 15.3 The ESP32 CAM Wi-Fi bluetooth development board with OV2640 camera module.

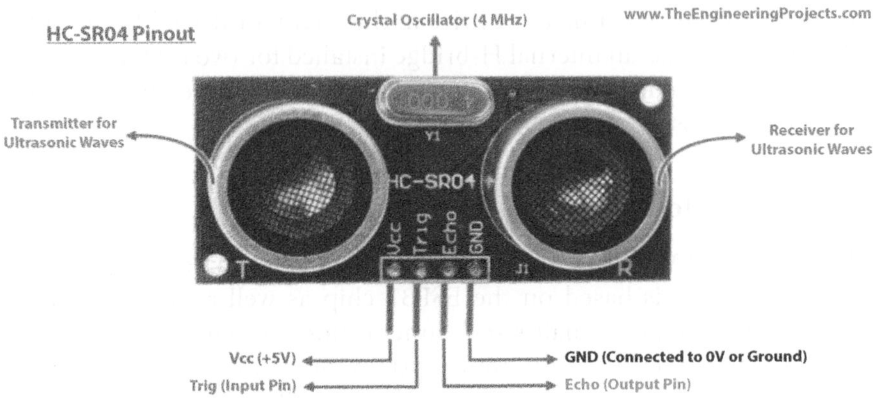

Figure 15.4 Ultrasonic sensor, used to detect the distance of an object.

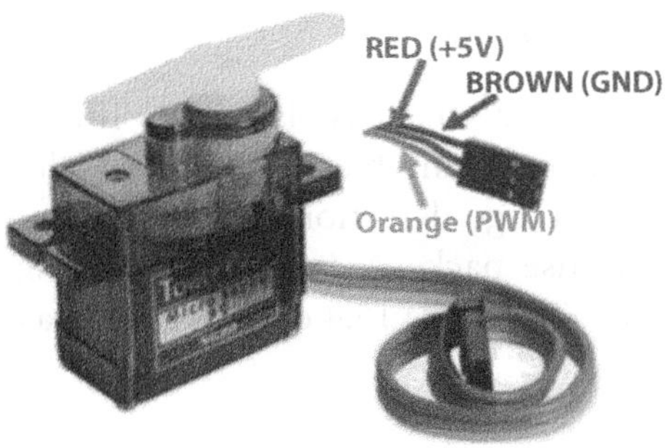

Figure 15.5 Servomotor.

15.2.5 Servomotor

In this work, only a single ultrasonic sensor is used. However, with the introduction of servo motor, coverage range is increased. It is helpful for the coverage of a wide range of environments. This motor provides us with the feedback on the current position of the motor shaft; thus, the feedback allows this motor to rotate with great precision. Figure 15.5 displays the servomotor [59].

15.2.6 PIR Sensor

In this work, a PIR sensor is also installed to detect the movement [60]. It is small, has low consumption, and also provides us with accurate detection. Figure 15.6 shows a PIR sensor.

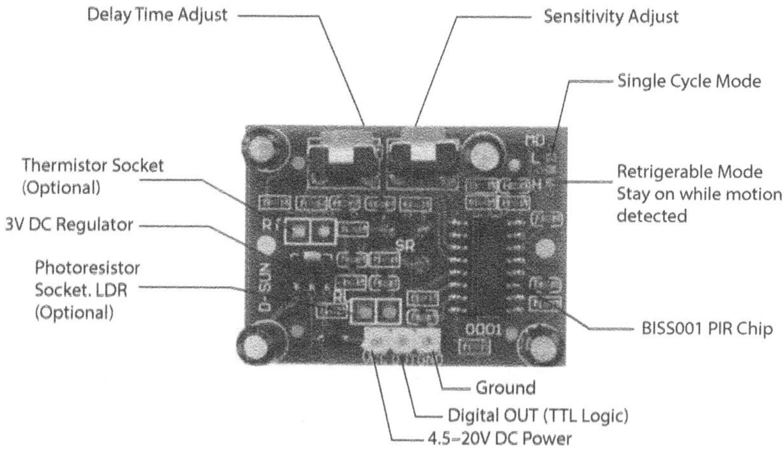

Figure 15.6 PIR sensor.

15.2.7 Components Specifications

- o Arduino UNO R3
- o FT232RL USB to TTL 3.3V/5V FTDI Serial Adapter
- o ESP32 Camera Development Board Wi-Fi + Bluetooth Module with OV2640 Camera Module
- o Tower Pro SG90 Servo Motor - 9 gms Mini/Micro Servo Motor
- o HC-SR04 Ultrasonic Sensor Module
- o HC SR501-PIR Motion Detection Sensor
- o L293D Motor Driver IC
- o Slide Switch
- o Jumper Wires

Apart from the above component used, Tinker CAD software is used for the simulation purpose.

15.3 Proposed Model

This section gives the explanation of the working of the Arduino-based multifeatured robot. It gives the information about the flow of data as shown in Figure 15.7, Figure 15.8 and Figure 15.9. Table 15.1 depicts the list of components used in this work.

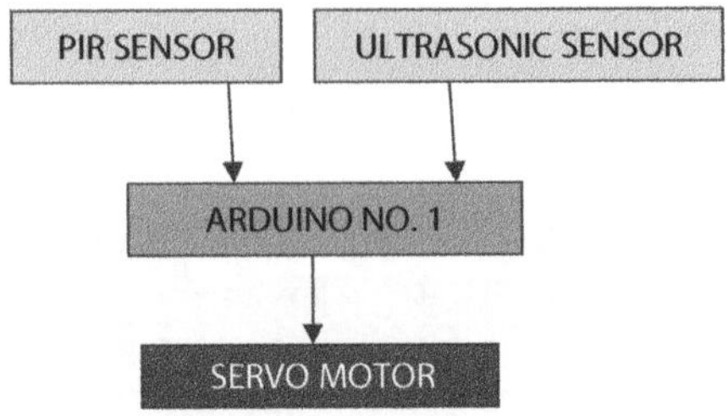

Figure 15.7 Block diagram of sensor.

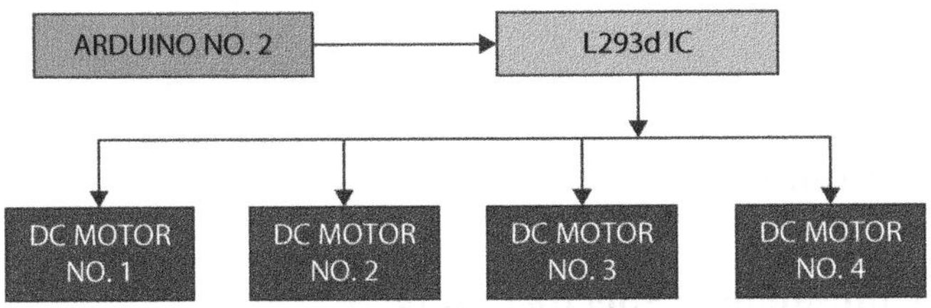

Figure 15.8 Block diagram of IC motor.

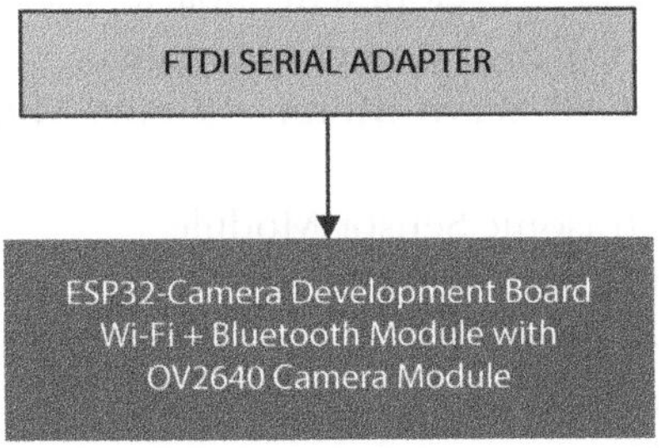

Figure 15.9 Block diagram of ESP32-CAM.

15.4 Process and Working Methodology

The COVID-19 global pandemic showed us how humans have to stay away from each other in any situation not only for theirs but also for the safety of others. The work has developed keeping in mind the same situation in which everyone is working in, in the current days. The work goes around a very popular and widely used development board Arduino UNO. We chose this board for not one but many reasons as it is easy to work on; libraries are well built, and enormous data are available. We started the design with

Table 15.1 Components used in the multifeatured robot.

Component	Power supply	Measurements	Size (mm²)
Arduino Uno	3.3 V/5 V	-----	68.6 × 53.4
Ultrasonic Sensor	5-V DC	2 cm to 400 cm	45.7 × 21.3
PIR Sensor	5 V	300 cm to 700 cm	24.03 × 32.34
Servo Motor	5 V	180° rotation	40.6 × 20.5 × 38.0
DC Motor	6 V	-----	25 × 15 × 20
L293d Motor Driver IC	5 V	-----	48 × 34 × 14
ESP32 Camera Development Board Wi-Fi + Bluetooth Module with OV2640 Camera Module	5 V	----	40 × 27
FTDI Serial Adapter Module	3.3 V/5.5 V	----	36 × 17.5

pretty simple yet very practical approach. We made novel multi-purpose robot in such a way that it can be modified in any way or form with just a little effort, which also makes it one of the only bots here in the industries that are capable of these kinds of manipulations.

The sensors that we used are the best possible that we could get in the budget and in the difficult situations due to the global pandemic. All the components that we used such as Arduino UNO, HW-201 infrared proximity sensors module, HC-05 Bluetooth module, HC-SR04 ultrasonic sensor module, L293D motor driver shield, BO 300 rpm motors, B.O. Motor wheels (diameter: 7.5cm; width: 1.5cm), jumper wires, and 12-V lithium-ion battery (2200 mAh) are the best that we could find in the market.

The calibration of these many chips together is not an easy task; one has to put in each and every part of their knowledge in order to calibrate all the sensors together to work as a team so as to provide a smooth output. Not only the sensors but the calibration of all the four motors together is a job in itself. The calibration is tough because of many factors. Conditions in which they are being calibrated are subjected to drastic changes; the quality of chips is not as great as they should be, and, hence, the values are never

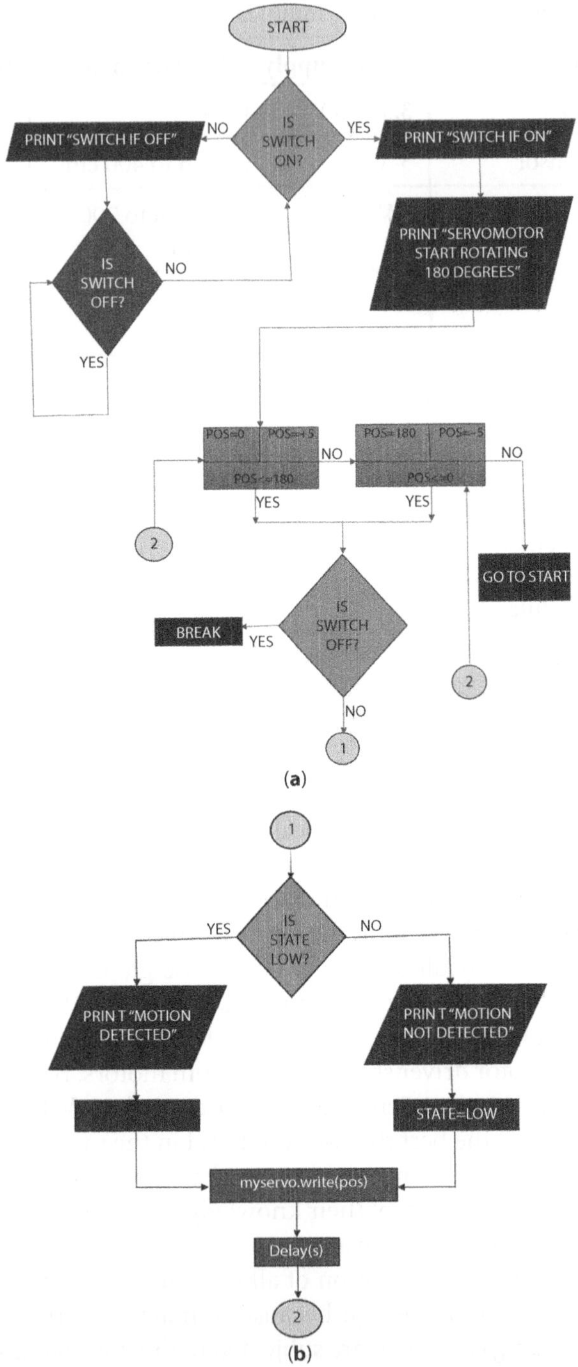

Figure 15.10 (a, b) The flowchart of the working of robot.

accurate enough, and the sensors cannot be calibrated for the predefined values because they do not output the same results for every unit change in the given inputs [61].

The flowchart for the working of the robot is shown in Figure 15.10 (a) and (b). The working methodology is explained with the help of both the flow charts.

The core of the robot which is the Arduino Uno R3 microcontroller is coded using Arduino IDE. A slide switch is used to power ON/OFF. When the switch is turned ON, the Arduino program runs; else, it runs on an empty loop.

We have used two Arduino: one dedicated for driving and the other for sensors and servo. The main idea is to use a single ultrasonic sensor and to widen the coverage using a servo. More specifically, if we attach the ultrasonic sensor on top of the servo, then the servo rotates the sensor, collects data and finds any object in the way, and prints it on the serial monitor. PIR sensor in general is placed independently and works simultaneously with the other components. Figure 15.11 shows the circuit diagram of sensor.

In the proposed work, we have used two Arduino's as we cannot run two programs together even if combined as it consumes too much time.

We have mainly used L293D IC for direction control and speed control. The online simulation is shown using Tinker CAD. Figure 15.12 depicts the circuit diagram of L293D IC.

Using FTDI (Future Technology Devices International) serial data adapter, we configure the ESP32 cam module in Arduino IDE, and, hence, we can find a readymade program in Arduino IDE.

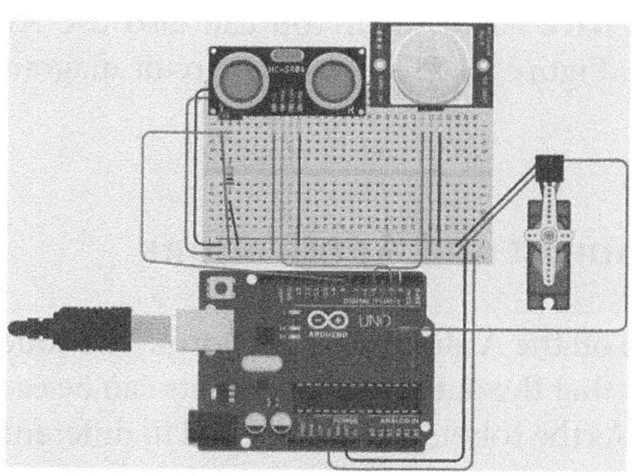

Figure 15.11 Circuit diagram of sensors.

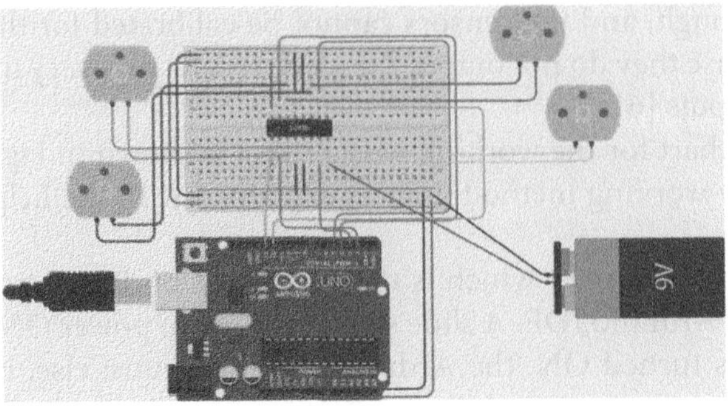

Figure 15.12 Circuit diagram of L293d IC.

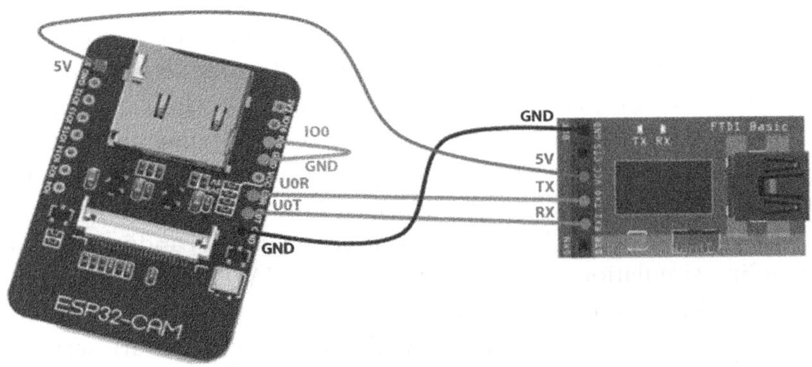

Figure 15.13 Circuit diagram of ESP32 CAM module.

Choose your SSID, PASSWORD, and MODULE and upload and run the program and then open the serial monitor. You will find an IP address, go to the IP address, and you will be able to see the camera running.

If you do not have FTDI, then you can also use Arduino to configure ESP32 CAM. Figure 15.13 shows the circuit diagram of ESP32 CAM Module.

15.5 Experiment and Applications

Coding was done on the Arduino IDE platform. The codes were developed in such a manner that the outputs of the robots can be easily modified with not much effort. As the robots have to be used in different states and forms, the coding has to be modified according to the needs as all the goods these robots are carrying will be of different weight ranges [62].

Working of both the master and slave bots are distinctive from each other. The master bot runs exclusively on Bluetooth sensors, whereas slave bots run on the cumulative data from the IR and ultrasonic sensors. Master bot uses the HC-05 Bluetooth module to communicate with the controlling device. The sensor is operated using four main signals that are left, right, forward, and backward. These four major signals are manipulated to produce other small adjusting movements. The signals received by the controller are processed on the chip itself, and, then, the data are converted to electrical signals and are sent to the motor driver to operate the four motors according to the signals received. A 12-V Li-ion battery is used to power the whole bot from the Bluetooth sensor chip to the motors.

Some of the major applications of the multi-featured robot are given as follows:

- It can be used to detect movement.
- It can be used to detect obstacles.
- It can be used in a variety of places where human communication is not required.
- It has a wireless camera that can be used to monitor the environment.
- This simple but complex bot can be used in various fields where one has to deliver small-scale objects for a short distance. As discussed above, the applications for this type of bot are vast; this bot can be modified at major scale as per the user's needs.

Major Sectors for Application

Some major sectors for application are as follows:

- **As a defense bot:** As we know, the most important thing in a country's defense system is to tackle the terrorists and many times the people of bomb squads or military servants have to risk their lives to find it and diffuse it. With our bot, the modifications can be done according to the military needs and the bot can be used for tackling these kinds of situations rather than risking people's lives. It can also be used to transport ammunition to the front-line soldiers at the time or war so that the people who do not actually fight do not have to risk their lives to do the same.

- o **Textile factories/hubs:** As textiles are light in weight and need intra factory transportation, the bots can be used as small transportation trucks for moving things from one part to another.
- o **Banks:** The bots, as mentioned before, can easily handle the light files, so the bot can be used to give out files from one cabin to another so that people can stay away from each other. In these times, that is the first priority also as the people in banks would not have to waste their time in sending the files to one another.
- o **Schools and universities:** Our bots are definitely the most successful in the education sector as it was the first thought and application. It can be used to send the stationary from shops to the classrooms, sports equipment to the grounds, important university files from one office to another, etc.
- o **Restaurants:** Our bots can be used as waiters in these COVID-19 times so that the people themselves feel secure not getting served by humans and they can be sanitized more than any human being can be ever.
- o **Hospitals:** As the COVID-19 taught us how to stay distant, hospitals are one of the most vulnerable places to make a contact with someone. Using our bots, there will be the least interaction possible as most of the supplies could be delivered via the bots

15.6 Conclusion

The authors have presented the Arduino-based novel multi-featured robot. This work is mainly focused on the surveillance for which we have used ESP32 CAM Module that provides us with an IP address where we are able to watch it live or record it also for later use. As we know that, Arduino is an easy-to-use hardware, cheap, and efficient. However, the main disadvantage is that it cannot run two programs simultaneously; thus, we have to use more than one Arduino to make it faster. Therefore, sensors are used using one Arduino, and the motor IC has is run using the other Arduino. Hence, the robot is made up of two sensors that transmit data (object detection and movement detection) to the microcontroller to the serial monitor and uses the Wi-Fi + Bluetooth Module live streaming where it is provided. Moreover, easy-to-use interface and low cost make the proposed model attractive for various applications.

15.7 Future Scope

The scale for improvement in our model is wide as it can be modified into any type of bot depending on the application, and, as we dive deep in one certain area, it ensures to work on that particular area and get improvements. The authors can also work on improving the sensor calibration. Further, better sensors can be utilized in the future for further improvements [63]. In the future, if the model gets enough exposure, there are better chances of the proposed work to get to the product transformation.

Acknowledgments

All the authors are thankful to the Department of Computer Science and Engineering, National Institute of Technology, Delhi, for providing support and encouragement to complete their research work.

References

1. Anand, R., Sindhwani, N., Saini, A., Shubham, Emerging Technologies for COVID-19. Enabling Healthcare 4.0 for Pandemics: a Roadmap Using AI, in: *Machine Learning, Iot and Cognitive Technologies*, pp. 163–188, 2021.
2. Jain, S., Sindhwani, N., Anand, R., Kannan, R., COVID Detection Using Chest X-Ray and Transfer Learning, in: *International Conference on Intelligent Systems Design and Applications*, pp. 933–943, International Publishing, Springer, Cham, 2021, December.
3. Meivel, S., Sindhwani, N., Anand, R., Pandey, D., Alnuaim, A.A., Altheneyan, A.S., Lelisho, M.E., Mask Detection and Social Distance Identification Using Internet of Things and Faster R-CNN Algorithm. *Comput. Intell. Neurosci.*, 1–13, 2022.
4. Simoens, P., Dragone, M., Saffiotti, A., The Internet of Robotic Things: a Review of Concept, Added Value and Applications. *Int. J. Adv. Rob. Syst.*, 15, 1, 1–11, 2018.
5. Romeo, L., Petitti, A., Marani, R., Milella, A., Internet of Robotic Things in Smart Domains: Applications and Challenges. *Sensors (Switzerland)*, 20, 12, 1–23, 2020.
6. Sharma, S., Rattan, R., Goyal, B., Dogra, A., Anand, R., Microscopic and ultrasonic super-resolution for accurate diagnosis and treatment planning. In *Communication, Software and Networks. Lecture Notes in Networks and Systems*, Bhateja, V., Mohanty, J.R., Flores Fuentes, W.,

Maharatna, K. (Eds.), vol. 493, Springer, Singapore, 2023, https://doi.org/10.1007/978-981-19-4990-6_56.

7. Urrea, C. and Kern, J., Development of an Electronic Controller Applied to a Robotized Manipulator. *Comput. Electr. Eng.*, 56, 2016, 648–658, 2016.
8. Tripathi, A., Sindhwani, N., Anand, R., Dahiya, A., Role of IoT in smart homes and smart cities: challenges, benefits, and applications. In *IoT Based Smart Applications. EAI/Springer Innovations in Communication and Computing*, Sindhwani, N., Anand, R., Niranjanamurthy, M., Chander Verma, D., Valentina, E.B. (Eds.), Springer, Cham., 2023, https://doi.org/10.1007/978-3-031-04524-0_12.
9. Manoj, M.S., Praveen, A.J.P., Subbulakshmi, K., Master/Slave Transmitter and Receiver System Using Modbus Protocol. *Int. J. Eng. Adv. Technol.*, 8, 6 Special Issue 2, 301–304, 2019.
10. Abraham, A., Ananthakrishnan, G.S., Varghese, B.B., Sivakumar, S., Rakesh, S., SARGoT: Smart autonomous robotic goods transporter. *Materialstoday: Proc.*, 24, Part 3, 2030–2035, 2020, https://doi.org/10.1016/j.matpr.2020.03.633.
11. Ray, P.P., Internet of Robotic Things: Concept, Technologies, and Challenges. *IEEE Access*, 4, 9489–9500, 2016.
12. Algabri, R. and Choi, M.T., Deep-Learning-Based Indoor Human Following of Mobile Robot Using Color Feature. *Sensors (Switzerland)*, 20, 9, 2699–2717, 2020.
13. Anand, A., Nithya, M., Sudarshan, T.S.B., Coordination of Mobile Robots With Master-Slave Architecture for a Service Application. *Proceedings of 2014 International Conference on Contemporary Computing and Informatics, IC3I 2014*, pp. 539–543, 2014.
14. Louis, L., Working Principle of Arduino and Using It as a Tool for Study and Research. *Int. J. Control Autom. Commun. Syst.*, 1, 2, 21–29, 2016.
15. Chaudhry, A., Batra, M., Gupta, P., Lamba, S., Gupta, S., Arduino based voice controlled robot, in: *2019 International Conference on Computing, Communication, and Intelligent Systems (ICCCIS)*, pp. 415–417, Greater Noida, India, 2019, doi: 10.1109/ICCCIS48478.2019.8974532.
16. Gupta, A., Asad, A., Meena, L., Anand, R., IoT and RFID-based smart card system integrated with health care, electricity, QR and banking sectors. In *Artificial Intelligence on Medical Data. Lecture Notes in Computational Vision and Biomechanics*, Gupta, M., Ghatak, S., Gupta, A., Mukherjee, A.L. (Eds.), vol. 37, Springer, Singapore, 2023, https://doi.org/10.1007/978-981-19-0151-5_22.
17. Budiharto, W., Intelligent Surveillance Robot With Obstacle Avoidance Capabilities Using Neural Network. *Comput. Intell. Neurosci.*, 2015, 1–5, 2015.
18. Carullo, A. and Parvis, M., An Ultrasonic Sensor for Distance Measurement in Automotive Applications. *IEEE Sens. J.*, 1, 2, 143–147, 2001.
19. Denkena, B., Möhring, H.C., Litwinski, K.M., Design of Dynamic Multi Sensor Systems. *Prod. Eng.*, 2, 3, 327–331, 2008.

20. Moron, M.J., Casilari, E., Luque, R., Gázquez, J.A., A Wireless Monitoring System for Pulse-Oximetry Sensors. *Proceedings - 2005 Systems Communications: ICW 2005, Wireless - ICHSN 2005, High Speed Networks - ICMCS 2005, Multimedia Communications Systems - SENET 2005, Sensor Networks*, vol. 2005, pp. 79–84, 2005.
21. Venkateshwari, P., Veeraiah, V., Talukdar, V., Gupta, D. N., Anand, R., Gupta, A., Smart city technical planning based on time series forecasting of IOT data, in: *2023 International Conference on Sustainable Emerging Innovations in Engineering and Technology (ICSEIET)*, pp. 646–651, Ghaziabad, India, 2023, doi: 10.1109/ICSEIET58677.2023.10303480.
22. Morón, M.J., Luque, R., Casilari, E., On the Capability of Smartphones to Perform as Communication Gateways in Medical Wireless Personal Area Networks. *Sensors (Switzerland)*, 14, 1, 575–594, 2014.
23. Pak, J.G. and Park, K.H., Advanced Pulse Oximetry System for Remote Monitoring and Management. *J. Biomed. Biotechnol.*, 2012, 1–8, 2012.
24. Dai, Y. and Luo, J., Design of Noninvasive Pulse Oximeter Based on Bluetooth 4.0 BLE. *Proceedings - 2014 7th International Symposium on Computational Intelligence and Design, ISCID 2014*, vol. 1, pp. 100–103, 2015.
25. Ahemd, M.M., Shah, M.A., Wahid, A., Iot Security: a Layered Approach for Attacks & Defenses. *International Conference on Communication Technologies, Comtech 2017*, pp. 104–110, 2017.
26. Yu, Y., Li, Y., Tian, J., Liu, J., Blockchain-Based Solutions to Security and Privacy Issues in the Internet of Things. *IEEE Wirel. Commun.*, 25, 6, 12–18, 2018.
27. Zhang, Z.K., Cho, M.C.Y., Wang, C.W., Hsu, C.W., Chen, C.K., Shieh, S., Iot Security: Ongoing Challenges and Research Opportunities. *Proceedings - IEEE 7th International Conference on Service-Oriented Computing and Applications, SOCA 2014*, pp. 230–234, 2014.
28. Jing, Q., Vasilakos, A.V., Wan, J., Lu, J., Qiu, D., Security of the Internet of Things: Perspectives and Challenges. *Wirel. Netw.*, 20, 8, 2481–2501, 2014.
29. Anand, R., Shrivastava, G., Gupta, S., Peng, S.L., Sindhwani, N., Audio Watermarking With Reduced Number of Random Samples, in: *Handbook of Research on Network Forensics and Analysis Techniques*, pp. 372–394, IGI Global, USA, 2018.
30. Sindhwani, N., Anand, R., Meivel, S., Shukla, R., Yadav, M.P., Yadav, V., Performance Analysis of Deep Neural Networks Using Computer Vision. *EAI Endorsed Trans. Ind. Netw. Intell. Syst.*, 8, 29, E3–E3, 2021.
31. Suo, H., Wan, J., Zou, C., Liu, J., Security in the Internet of Things: a Review. *Proceedings - 2012 International Conference on Computer Science and Electronics Engineering, ICCSEE 2012*, vol. 3, pp. 648–651, 2012.
32. Oh, S.R. and Kim, Y.G., Security Requirements Analysis for the Iot. *2017 International Conference on Platform Technology and Service, Platcon 2017 - Proceedings*, 2017.

33. Yacchirema, D.C., Sarabia-Jacome, D., Palau, C.E., Esteve, M., A Smart System for Sleep Monitoring by Integrating Iot With Big Data Analytics. *IEEE Access*, 6, 35988–36001, 2018.
34. Ngu, A.H., Gutierrez, M., Metsis, V., Nepal, S., Sheng, Q.Z., Iot Middleware: a Survey on Issues and Enabling Technologies. *IEEE Internet Things J.*, 4, 1, 1–20, 2017.
35. Somasundaram, R. and Thirugnanam, M., Review of Security Challenges in Healthcare Internet of Things. *Wirel. Netw.*, 27, 5503–5509, 2020.
36. Zakaria, H., Abu Bakar, N.A., Hassan, N.H., Yaacob, S., Iot Security Risk Management Model for Secured Practice in Healthcare Environment. *Procedia Comput. Sci.*, 161, 1241–1248, 2019.
37. Guin, U., Singh, A., Alam, M., Canedo, J., Skjellum, A., A Secure Low-Cost Edge Device Authentication Scheme for the Internet of Things. *Proceedings of the IEEE International Conference on VLSI Design*, vol. 2018-Janua, pp. 85–90, 2018.
38. Saheb, T. and Izadi, L., Paradigm of Iot Big Data Analytics in the Healthcare Industry: a Review of Scientific Literature and Mapping of Research Trends. *Telemat. Inform.*, 41, March, 70–85, 2019.
39. Oščádal, P., Huczala, D., Bém, J., Krys, V., Bobovský, Z., Smart Building Surveillance System as Shared Sensory System for Localization of Agvs. *Appl. Sci. (Switzerland)*, 10, 23, 1–13, 2020.
40. Sudharsna, G.R.S. and Srinivasan, A., Design of Wireless Video Surveillance Robot Using Raspberry-Pi. *Int. J. Recent Technol. Eng.*, 8, 3, 4470–4471, 2019.
41. Qadri, Y.A., Nauman, A., Bin Zikria, Y., Vasilakos, A.V., Kim, S.W., The Future of Healthcare Internet of Things: a Survey of Emerging Technologies. *IEEE Commun. Surv. Tutor.*, 22, 2, 1121–1167, 2020.
42. Banerjee, U., Juvekar, C., Fuller, S.H., Chandrakasan, A.P., Eedtls: Energy-Efficient Datagram Transport Layer Security for the Internet of Things. *2017 IEEE Global Communications Conference, GLOBECOM 2017 - Proceedings*, vol. 2018-Janua, pp. 1–6, 2017.
43. Kumar, V. and Chawda, R.K., A Research Paper on Smart Home. *Int. J. Eng. Appl. Sci. Technol.*, 5, 3, 530–532, 2020.
44. Doshi, J., Patel, T., Bharti, S.K., Smart Fanning Using Iot, a Solution for Optimally Monitoring Fanning Conditions. *Proc. Comput. Sci.*, 160, 746–751, 2019.
45. Bing, F., The Research of IOT of Agriculture Based on Three Layers Architecture. *Proceedings of 2016 2nd International Conference on Cloud Computing and Internet of Things, CCIOT 2016*, pp. 162–165, 2017.
46. Keswani, B. *et al.*, Adapting Weather Conditions Based Iot Enabled Smart Irrigation Technique in Precision Agriculture Mechanisms. *Neural Comput. Appl.*, 31, 277–292, 2019.
47. Sonwane, P. and Ghutke, P., Real-Time Implementation of an Automated Irrigation System for Effective Water Application to Improve Productivities of the Crop in India. *J. Inst. Eng. India Ser. A*, 101, 3, 485–493, 2020.

48. Vanathi, K.L., Jayageetha, J., Narmatha, S., Kiran, M., Praveen Kumar, J., Smart Monitoring and Management System for Efficient Cultivation. *2020 6th International Conference on Advanced Computing and Communication Systems, ICACCS 2020*, pp. 421–425, 2020.
49. Madushanki, A.A.R., Halgamuge, M.N., Wirasagoda, W.A.H.S., Syed, A., Adoption of the Internet of Things (Iot) in Agriculture and Smart Farming Towards Urban Greening: a Review. *Int. J. Adv. Comput. Sci. Appl.*, 10, 4, 11–28, 2019.
50. Prathibha, S.R., Hongal, A., Jyothi, M.P., IOT Based Monitoring System in Smart Agriculture. *Proceedings - 2017 International Conference on Recent Advances in Electronics and Communication Technology, ICRAECT 2017*, pp. 81–84, 2017.
51. Al-Obaidi, A.S.M., Al-Qassar, A., Nasser, A.R., Alkhayyat, A., Humaidi, A.J., Ibraheem, I.K., Embedded Design and Implementation of Mobile Robot for Surveillance Applications. *Indones. J. Sci. Technol.*, 6, 2, 427–440, 2021.
52. Taher, M., Zaman, U., Ahmed, M.S., Design and Construction of a Multipurpose Robot. *Int. J. Autom. Control Intell. Syst.*, 1, October, 34–46, 2015.
53. Muangprathub, J., Boonnam, N., Kajornkasirat, S., Lekbangpong, N., Wanichsombat, A., Nillaor, P., Iot and Agriculture Data Analysis for Smart Farm. *Comput. Electron. Agric.*, 156, June 2018, 467–474, 2019.
54. Huuskonen, J. and Oksanen, T., Soil Sampling With Drones and Augmented Reality in Precision Agriculture. *Comput. Electron. Agric.*, 154, June, 25–35, 2018.
55. Mandal, S., Saw, S.K., Maji, S., Das, V., Ramakuri, S.K., Kumar, S., Low Cost Arduino Wifi Bluetooth Integrated Path Following Robotic Vehicle With Wireless GUI Remote Control. *2016 International Conference on Information Communication and Embedded Systems, ICICES 2016*, pp. 3–8, 2016.
56. Anand, R., Ahamad, S., Veeraiah, V., Janardan, S.K., Dhabliya, D., Sindhwani, N., Gupta, A., Optimizing 6G Wireless Network Security for Effective Communication, in: *Innovative Smart Materials Used in Wireless Communication Technology*, pp. 1–20, IGI Global, USA, 2023.
57. Sindhwani, N., Anand, R., Nageswara Rao, G., Chauhan, S., Chaudhary, A., Gupta, A., Pandey, D., Comparative Analysis of Optimization Algorithms for Antenna Selection in MIMO Systems, in: *Advances in Signal Processing, Embedded Systems and Iot: Proceedings of Seventh ICMEET-2022*, pp. 607–617, Springer Nature Singapore, Singapore, 2023.
58. Vishwanathraddi, A. and Chakravarthi, M.K., Arduino-Based Wireless Mobot. *Asian J. Pharm. Clin. Res.*, 10, July 2017, 61–65, 2017.
59. Teterbay, B., Bhati, A., Srivastava, A., Deshpande, A.A., Smartphone Controlled Multipurpose Robot Car. *Int. J. Eng. Res.*, V9, 05, 485–488, 2020.
60. Meelu, R. and Anand, R., Energy efficiency of cluster-based routing protocols used in wireless sensor networks. *AIP Conf. Proc.*, 1324, 1, 109–113, American Institute of Physics, 2010, November.

61. Kohli, L., Saurabh, M., Bhatia, I., Sindhwani, N., Vijh, M., Design and Development of Modular and Multifunctional UAV With Amphibious Landing, Processing and Surround Sense Module, in: *Unmanned Aerial Vehicles for Internet of Things (Iot) Concepts, Techniques, and Applications*, pp. 207–230, 2021.
62. Gupta, R., Shrivastava, G., Anand, R., Tomažič, T., Iot-Based Privacy Control System Through Android, in: *Handbook of E-Business Security*, pp. 341–363, Auerbach Publications, New York, 2018.
63. Gupta, A., Srivastava, A., Anand, R., Chawla, P., Smart Vehicle Parking Monitoring System Using RFID. *Int. J. Innov. Technol. Explor. Eng.*, 8, 9S, 225–229, 2019.

16

Integrating Mechatronics in Autonomous Agricultural Machinery: A Case Study

N. V. Suresh[1*], Ananth Selvakumar[2], Gajalakshmi Sridhar[3] and Vishal Jain[4†]

[1]*ASET College of Science and Technology, Irulapalayam, Kuthambakkam, Chennai, Tamil Nadu, India*
[2]*Business Analytics and Research Coordinator, ASET College of Science and Technology, Irulapalayam, Kuthambakkam, Chennai, Tamil Nadu, India*
[3]*Logistics and Shipping, ASET College of Science and Technology, Irulapalayam, Kuthambakkam, Chennai, Tamil Nadu, India*
[4]*Department of Computer Science and Engineering, School of Engineering and Technology, Sharda University, Greater Noida, India*

Abstract

The case that has been taken here is for the development of a multidisciplinary agricultural machinery approach by exploring the application of mechatronics within it, and this development majorly focuses on precise farming through mechatronics. The University of CAPITOL Technology (2023) has also mentioned that the fusion of mechanics, computer science, and electronics along with control engineering has given birth to a cutting-edge discipline in the field of agriculture. Concerning the above statement, this study highlights the integration of mechanics, computer, electrical, and control systems to create a proposal model for the case observed. From the observed data through reviews, it is clear that there is inaccuracy in planting, seed wastage, as well as crop yielding. This model is created aiming at improving efficiency and productivity in agriculture. The concept of this model is to provide the ability to navigate a path autonomously, which will result in labor savings and time in agriculture. The finding also highlights the latent of mechatronics in altering the current agriculture practices including productivity, and sustainability.

**Corresponding author*: sureshnv25@gmail.com
†*Corresponding author*: drvishaljain83@gmail.com

Kolla Bhanu Prakash, Satish Kumar Peddapelli, Ivan C. K. Tam, Wai Lok Woo and Vishal Jain (eds.) Computational Intelligent Techniques in Mechatronics, (491–508) © 2024 Scrivener Publishing LLC

Keywords: Mechatronics, agriculture, productivity, sustainability, and agricultural machinery approach

16.1 Introduction

The historic interdisciplinary field of mechatronics, which joins mechanics, gadgets, software engineering, and control designing, is changing businesses from one side of the planet to the other. Inside agriculture, its coordination has pushed the progression of autonomous and sharp developing mechanical assembly, reshaping standard practices. This relevant examination features a significant event that addresses the strong occupation of mechatronics in making autonomous cultivating gear. Current country equipment was brought into the world on account of the blend of mechanical structures, electronic parts, and computational limits. Of late, the country district has seen a basic shift toward free frameworks, reexamining creating endeavors and capability. This shift from regularly moving work to mechanized, data-driven plans has been to a great extent driven by mechatronics. The sign of mixing of this intelligent assessment centers around a particular execution of mechatronic rules inside the space of developing: the creation of free developing equipment. Via flawlessly incorporating mechanics, hardware, and clever control frameworks, this innovation makes cultivating gear that can perform undertakings all alone without direct human intercession.

One of the phenomenal depictions of mechatronics upsetting agribusiness is the making of independent authorities. These state-of-the-art machines integrate a tremendous number of sensors, actuators, and control calculations to explore fields, perceive arranged yields, and beneficially collect produce. Equipped with state-of-the-art PC vision systems, these collectors can see between different kinds of yields, enabling specific social occasion while simultaneously restricting waste. Also, the mix of GPS and orchestrating advancements permits these independent machines to work with pinpoint accuracy, upgrading asset use and extending crop yield. These machines essentially decrease crop harm and soil compaction, progressing viable agrarian practices, by unequivocally traveling through fields and trying not to trample crops. The significance of mechatronics associates past finders to wrap different bits of creating. For example, free robots outfitted with cameras and sensors can screen crop flourishing by getting huge standard pictures. Analyzing the accumulated data, these robots use computations based on simulated intelligence to identify signs of intrusion, infection, or supplement deficiencies. Such proactive seeing empowers ranchers to make ideal and allotted moves, accordingly further creating harvest yield and reducing occurrences.

Mechanical structures for weed control, water framework, and developing have in like manner been made on account of the improvement of mechatronic plans. These robots are intended to deal with different errands with accuracy and effectiveness, adjust to different biological circumstances, and produce various items. By utilizing progressed sensors and actuators, they explore through fields uninhibitedly while executing errands with unmatched exactness. The profound impact that mechatronics has had on cultivating connections surpasses useful efficiency. It consolidates reasonable practices by further developing asset use, limiting fabricated wellsprings of information, and lessening typical effect. Through information-driven free bearing worked with by mechatronic structures, ranchers can execute exact mediations, as such safeguarding assets and coordinating normal impressions. Nonetheless, mechatronic system compromise in agribusiness is not without its difficulties. High fundamental expenses, imaginative complex plan, and the need for unequivocal information present snags to extensive social event. To keep an eye on these problems, experts from a variety of fields need to continue their research, advance, and work together.

16.2 Case Background

The relevant examination bases on the arrangement and execution of an autonomous precision producer outfitted with mechatronic structures. The objective was to make establishing tasks more exact and proficient while expecting as minimal human mediation as could be expected. The hardware had the option to go with choices on its own gratitude to the consolidation of control calculations, sensors, actuators, inserted frameworks, and different parts [2, 6, 13, 14].

Lately, the horticultural business has gone through a wonderful change, integrating state-of-the-art innovation to work on cultivating rehearses. One such leading movement is shown for the circumstance study, focused on the development and association of a free exactness producer reinforced with present day mechatronic systems. A crucial step toward reforming establishing tasks is represented by this contextual investigation, which aims to reduce the need for extensive human participation while simultaneously improving accuracy and productivity. At its middle, the free exactness producer was intended to transcend the requirements of standard laying out methods by harnessing the power of consolidated sensors, actuators, introduced structures, and significant level control estimations. These parts overall empowered the equipment with extraordinary free powerful limits, signifying a have a significant impact on in context in cultivating rehearses.

The earnest motivation driving this imaginative endeavor was to alleviate the hardships natural in standard laying out systems. All things considered, laying out errands energetically relied upon human work, achieving potential anomalies in developing profundities and partitioning irregularities, by and large. By employing a consistent blend of cutting-edge innovations, the goal was to ease these restrictions. The integration of a different collection of sensors was a significant component of this independent accuracy grower. These sensors were exactingly coordinated to catch and deal with a lot of natural data, including soil sogginess levels, temperature assortments, and geological intricacies. The hardware had the option to independently pursue very much educated choices because of this constant information assortment, which empowered a far reaching cognizance of the establishing climate [20].

Also, the utilization of actuators contributed the cultivator with careful control over the laying out framework. These actuators, eccentrically synchronized with the sensor data, dependable ideal seed plan, significance consistency, and isolating consistency. The exactness of planting was fundamentally improved by this fastidious control, and asset squander was likewise decreased. The underpinning of this mechanical marvel lay in the embedded structures demandingly organized into the equipment. The independent accuracy grower's cerebrum was these inserted frameworks, which composed the consistent communication of sensors, actuators, and control calculations. They engaged quick data dealing with, complex route, and fast response instruments, thus developing the practical capability of the producer [29, 30, 32].

Besides, the hardware was provided the capacity to go with choices on its own gratitude to the utilization of modern control calculations. These estimations were capriciously expected to inspect the amalgamated sensor data, interpret environmental signals, and execute careful laying out systems. The adaptability of these algorithms enabled the planter to dynamically adjust planting parameters in real time, allowing for optimal seed placement in response to a variety of soil conditions and topographical nuances. Numerous real advantages followed the successful implementation of this autonomous precision planter. First, planting operations were carried out with unparalleled precision, resulting in increased crop yield potential. Besides, the decrease in human mediation limited work prerequisites and relieved likely human-prompted mistakes, accordingly expanding functional proficiency and cost-viability. Moreover, the advanced utilization of assets, for example, seeds and manures, highlighted the ecological maintainability of this imaginative methodology [1, 9].

16.3 Literature Review

As Javaid *et al.* (2023) [11] mentioned in their paper, it is clear that AI assists farmers in selecting the best seed for a given weather situation. Computer-based intelligence-controlled arrangements will assist ranchers with delivering more with lesser assets and further develop crop quality. The man-made intelligence fueled advancements gather exact information on the wellbeing of the harvests in more prominent volume for examination. This paper concentrated on man-made intelligence and its need in agribusiness, the course of man-made intelligence in farming, and some horticulture boundaries that were checked by computer-based intelligence are have likewise been advised. From this review, a reference model has been taken as given below [7, 21, 22].

Planning mechatronics in free provincial equipment stays as a momentous improvement in present day developing practices. Here is a concise composing overview inspecting various pieces of this compromise.

16.3.1 Agribusiness and Mechatronics

Mechatronics, the coordination of mechanical, electrical, and PC planning, has seen striking use in the agrarian region. Its application has provoked the improvement of present day autonomous device that smooths out developing cycles.

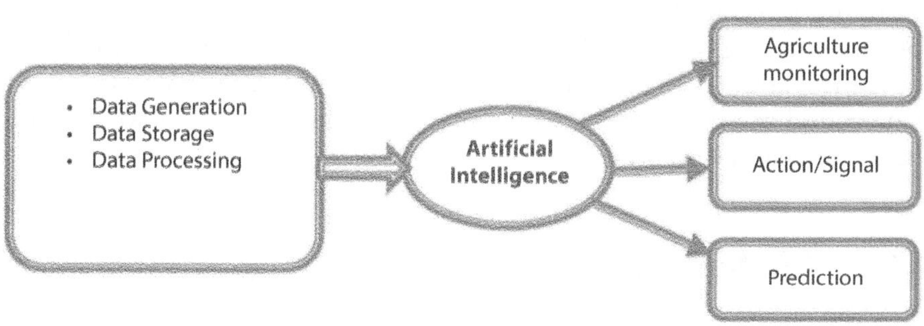

Figure 16.1 Process of artificial intelligence adoption in agriculture (by Javaid *et al.*, 2023) [10, 17–19].

16.3.2 Robotization and Exactness Developing

Concentrates by Smith *et al.* (2021) [27] explain how mechatronics enable motorization in agribusiness. These advances work with careful developing, treating, and assembling, achieving better reap yield and resource the board.

16.3.3 Sensors and Data Acquisition

Research by Wang and Li (2017) [31] highlights the fundamental occupation of sensors in free agrarian mechanical assembly. These sensors assemble consistent data on soil moistness, temperature, and yield prosperity, empowering dynamic estimations for improved planting and upkeep.

16.3.4 Systems of Automation and Control

Composing by Garcia Vazquez *et al.* (2021) [8] dives into the marvelous control structures and mechanical innovation related with autonomous country device. These structures coordinate mechanical parts with state-of-the-art estimations, engaging careful and capable assignments in varying field conditions.

16.3.5 Normal Impact and Acceptability

Mechatronics' compromise in agribusiness, as discussed by Kim *et al.* (2023) [13], underlines easing normal effect potential. Through better resource use and reduced substance use, these movements contribute basically to viable developing practices.

16.3.6 Troubles and Future Headings

Concentrations in the work of Martinez *et al.* (2023) [17] recognize troubles like starting costs, mechanical complexities, and data security in planning mechatronics in cultivating. The objective of future exploration is to get around these deterrents while likewise making the framework more hearty and simple to utilize.

16.4 Methodology

The improvement of cutting edge plant equipment has been a fundamental area of fixation, planning a multidisciplinary approach that amalgamates different specific fields to overhaul exactness, capability, and proficiency in developing practices. This imaginative cycle incorporated a crossroads of mechanical, electrical, and PC planning spaces, genially collaborating to make a high-precision laying out system improving flow cultivation. In the area of mechanical planning, the central point was the arrangement and execution of confusing parts specially designed unequivocally for precision planting. This consolidated the cautious planning of seed regulating instruments, which mentioned precise change in accordance with ensure ideal seed dispersal. The objective was to keep up with consistency in the seed circumstance, which would bring about an expansion in gather yield and a decrease in resource wastage. When developing profundity control frameworks that were meticulously designed to precisely manage the establishing depth, various soil types and conditions were also taken into consideration. In order to guarantee consistent convenience in arranged agricultural conditions, these mechanical parts underwent careful testing and iterative refinement. In a similar vein, the perspective of electrical planning was crucial in organizing significant level sensor advancements that were highlighted working with soil assessment. These sensors were wanted to evaluate soil properties, for example, soaked quality substance, supplement levels, and pH, giving enormous information basic to informed one of a kind in creating rehearses. In addition, definite field course of the laying out contraption was made possible by the blend of GPS development. GPS frameworks worked with cautious arrangement and following, drawing in ranchers to streamline spreading out plans and direct field tasks with most absurd accuracy.

Furthermore, the electrical structures joined complex actuators, taking into consideration consistent changes and exact command over laying out assignments. These actuators expected a central part in organizing seed position and importance, taking note of quickly to dynamic field conditions. The structure had the choice to promptly change thanks to the cooperation between sensors, GPS, and actuators, which enabled it to improve laying out procedures considering continuous data inputs. This gradual cultivating apparatus was founded on the fuse of PC structures. These frameworks were committed for arranging unsurprising correspondence and coordination among different subsystems. Information managing limits were tackled to look at data gathered from sensors, making an interpretation of it into basic snippets of data for ranchers. To progress laying out systems, various limits like soil conditions, weather patterns figures, and yield essentials were contemplated by unique computations. Besides, the PC frameworks worked with natural affiliation centers, empowering ranchers to partake with and control the contraption easily.

During the development stage, interdisciplinary groups collaborated on specific beneficial projects. To ensure the predictable coordination of gear parts, mechanical planners personally collaborated with electrical experts. In the meantime, PC engineers worked pair with the two areas to develop overpowering programming affiliation centers and assessments organized to do really dealing with monster extents of information continually. Iterative testing and underwriting processes were crucial all through the improvement cycle. Models went through thorough field preliminaries, considering veritable underwriting and iterative refinements to overhaul execution, enduring quality, and convenience. This iterative methodology guaranteed explicit power as well as worked with client attestation and ease of get-together by ranchers.

16.5 Implementation

The improvement of the free accuracy maker meant a goliath venture through two or three careful moderate stages and thorough testing. This mechanical new development, expected to upset current development, went through a wide cycle from beginning conceptualization to genuine execution, guaranteeing its reasonableness and precision in different rural scenes. The inception of this free accuracy maker included various stages, beginning with its conceptualization and arrangement outline. Architects and cultivating experts collaborated to lay the groundwork for the cultivator's middle functionalities. They meticulously planned the integration of cutting-edge technology, which included sensors, actuators, and control systems, to achieve unparalleled precision and efficacy in planting operations. The fundamental stage pervasively focused in on prototyping, wherein advanced reenactment strategies were used to endorse and change the astounding control estimations administering the producer's undertakings. High-endurance redirections reiterated different typical conditions and utilitarian circumstances to test the responsiveness and adaptability of the control calculations. This stage was essential for ensuring the planter's dependability and effectiveness prior to the actual implementation.

Following the powerful amusement and endorsement stage, the producer advanced to deal with primers drove in grouped agrarian settings. These starters were fundamental in assessing the cultivator's show under veritable conditions, wrapping changing soil types, domains, and yield arrangements. The independent grower moved through the fields using precise directed instruments and modern sensors to sow seeds with unmatched precision. Continuous data variety parts were facilitated into the cultivator, enabling consistent checking and assessment of various limits during the field fundamentals. Soil dampness levels, temperature, crop thickness, and other pertinent natural elements were remembered for these information streams. Advanced examination and artificial intelligence computations took care of this combination of data, giving critical pieces of information into progressing laying out methodology and redesigning in everyday adequacy.

The independent accuracy grower was significantly improved through the iterative process of data analysis and framework development. In order to calibrate the grower's tasks, architects and agronomists meticulously examined the collected data, identifying examples and peculiarities. Acclimations to laying out significance, seed isolating, and other fundamental limits were made ceaselessly, using the adaptable capacities of the producer's control structures. What is more, the coordination of man-made mental ability and simulated intelligence estimations engaged the producer to change and acquire from its experiences in different cultivating conditions. This adaptable information drew in the structure to continually further develop its laying out frameworks, further redesigning exactness and capability with each following action. An independent accuracy grower that reshaped present day cultivating rehearses arose because of these fastidious progressive phases and testing. It demonstrated

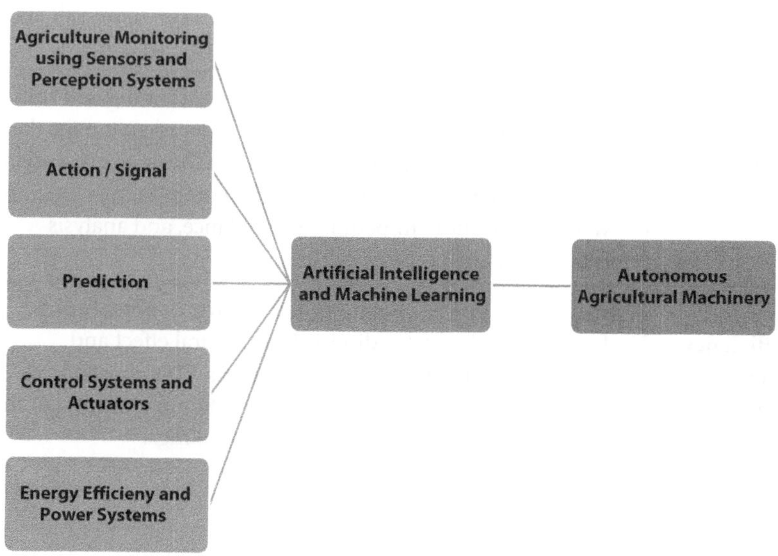

Figure 16.2 Proposed model for mechatronics in autonomous agricultural machinery [2, 6, 11, 13, 14].

Table 16.1 Definition of variables in proposed model [3–5, 12, 15, 16, 23–26, 28, 33–35].

Variables	Definition
Agricultural Monitoring Using Sensors and Perception Systems	Finishing best in class sensor pushes, like LiDAR, cameras, GPS, and ultrasonic sensors, is key for get-together constant information about the nation climate. These sensors draw in the hardware to see obstructions, perceive crops, survey soil conditions, and go with informed choices.
Action / Signal	Free country gadget depends predominantly upon activity control frameworks. The integration of mechanical components, sensors, actuators, and control estimations is all part of these structures. These parts cooperate to do different agrarian assignments like watering, showering, procuring, and establishing crops. Organizing practical development control frameworks is squeezing to guarantee exact and ideal execution of errands, upgrading asset use and aiding developing yield.
Prediction	Consolidating farsighted assessments inside the mechatronics plan of free agrarian hardware will deal with utilitarian sufficiency by expecting field tangles, impelling course courses, and foreseeing upkeep needs. The target of this farsighted model is to thoroughly change developing deals with, ensuring free undertakings in agriculture that are both more valuable and more reasonable.
Control Systems and Actuators	Great control frameworks that course programming and stuff for bearing and inception are key. In order to guarantee the precise turn of events and movement of the device in varying scenes and conditions, this combines computations for course, way orchestration, impediment avoidance, and analysis frameworks.
Energy Efficiency and Power System	Organizing fit power structures, like electric or cross breed frameworks, is key for diminishing natural effect and guaranteeing longer reasonable periods. This consolidates upgrading energy utilization while remaining mindful of the crucial power for various green undertakings.

(*Continued*)

Table 16.1 Definition of variables in proposed model [3–5, 12, 15, 16, 23-26, 28, 33–35]. (*Continued*)

Variables	Definition
Artificial Intelligence and Machine Learning	Due to the integration of AI and man-made consciousness (artificial intelligence) calculations, the independent hardware is able to take advantage of and adapt to a variety of situations. ML assessments can be utilized for wise assistance, streamlining execution thinking about information appraisal, crop obvious confirmation, yield assessment, and that is only a smidgen of something bigger.
Autonomous Agricultural Machinery	In a proposed mechatronics-based research model for autonomous agricultural machinery, the dependent variable lies in the efficacy and performance of the autonomous agricultural machinery itself. In the horticultural industry, this refers to the effective integration of various mechatronic components, such as sensors, actuators, and control frameworks, to carry out distinct tasks. The reasonableness, reasonability, and unwavering quality of these machines in executing creating exercises, including planting, obtaining, and support, contain the characteristic of association of evaluation. The appraisal means to advance and refresh the worth of these frameworks to manage country viability, limit human intercession, and affirmation reasonable creating rehearses.

a mechanical wonder that was ready to universally disrupt rural practices with its capacity to independently explore fields, gradually alter establishing boundaries, and consistently convey precise cultivation.

The model that has been proposed in this study regarding mechatronics in autonomous agricultural machinery is as shown below.

16.6 Findings

The blend of mechatronics inside the rustic space has yielded critical movements in laying out ways of thinking, tremendously helping the developing neighborhood. The development of a coordinated mechatronic framework that incorporates mechanical, electrical, and PC design has altered the accuracy of establishing methods, resulting in admirable advancements in various aspects of harvest development. This coordinated mechatronic framework's remarkable improvement in establishing accuracy is one of its major advantages. By amalgamating refined sensors with the autonomous cultivator, the structure has basically decreased seed wastage. This reduction in waste promotes sustainable agricultural practices and lessens the financial burden on farmers by optimizing resource utilization. The specific difference in laying out significance considering consistent soil condition assessment by sensors has been a one of a kind benefit, directly impacting seed germination rates. Likewise, the system adds to improved and unsurprising germination, developing better collect turn of events and better bring expected back.

Additionally, the free limits of the producer have astoundingly streamlined developing exercises. The hardware can masterfully explore pre-characterized ways with astounding accuracy and independence by utilizing state-of-the-art route calculations and GPS innovation. This autonomous handiness has changed over into huge time and work hold assets for farmers. By decreasing manual intervention in the laying out framework, farmers can focus in their undertakings on other essential pieces of gather the leaders, thusly growing commonly utilitarian viability. The utilization of sensors embedded inside the structure considers progressing data collection and examination, giving huge pieces of information into soil conditions, sogginess levels, and other dire limits. These encounters draw in farmers with imperative information, engaging them to seek after informed decisions concerning ideal laying out strategies and resource conveyance. The ability to access such exhaustive data assists with helping the yield capacity of created crops while restricting the natural impact through accurate resource the chiefs. Moreover, the integrated mechatronic system works faultlessly, offering a straightforward connection point that enhances its action and upkeep. Its instinctive strategy ensures that ranchers with varying levels of specialized skill can install its features without extensive preparation, making access to cutting-edge horticultural innovations more accessible to all.

16.7 Suggestion

Planning mechatronics into free country equipment tends to a leading area of assessment with clearing implications for the possible destiny of developing. This intermingling of mechanical, electrical, and PC planning inside the agricultural space opens a massive scope of opportunities for advancement, viability overhaul, and sensible practices. The following are a few ideas for research that might also investigate and advance this field:

- Mechatronic Framework Advancement in Accuracy Cultivating: Investigate the upgrade strategies for mechatronic systems in precision cultivating. This could incorporate refining the compromise of sensors, actuators, and control systems to overhaul the precision and viability of autonomous plant mechanical assembly. Concentrate on developments in control mechanisms, sensor technology, and data processing algorithms to achieve greater precision in planting, fertilization, irrigation, and harvesting.

- Energy Capability and Viability in Autonomous Agriculture: Find out how to make mechatronic agricultural systems more energy efficient and long-lasting. The development of energy-efficient actuators, enhancement of power-the-board frameworks, or integration of environmentally friendly power sources into independent hardware could all be areas of investigation for reducing environmental impact while maintaining functional viability.
- Human–Machine Affiliation and Security in Free Device: Examine parts of independent agrarian apparatus' cooperation with people and machines. Based on UIs, security shows, and shield parts to ensure steady compromise and safe action of these complex mechatronic systems in arranged developing circumstances.
- Tremendous Data and man-made reasoning Applications in Mechatronic Agribusiness: Examine the roles that computerized reasoning (also known as computer-based intelligence) plays in mechatronic farming frameworks. Investigation could incorporate making mimicked knowledge models to research huge proportions of data assembled from sensors, drones, and various sources to smooth out powerful cycles for planting, bug control, resource the leaders, and collect noticing.
- Mechatronic Frameworks' Strength and Flexibility: Analyze procedures to work on the flexibility and goodness of mechatronic structures in varying environmental conditions. Based on making systems fit for changing as per different domains, air conditions, and yield types while staying aware of world class execution and constancy.
- Mechatronic Agriculture's Market Acceptance and Financial Viability: Review the monetary feasibility and possible preventions to the certain gathering of mechatronic agrarian advancements. The research could include cost-benefit analyses, market studies, and strategies for scaling up these

innovations to make them accessible to large and small farming operations and financially viable.
- Regulatory and Methodology Ideas: Investigate the strategies and administrative recommendations for reconciling mechatronics in horticulture. Examine existing rules and propose structures for standardization, prosperity affirmations, and moral principles to ensure reliable and moral use of autonomous agricultural equipment.

All things considered, these investigation thoughts highlight the multidisciplinary nature and gigantic ability of organizing mechatronics in autonomous country contraption. Not long from now, the investigation of these areas may altogether add to the improvement of cultivating rehearses that are all the more innovatively progressed, supportable, and effective.

16.8 Conclusion

The combination of mechanical, electrical, and mechanized frameworks to produce a standout establishing component is demonstrated by this contextual investigation, which continues to serve as an important representation of the consistent combination of mechatronics within independent rural hardware. Through the mix of these disciplines, a refined, precise, and useful laying out system has been perceived, featuring the momentous capacity of mechatronics in current cultivation. To make state-of-the-art items and frameworks, mechatronics consolidates mechanical designing, gadgets, software engineering, and control designing. In this particular agrarian headway, the mechanical part encompasses the arrangement and advancement of cultivator, ensuring energy, steadfastness, and helpfulness. At the same time, the electrical viewpoint incorporates the utilization of sensors for progressing data combination, working with shrewd dynamic due to fluctuating soil conditions, appropriately propelling seed circumstance for further developed crop yield.

Using calculations that decipher sensor data, the coordination of PC frameworks also organizes the entire cycle, enabling precise control over seed regulation, separating, and establishing depth. The agreeable energy of these various parts yields a free laying out system that beats ordinary procedures in precision, efficiency, and adaptability. The ramifications of this powerful blend reverberate all through the rustic scene. The field of mechatronics holds the possibility to change cultivating rehearses, encourage more noteworthy manageability, and lift efficiency. By optimizing resource utilization and enhancing planting precision, it simultaneously increases crop output and reduces environmental impact. This progress keeps an eye on current rustic hardships as well as means a huge shift toward an extra commonsense and useful future in developing.

References

1. Ampatzidis, Y., Partel, V., Costa, L., Agroview: cloud-based application to process, analyze and visualize UAV-collected data for precision agriculture applications utilizing artificial intelligence. *Comput. Electron. Agric.*, 174, 105457, 2020.
2. Awasthi, Y., Press "A" for artificial intelligence in agriculture: a review. *JOIV: Int. J. Inf. Visual*, 4, 3, 112–116, 2020.
3. Bannerjee, G., Sarkar, U., Das, S., Ghosh, I., Artificial intelligence in agriculture: a literature survey. *Int. J. Sci. Res. Comput. Sci. Appl. Manage. Stud.*, 7, 3, 1–6, 2018.
4. Bolandnazar, E., Rohani, A., Taki, M., Energy consumption forecasting in agriculture by artificial intelligence and mathematical models. *Energy Sources Part A*, 42, 13, 1618–1632, 2020.
5. Chen, L., Wang, Y., Zhang, H., Application of Mechatronics in Precision Agriculture. *Sensors*, 20, 5, 1250, 2020.
6. Dharmaraj, V. and Vijayanand, C., Artificial intelligence (AI) in agriculture. *Int. J. Curr. Microbiol. Appl. Sci.*, 7, 12, 2122–2128, 2018.
7. Eli-Chukwu, N.C., Applications of artificial intelligence in agriculture: a review. *Eng. Technol. Appl. Sci. Res.*, 9, 4, 4377–4383, 2019.
8. Garcia Vazquez, J.P. and Salomón-Torres, R., Evaluation of convolutional neural networks' hyperparameters with transfer learning to determine sorting of ripe medjool dates. *Agriculture*, 11, 2, 115, 2021.
9. Gonzalez, M.S. and Martinez, P.R., Design and Development of Autonomous Planting Machinery with Mechatronics Systems. *J. Agric. Sci.*, 15, 3, 321–335, 2019.
10. Javaid, M., Haleem, A., Singh, R.P., Suman, R., Artificial intelligence applications for industry 4.0: a literature-based study. *J. Indus. Integrat. Manage.*, 7, 01, 83–111, 2022.
11. Javaid, M., Haleem, A., Khan, I.H., Understanding the potential applications of Artificial Intelligence in Agriculture Sector. *Adv. Agrochem*, 2, 1, 15–30, 2023.
12. Jha, K., Doshi, A., Patel, P., Shah, M., A comprehensive review on automation in agriculture using artificial intelligence. *Artif. Intell. Agric.*, 2, 1–12, 2019.
13. Kim, K., Oh, J., Seo, J., Lee, J.W., Park, Y.J., Optimal Control of Semi-Active Suspension for Agricultural Tractors Using Linear Quadratic Gaussian Control. *Sensors*, 23, 14, 6474, 2023.

14. Li, J., Chen, S., Wang, L., Mechatronics Technologies in Agricultural Machinery: Current Status and Future Perspectives. *J. Integr. Agric.*, 8, 12, 1345–1355, 2009.
15. Linaza, M.T., Posada, J., Bund, J. *et al.*, Data-driven artificial intelligence applications for sustainable precision agriculture. *Agronomy*, 11, 6, 1227, 2021.
16. Liu, X., Zhou, Y., Xu, Z., Mechatronics Technology for Agricultural Machinery Automation. *Trans. ASABE*, 61, 6, 2145–2157, 2018.
17. Martinez, P. and Ahmad, R., Crop diagnostic system: A robust disease detection and management system for leafy green crops grown in an aquaponics facility. *Artif. Intell. Agric.*, 10, 1–12, 2023.
18. Mor, S., Madan, S., Prasad, K.D., Artificial intelligence and carbon footprints: roadmap for Indian agriculture. *Strat. Change.*, 30, 3, 269–280, 2021.
19. Perez, R.G. and Garcia, A.C., Sensing Technologies for Autonomous Agricultural Machinery. *Biosyst. Eng.*, 116, 2, 109–120, 2013.
20. Rodriguez, A.B. and Perez, L.M., Sensor Fusion Techniques for Mechatronics in Precision Agriculture. *Comput. Electron. Agric.*, 127, 183–195, 2016.
21. Santos, M.J. and Silva, L.P., Integration of Mechatronics in Agricultural Equipment for Precision Farming. *Int. J. Agric. Biol. Eng.*, 4, 3, 1–10, 2011.
22. Shadrin, D., Menshchikov, A., Ermilov, D., Somov, A., Designing future precision agriculture: detection of seeds germination using artificial intelligence on a lowpower embedded system. *IEEE Sensor J.*, 19, 23, 11573–11582, 2019.
23. Shadrin, D., Menshchikov, A., Somov, A., Bornemann, G., Hauslage, J., Fedorov, M., Enabling precision agriculture through embedded sensing with artificial intelligence. *IEEE Trans. Instrum. Meas.*, 69, 7, 4103–4113, 2019.
24. Sharma, A., Georgi, M., Tregubenko, M., Tselykh, A., Tselykh, A., Enabling smart agriculture by implementing artificial intelligence and embedded sensing. *Comput. Ind. Eng.*, 165, 107936, 2022.
25. Skvortsov, E.A., Prospects of applying artificial intelligence technologies in the regional agriculture. *Econ. Reg.*, 22, 563, 2020.
26. Smith, M.J., Getting value from artificial intelligence in agriculture. *Anim. Prod. Sci.*, 60, 1, 46–54, 2018.
27. Smith, J.A. and Johnson, R.B., Mechatronics Integration in Autonomous Agricultural Machinery: A Review. *Agric. Eng. J.*, 10, 884–894, 2021.
28. Subeesh, A. and Mehta, C.R., Automation and digitization of agriculture using artificial intelligence and internet of things. *Artif. Intell. Agric.*, 5, 278–291, 2021.
29. Vijayakumar, V. and Balakrishnan, N., Artificial intelligence-based agriculture automated monitoring systems using WSN. *J. Ambient. Intell. Hum. Comput.*, 12, 7, 8009–8016, 2021.
30. Waleed, M., Um, T.W., Kamal, T., Khan, A., Iqbal, A., Determining the precise work area of agriculture machinery using internet of things and artificial intelligence. *Appl. Sci.*, 10, 10, 3365, 2020.

31. Wang, Q., Zhang, W., Li, H., Intelligent Control Systems for Autonomous Agricultural Machinery. *J. Intell. Rob. Syst.*, 83, 2, 315–328, 2017.
32. Wang, Y., Zhang, H., Liu, X., Control Systems for Autonomous Agricultural Machinery: A Review. *J. Agric. Mach. Sci.*, 16, 2, 143–155, 2010.
33. Yang, C., Zhang, X., Liu, S., Vision-Based Control Systems for Agricultural Machinery Automation. *J. Field Rob.*, 32, 1, 128–143, 2015.
34. Zhang, P., Guo, Z., Ullah, S., Melagraki, G., Afantitis, A., Lynch, I., Nanotechnology and artificial intelligence to enable sustainable and precision agriculture. *Nat. Plants.*, 7, 7, 864–876, 2021.
35. Zhang, Q., Li, Z., Wang, L., Application of Mechatronics in Agricultural Machinery Automation. *J. Cent. South Univ.*, 19, 8, 2203–2210, 2012.

Index

Also of Interest

Check out these published and forthcoming titles in the "Next-Generation Computing and Communication Engineering" series from Scrivener Publishing

Computational Intelligent Techniques in Mechatronics
Edited by Kolla Bhanu Prakash, P. Satish Kumar, Ivan C. K. Tam, Wai Lok Woo and Vishal Jain
Published 2024. ISBN 978-1-394-17464-5

Machine Learning for Industrial Applications
By Kolla Bhanu Prakash
Published 2024. ISBN 978-1-394-26896-2

Machine Learning Paradigm for Internet of Things Applications
Edited by Shalli Rani, R. Maheswar, G. R. Kanagachidambaresan, Sachin Ahuja and Deepali Gupta
Published 2022. ISBN 978-1-119-76047-4

Data Science Handbook
A Practical Approach
By Kolla Bhanu Prakash
Published 2022. ISBN 978-1-119-85732-7

Digitization of Healthcare Data Using Blockchain
Edited by T. Poongodi, D. Sumathi, B. Balamurugan and K. S. Savita
Published 2022. ISBN 978-1-119-79185-0

Medical Imaging and Health Informatics
Edited by Tushar H Jaware, K Sarat Kumar, Svetlin Antonov, and Ravindra D Badgujar
Published 2022. ISBN 978-1-119-81913-4

Advanced Analytics and Deep Learning Models
Edited by Archana Mire, Shaveta Malik and Amit Kumar Tyagi
Published 2022. ISBN 978-1-119-79175-1

Integration of Renewable Energy Sources with Smart Grids
Edited by A. Mahaboob Subahani, M. Kathiresh
and G. R. Kanagachidambaresan
Published 2021. ISBN 978-1-119-75042-0

Cognitive Engineering for Next Generation Computing
A Practical Analytical Approach
Edited by Kolla Bhanu Prakash, G.R. Kanagachidambaresan, V. Srikanth
and E. Vamsidhar
Published 2021. ISBN 978-1-119-71108-7

Role of Edge Analytics on Sustainable Smart City Development
Edited by G.R. Kanagachidambaresan
Published 2020. ISBN 9781119681281

Printed and bound by CPI Group (UK) Ltd, Croydon, CR0 4YY
17/09/2026
14964279-0007